The IMA Volumes in Mathematics and its Applications

Volume 138

Series Editors
Douglas N. Arnold   Fadil Santosa

Springer
New York
Berlin
Heidelberg
Hong Kong
London
Milan
Paris
Tokyo

Institute for Mathematics and its Applications (IMA)

The **Institute for Mathematics and its Applications** was established by a grant from the National Science Foundation to the University of Minnesota in 1982. The primary mission of the IMA is to foster research of a truly interdisciplinary nature, establishing links between mathematics of the highest caliber and important scientific and technological problems from other disciplines and industry. To this end, the IMA organizes a wide variety of programs, ranging from short intense workshops in areas of exceptional interest and opportunity to extensive thematic programs lasting a year. IMA Volumes are used to communicate results of these programs that we believe are of particular value to the broader scientific community.

The full list of IMA books can be found at the Web site of the Institute for Mathematics and its Applications:

http://www.ima.umn.edu/springer/full-list-volumes.html.

Douglas N. Arnold, Director of the IMA

* * * * * * * * * * *

IMA ANNUAL PROGRAMS

1982–1983	Statistical and Continuum Approaches to Phase Transition
1983–1984	Mathematical Models for the Economics of Decentralized Resource Allocation
1984–1985	Continuum Physics and Partial Differential Equations
1985–1986	Stochastic Differential Equations and Their Applications
1986–1987	Scientific Computation
1987–1988	Applied Combinatorics
1988–1989	Nonlinear Waves
1989–1990	Dynamical Systems and Their Applications
1990–1991	Phase Transitions and Free Boundaries
1991–1992	Applied Linear Algebra
1992–1993	Control Theory and its Applications
1993–1994	Emerging Applications of Probability
1994–1995	Waves and Scattering
1995–1996	Mathematical Methods in Material Science
1996–1997	Mathematics of High Performance Computing
1997–1998	Emerging Applications of Dynamical Systems
1998–1999	Mathematics in Biology

Continued at the back

Mark Johnson Sanjeev P. Khudanpur
Mari Ostendorf Roni Rosenfeld
Editors

Mathematical Foundations of Speech and Language Processing

With 56 Illustrations

Springer

Mark Johnson
Dept. of Cognitive and
Linguistic Studies
Brown University
Providence, RI 02912
USA

Sanjeev P. Khudanpur
Dept. of ECE and Dept. of
Computer Science
Johns Hopkins University
Baltimore, MD 21218
USA

Mari Ostendorf
Dept. of Electrical Engineering
University of Washington
Seattle, WA 98195
USA

Roni Rosenfeld
School of Computer Science
Carnegie Mellon University
Pittsburgh, PA 15213
USA

Series Editors:
Douglas N. Arnold
Fadil Santosa
Institute for Mathematics and
its Applications
University of Minnesota
Minneapolis, MN 55455
USA
http://www.ima.umn.edu

Mathematics Subject Classification (2000): 68T10, 68T50, 94A99, 68-06, 94A12, 94A40, 60J22, 60J20, 68U99, 94-06

Library of Congress Cataloging-in-Publication Data
Mathematical foundations of speech and language processing / Mark Johnson . . . [et al.]
p. cm. — (IMA volumes in mathematics and its applications ; v. 138)
Includes bibliographical references.
ISBN 0-387-20326-5 (alk. paper)
1. Speech processing systems—Mathematical models. I. Johnson, Mark Edward, 1970–
II. Series.
TK7882.S65.M38 2004
006.4′54—dc22 2003065729

ISBN 0-387-20326-5 Printed on acid-free paper.

Printed in the United States of America. (SBA)

9 8 7 6 5 4 3 2 1 SPIN 10951453

Springer-Verlag is part of *Springer Science+Business Media*

springeronline.com

FOREWORD

This IMA Volume in Mathematics and its Applications

MATHEMATICAL FOUNDATIONS OF SPEECH AND LANGUAGE PROCESSING

contains papers presented at two successful one-week workshops: Mathematical Foundations of Speech Processing and Recognition and Mathematical Foundations of Natural Language Modeling. Both workshops were integral to the 2000–2001 IMA annual program on Mathematics in Multimedia.

Sanjeev Khudanpur (Department of Electrical and Computer Engineering and Department of Computer Science, Johns Hopkins University), Mari Ostendorf (Signal and Image Processing, University of Washington), and Roni Rosenfeld (School of Computer Science, Carnegie Mellon University) were the organizers of the first workshop held on September 18–22, 2000. The second workshop which took place on October 30–November 3, 2000 was also organized by Khudanpur and Rosenfeld. They were joined by Mark Johnson (Department of Cognitive and Linguistic Sciences, Brown University) and Frederick Jelinek (Center for Language and Speech Processing, Johns Hopkins University)

We are grateful to all the organizers for making the events successful. We further thank Mark Johnson, Sanjeev P. Khudanpur, Mari Ostendorf, and Roni Rosenfeld for their superb role in editing the proceedings.

We take this opportunity to thank the National Science Foundation for its support of the IMA. We are also grateful to the Office of Naval Research for providing additional funds to support the Multimedia annual program.

Series Editors

Douglas N. Arnold, Director of the IMA

Fadil Santosa, Deputy Director of the IMA

PREFACE

The importance of speech and language technologies continues to grow as information, and information needs, pervade every aspect of our lives and every corner of the globe. Speech and language technologies are used to automatically transcribe, analyze, route and extract information from high-volume streams of spoken and written information. Equally important, these technologies are also used to create natural and efficient interfaces between people and machines.

The workshop on Mathematical Foundations of Speech Processing and Recognition, (September 18-22, 2000), and the one on Mathematical Foundations of Natural Language Modeling (October 30–November 3, 2000), were held at the University of Minnesota's NSF-sponsored Institute for Mathematics and Its Applications (IMA), as part of the "Mathematics in Multimedia" year-long program. These workshops brought together practitioners in the respective technologies on one hand, and mathematicians and statisticians on the other hand, for an intensive week of introduction and cross-fertilization. The intent of these workshops was (1) to provide the mathematicians and statisticians with an accelerated introduction to the state-of-the-art in the aforementioned technologies, and the mathematical challenges lying therein; (2) to expose the practitioners to the state-of-the-art in various mathematical and statistical disciplines of potential relevance to their field; (3) to create an environment for the emergence of cross-fertilization and break-through ideas; and (4) to encourage and facilitate new long-term collaborations. Judging from the level of enthusiasm during the workshops and the long off-hours discussions, the first three goals achieved unqualified success. As for the fourth goal, some collaboration between practitioners and mathematicians had already begun during the workshop planning; only time will tell the long-term effects of such beginnings.

There is a long history of benefit from introducing mathematical techniques and ideas to speech and language technologies. Examples include applying the source-channel paradigm from information theory to automatic speech recognition (and later also to machine translation and information retrieval); applying hidden Markov models to acoustic modeling and hidden variables to speech and language modeling more generally; applying decision trees, singular value decomposition and exponential models to the modeling of natural language; and applying formal languages theory to parsing. It is likely that new mathematical techniques, or novel applications of existing techniques, will once again prove pivotal for moving the field forward. For example, recent work on making Monte Carlo Markov Chain techniques more computationally feasible holds promise for breaking away from point estimation (e.g. maximum likelihood and discriminative criteria) towards full Bayesian modeling in both the acoustic and linguistic domains.

The role of mathematics and statistics in speech and language technologies cannot be overestimated. The rate at which we continue to accumulate speech and language training data is far greater than the rate at which our understanding of the speech and language phenomena grows. As a result, the relative advantage of data driven techniques continues to grow with time, and with it, the importance of mathematical and statistical methods that make use of such data.

In this volume, we have compiled papers representing some original contributions presented by participants during the two workshops. More information about the various workshop presentations and discussions can be found online, at http://www.ima.umn.edu/multimedia/. In this volume, chapters are organized starting with four contributions related to language processing, moving from more general work to specific advances in structure and topic representations in language modeling. The fifth paper on prosody modeling provides a nice transition, since prosody can be seen as an important link between acoustic and language modeling. The next five papers relate primarily to acoustic modeling, starting with work that is motivated by speech production models and acoustic-phonetic studies, and then moving toward more general work on new models. The book concludes with two contributions from the statistics community that we believe will impact speech and language processing in the future.

Finally, we would like to express our gratitude to the National Science Foundation for making these workshops possible via its funding of the IMA and its activities; to the IMA's staff for so ably organizing and administering the workshops and related events; and to all the participants for contributing to the success of these workshops in particular and "the Year of Mathematics in Multimedia" in general.

Mark Johnson
Department of Cognitive and Linguistic Sciences
Brown University

Sanjeev P. Khudanpur
Department of Electrical and Computer Engineering and Department of Computer Science
Johns Hopkins University

Mari Ostendorf
Signal and Image Processing
University of Washington

Roni Rosenfeld
School of Computer Science
Carnegie Mellon University

CONTENTS

PROBABILITY AND STATISTICS IN COMPUTATIONAL LINGUISTICS, A BRIEF REVIEW

STUART GEMAN* AND MARK JOHNSON*

1. Introduction. Computational linguistics studies the computational processes involved in language learning, production, and comprehension. Computational linguists believe that the essence of these processes (in humans and machines) is a computational manipulation of information. Computational *psycholinguistics* studies psychological aspects of human language (e.g., the time course of sentence comprehension) in terms of such computational processes.

Natural language processing is the use of computers for processing natural language text or speech. Machine translation (the automatic translation of text or speech from one language to another) began with the very earliest computers [Kay et al., 1994]. Natural language interfaces permit computers to interact with humans using natural language, e.g., to query databases. Coupled with speech recognition and speech synthesis, these capabilities will become more important with the growing popularity of portable computers that lack keyboards and large display screens. Other applications include spell and grammar checking and document summarization. Applications outside of natural language include compilers, which translate source code into lower-level machine code, and computer vision [Foo, 1974, Foo, 1982].

The notion of a grammar is central to most work in computational linguistics and natural language processing. A *grammar* is a description of a language; usually it identifies the sentences of the language and provides descriptions of them, e.g., by defining the phrases of a sentence, their inter-relationships, and perhaps also aspects of their meanings. *Parsing* is the process of recovering a sentence's description from its words, while *generation* is the process of translating a meaning or some other part of a sentence's description into a *grammatical* or well-formed sentence. Parsing and generation are major research topics in their own right. Evidently, human use of language involves some kind of parsing and generation process, as do many natural language processing applications. For example, a machine translation program may parse an input language sentence into a (partial) representation of its meaning, and then generate an output language sentence from that representation.

Although the intellectual roots of modern linguistics go back thousands of years, by the 1950s there was considerable interest in applying the then newly developing ideas about finite-state machines and other kinds of automata, both deterministic and stochastic, to natural language. Automata

*Department of Cognitive and Linguistic Sciences, Brown University, Providence, RI 02912, USA.

are Markov-like machines consisting of a set of states and a set of allowed state-to-state transitions. An input sequence, selected from a finite input alphabet, moves the machine from state to state along allowed transitions. [Chomsky, 1957] pointed out clearly the inadequacies of finite-state machines for modelling English syntax. An effect of Chomsky's observations, perhaps unintended, was to discourage further research into probabilistic and statistical methods in linguistics. In particular, stochastic grammars were largely ignored. Instead, there was a shift away from simple automata, both deterministic and stochastic, towards more complex non-stochastic grammars, most notably "transformational" grammars. These grammars involved two levels of analyses, a "deep structure" meant to capture more-or-less simply the meaning of a sentence, and a "surface structure" which reflects the actual way in which the sentence was constructed. The deep structure might be a clause in the active voice, "Sandy saw Sam," whereas the surface structure might involve the more complex passive voice, "Sam was seen by Sandy."

Transformational grammars are computationally complex, and in the 1980s several linguists came to the conclusion that much simpler kinds of grammars could describe most syntactic phenomena, developing Generalized Phrase-Structure Grammars [Gazdar et al., 1985] and Unification-based Grammars [Kaplan and Bresnan, 1982, Pollard and Sag, 1987], [Shieber, 1986]. These grammars generate surface structures directly; there is no separate deep structure and therefore no transformations. These kinds of grammars can provide very detailed syntactic and semantic analyses of sentences, but as explained below, even today there are no comprehensive grammars of this kind that fully accommodate English or any other natural language.

Natural language processing using hand-crafted non-stochastic grammars suffers from two major drawbacks. First, the syntactic coverage offered by any available grammar is incomplete, reflecting both our lack of understanding of even relatively frequently occuring syntactic constructions and the organizational difficulty of manually constructing any artifact as complex as a grammar of a natural language. Second, such grammars almost always permit a large number of *spurious ambiguities*, i.e., parses which are permitted by the rules of syntax but have unusual or unlikely semantic interpretations. For example, in the sentence *I saw the boat with the telescope*, the prepositional phrase *with the telescope* is most easily interpreted as the instrument used in seeing, while in *I saw the policeman with the rifle*, the prepositional phrase usually receives a different interpretation in which the policeman has the rifle. Note that the corresponding alternative interpretation is marginally accessible for each of these sentences: in the first sentence one can imagine that the telescope is on the boat, and in the second, that the rifle (say, with a viewing scope) was used to view the policeman.

In effect, there is a dilemma of coverage. A grammar rich enough to accommodate natural language, including rare and sometimes even "ungrammatical" constructions, fails to distinguish natural from unnatural interpretations. But a grammar sufficiently restricted so as to exclude what is unnatural fails to accommodate the scope of real language. These observations lead, in the 1980's, to a renewed interest in stochastic approaches to natural language, particularly to speech. Stochastic finite-state automata became the basis of speech recognition systems by out-performing the best of the systems based on deterministic hand-crafted grammars. Largely inspired by the success of stochastic approaches in speech recognition, computational linguists began applying them to other natural language processing applications. Usually, the architecture of such a stochastic model is specified manually (e.g., the possible states of a stochastic finite-state automaton and the allowed transitions between them), while the model's parameters are estimated from a *training corpus*, i.e., a large representative sample of sentences.

As explained in the body of this paper, stochastic approaches replace the binary distinctions (grammatical versus ungrammatical) of non-stochastic approaches with probability distributions. This provides a way of dealing with the two drawbacks of non-stochastic approaches. Ill-formed alternatives can be characterized as extremely low probability rather than ruled out as impossible, so even ungrammatical strings can be provided with an interpretation. Similarly, a stochastic model of possible interpretations of a sentence provides a method for distinguishing more plausible interpretations from less plausible one.

The next section, §2, introduces formally various classes of grammars and languages. Probabilistic grammars are introduced in §3, along with the basic issues of parametric representation, inference, and computation.

2. Grammars and languages. The formal framework, whether used in a transformational grammar, a generalized phrase-structure grammar, or a more traditionally styled context-free grammar, is due to [Chomsky, 1957] and his co-workers. In this section, we will present a brief introduction to this framework. But for a thorough (and very readable) presentation we highly recommend the book by [Hopcroft and Ullman, 1979].

If T is a finite set of symbols, let T^* be the set of all strings (i.e., finite sequences) of symbols of T, including the empty string, and let T^+ be the set of all nonempty strings of symbols of T. A *language* is a subset of T^*. A *rewrite grammar* G is a quadruple $G = (T, N, S, R)$, where T and N are disjoint finite sets of symbols (called the *terminal* and *non-terminal* symbols respectively), $S \in N$ is a distinguished non-terminal called the *start symbol*, and R is a finite set of productions. A *production* is a pair (α, β) where $\alpha \in N^+$ and $\beta \in (N \cup T)^*$; productions are usually written $\alpha \to \beta$. Productions of the form $\alpha \to \epsilon$, where ϵ is the empty string, are called *epsilon productions*. In this paper we will restrict attention to grammars without epsilon productions, i.e., $\beta \in (N \cup T)^+$, as this simplifies the mathematics considerably.

A rewrite grammar G defines a *rewriting relation* $\Rightarrow_G \subseteq (N \cup T)^* \times (N \cup T)^*$ over pairs of strings consisting of terminals and nonterminals as follows: $\gamma\alpha\delta \Rightarrow \gamma\beta\delta$ iff $\alpha \to \beta \in R$ and $\gamma, \delta \in (N \cup T)^*$ (the subscript G is dropped when clear from the context). The reflexive, transitive closure of $\Rightarrow$ is denoted $\Rightarrow^*$. Thus $\Rightarrow^*$ is the rewriting relation using arbitrary finite sequences of productions. (It is called "reflexive" because the identity rewrite, $\alpha \Rightarrow^* \alpha$, is included). The *language generated by* G, denoted L_G, is the set of all strings $w \in T^+$ such that $S \Rightarrow^* w$.

A terminal or nonterminal $X \in N \cup T$ is *useless* unless there are $\gamma, \delta \in (N \cup T)^*$ and $w \in T^*$ such that $S \Rightarrow^* \gamma X \delta \Rightarrow^* w$. A production $\alpha \to \beta \in R$ is useless unless there are $\gamma, \delta \in (N \cup T)^*$ and $w \in T^*$ such that $S \Rightarrow^* \gamma\alpha\delta \Rightarrow \gamma\beta\delta \Rightarrow^* w$. Informally, useless symbols or productions never appear in any sequence of productions rewriting the start symbol S to any sequence of terminal symbols, and the language generated by a grammar is not affected if useless symbols and productions are deleted from the grammar.

Example 1. Let the grammar $G_1 = (T_1, N_1, \mathrm{S}, R_1)$, where $T_1 = \{\mathrm{grows}, \mathrm{rice}, \ \mathrm{wheat}\}$, $N_1 = \{\mathrm{S}, \mathrm{NP}, \mathrm{VP}\}$ and $R_1 = \{\mathrm{S} \to \mathrm{NP}\,\mathrm{VP}, \mathrm{NP} \to \mathrm{rice}, \mathrm{NP} \to \mathrm{wheat}, \mathrm{VP} \to \mathrm{grows}\}$. Informally, the nonterminal S rewrites to sentences or clauses, NP rewrites to noun phrases and VP rewrites to verb phrases. Then $L_{G_1} = \{\mathrm{rice\ grows}, \ \mathrm{wheat\ grows}\}$. G_1 does not contain any useless symbols or productions.

Rewrite grammars are traditionally classified by the shapes of their productions. $G = (T, N, S, R)$ is a *context-sensitive grammar* iff for all productions $\alpha \to \beta \in R$, $|\alpha| \le |\beta|$, i.e., the right-hand side of each production is not shorter than its left-hand side. G is a *context-free grammar* iff $|\alpha| = 1$, i.e., the left-hand side of each production consists of a single non-terminal. G is a *left-linear grammar* iff G is context-free and β (the right-hand side of the production) is either of the form Aw or of the form w where $A \in N$ and $w \in T^*$; in a *right-linear grammar* β always is of the form wA or w. A right or left-linear grammar is called a *regular* grammar.

It is straight-forward to show that the classes of languages generated by these classes of grammars stand in equality or subset relationships. Specifically, the class of languages generated by right-linear grammars is the same as the class generated by left-linear grammars; this class is called the *regular languages*, and is a strict subset of the class of languages generated by context-free grammars, which is a strict subset of the class of languages generated by context-sensitive grammars, which in turn is a strict subset of the class of languages generated by rewrite grammars.

The computational complexity of deciding whether a string is generated by a rewrite grammar is determined by the class that the grammar belongs to. Specifically, the *recognition problem for grammar* G takes a string $w \in T^+$ as input and returns TRUE iff $w \in L_G$. Let $\mathcal{G}$ be a class of grammars. The *universal recognition problem for $\mathcal{G}$* takes as input a string $w \in T^+$ and a grammar $G \in \mathcal{G}$ and returns TRUE iff $w \in L_G$.

There are rewriting grammars that generate languages that are *recursively enumerable* but not *recursive.* In essence, a language L_G is recursively enumerable if there exists an algorithm that is guaranteed to halt and emit TRUE whenever $w \in L_G$, and *may* halt and emit NOT TRUE, *or may not halt*, whenever $w \notin L_G$. On the other hand, a language is recursive if there exists an algorithm that always halts, and emits TRUE or NOT TRUE depending on whether $w \in L_G$ or $w \notin L_G$, respectively.[1] Obviously, the set of recursive languages is a subset of the set of recursively enumerable languages. The recognition problem for a language that is recursively enumerable but not recursive is said to be *undecidable.* Since such languages do exist, generated by rewrite grammars, the universal recognition problem for rewrite grammars is undecidable.

The universal recognition problem for context-sensitive grammars *is* decidable, and furthermore is in PSPACE (space polynomial in the size of G and w), but there are context-sensitive grammars for which the recognition problem is PSPACE-complete [Garey and Johnson, 1979], so the universal recognition problem for context-sensitive grammars is PSPACE-complete also. Since NP$\subseteq$PSPACE, we should not expect to find a polynomial-time recognition algorithm for arbitrary context-sensitive grammars. The universal recognition problem for context-free grammars is decidable in time polynomial in the size of w and linear in the size of G; as far as we are aware a tight upper bound is not known. Finally, the universal recognition problem for regular grammars is decidable in time linear in w and G.

It turns out that context-sensitive grammars (where a production rewrites more than one nonterminal) have not had many applications in natural language processing, so from here on we will concentrate on context-free grammars, where all productions take the form $A \to \beta$, where $A \in N$ and $\beta \in (N \cup T)^+$.

An appealing property of grammars with productions in this form is that they induce tree structures on the strings that they generate. And, as we shall see shortly (§3), this is the basis for bringing in probability distributions and the theory of inference. We say that the context-free grammar $G = (T, N, S, R)$ generates the labelled, ordered tree ψ iff the root node of ψ is labelled S, and for each node n in ψ, either n has no children and its label is a member of T (i.e., it is labelled with a terminal) or else there is a production $A \to \beta \in R$ where the label of n is A and the left-to-right sequence of labels of n's immediate children is β. It is straight forward to show that w is in L_G iff G generates a tree ψ whose *yield* (i.e., the left-to-right sequence of terminal symbols labelling ψ's leaf nodes) is

[1] A rigorous definition requires a proper introduction to Turing machines. Again, we recommend [Hopcroft and Ullman, 1979].

w; ψ is called a *parse tree* of w (with respect to G). In what follows, we define Ψ_G to be the set of parse trees generated by G, and $\mathcal{Y}(\cdot)$ to be the function that maps trees to their yields.

Example 1 (continued). The grammar G_1 defined above generates the following two trees, ψ_1 and ψ_2.

$\psi_1 =$ 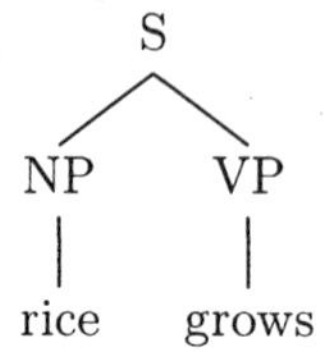

$\psi_2 =$ 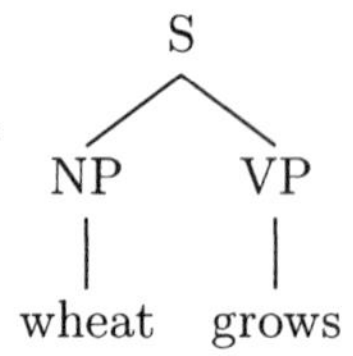

In this example, $\mathcal{Y}(\psi_1) = \textit{rice grows}$ and $\mathcal{Y}(\psi_2) = \textit{wheat grows}$

A string of terminals w is called *ambiguous* iff w has two or more parse trees. Linguistically, each parse tree of an ambiguous string usually corresponds to a distinct interpretation.

Example 2. Consider $G_2 = (T_2, N_2, \mathrm{S}, R_2)$, where $T_2 = \{\mathrm{I}, \mathrm{saw}, \mathrm{the}, \mathrm{man}, \mathrm{with}, \mathrm{telescope}\}$, $N_2 = \{\mathrm{S}, \mathrm{NP}, \mathrm{N}, \mathrm{Det}, \mathrm{VP}, \mathrm{V}, \mathrm{PP}, \mathrm{P}\}$ and $R_2 = \{\mathrm{S} \to \mathrm{NP\,VP}, \mathrm{NP} \to \mathrm{I}, \mathrm{NP} \to \mathrm{Det\,N}, \mathrm{Det} \to \mathrm{the}, \mathrm{NP} \to \mathrm{NP\,PP}, \mathrm{N} \to \mathrm{man}, \mathrm{N} \to \mathrm{telescope}, \mathrm{VP} \to \mathrm{V\,NP}, \mathrm{VP} \to \mathrm{VP\,PP}, \mathrm{PP} \to \mathrm{P\,NP}, \mathrm{V} \to \mathrm{saw}, \mathrm{P} \to \mathrm{with}\}$. Informally, N rewrites to nouns, Det to determiners, V to verbs, P to prepositions and PP to prepositional phrases. It is easy to check that the two trees ψ_3 and ψ_4 with the yields $\mathcal{Y}(\psi_3) = \mathcal{Y}(\psi_4) =$ *I saw the man with the telescope* are both generated by G_2. Linguistically, these two parse trees represent two different syntactic analyses of the sentence. The first analysis corresponds to the interpretation where the seeing is by means of a telescope, while the second corresponds to the interpretation where the man has a telescope.

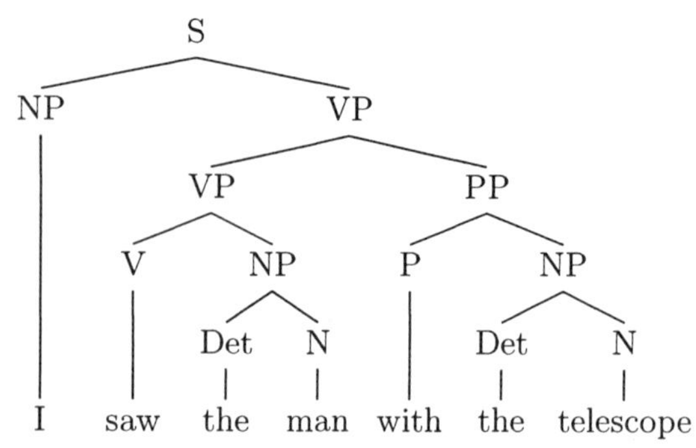

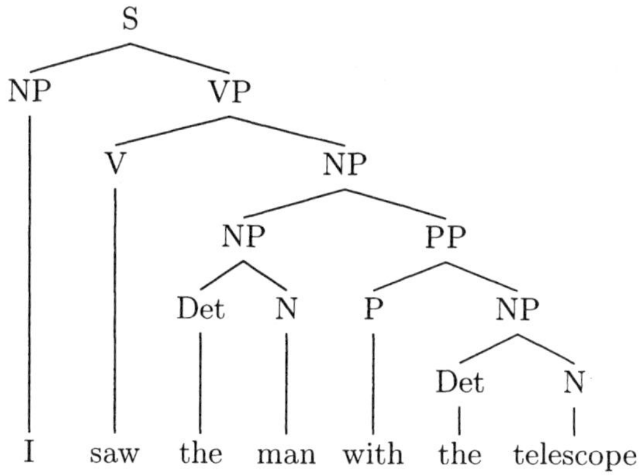

There is a close relationship between linear grammars and finite-state machines. A finite-state machine is a kind of automaton that makes state-to-state transitions driven by letters from an input alphabet (see [Hopcroft and Ullman, 1979]) for details). Each finite-state machine has a corresponding right-linear grammar which has the property that the set of strings accepted by the machine is the same as the set of strings generated by the grammar (modulo an endmarker, as discussed below), and the nonterminals of this grammar are exactly the set of states of the machine. Moreover, there is an isomorphism between accepting computations of the machine and parse trees generated by this grammar: for each sequence of states that the machine transitions through in an accepting computation there is a parse tree of the corresponding grammar containing exactly the same sequence of states (the example below clarifies this).

The grammar $G_M = (T, N, S, R)$ that corresponds to a finite-state machine M is one where the nonterminal symbols N are the states of M, the start symbol S is the start state of M, and the terminal symbols T are the input symbols to M together with a new symbol '\$', called the *endmarker*, that does not appear in the transition labels of M. The productions R of G_M come in two kinds. R can contain productions $A \to b\,B$, where $A, B \in N$ and $b \in T$, iff there is a transition in M from state A to state B on input symbol b. R contains the production $A \to \$$ iff A is a final state in M. Informally, M accepts a string $w \in T^*$ iff w is the sequence of inputs along a path from M's start state to some final state. It is easy to show that G_M generates the string $w\$$ iff M accepts the string w. (If we permitted epsilon productions then it would not be necessary to use an endmarker; R would contain a production $A \to \epsilon$, where ϵ is the empty string, iff A is a final state in M).

Example 3. Consider the right-linear grammar $G_3 = (T_3, N_3, \mathrm{S}, R_3)$, where $T_3 = \{\mathrm{a}, \mathrm{b}, \$\}$, $N_3 = \{\mathrm{S}, \mathrm{A}\}$ and $R_3 = \{\mathrm{S} \to \mathrm{b}, \mathrm{S} \to \mathrm{a\,S}, \mathrm{S} \to \mathrm{b\,A}, \mathrm{A} \to \mathrm{a\,A}, \mathrm{A} \to \$\}$. G_3 corresponds to the finite-state machine depicted below, where the '>' attached to the state labelled S indicates it is the start state, and the double circle indicates that the state labelled A is a final state. The parse tree for 'aaba\$' with respect to G_3 translates immediately into the sequence of states that the machine transitions through when accepting 'aaba'.

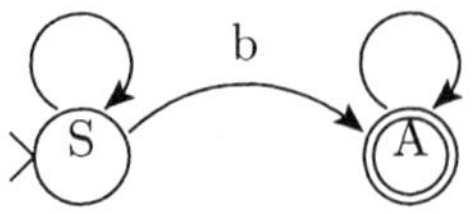

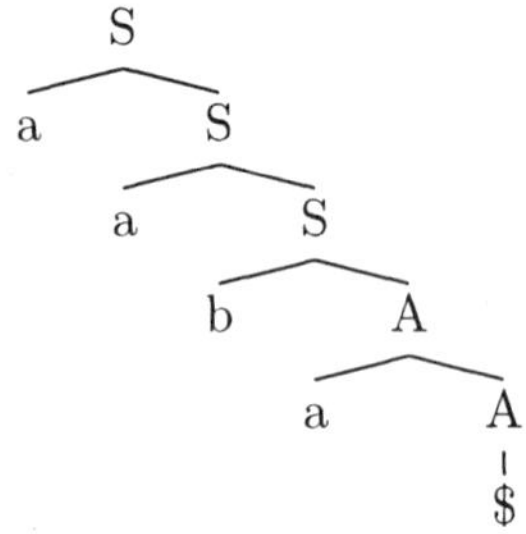

As remarked earlier, context-sensitive and unrestricted rewrite grammars do not seem to be useful in many natural language processing applications. On the other hand, the notation of context-free grammars is not ideally suited to formulating natural language grammars. Furthermore, it is possible to show that some natural languages are not context-free languages [Culy, 1985, Shieber, 1985]. These two factors have led to the development of a variety of different kinds of grammars. Many of these can be described as *annotated phrase structure grammars*, which are extensions of context-free grammars in which the set of nonterminals N is very large, possibly infinite, and N and R possess a linguistically motivated structure. In Generalized Phrase Structure Grammars [Gazdar et al., 1985] N is finite, so these grammars always generate context-free languages, but in unification grammars such as Lexical-Functional Grammar [Kaplan and Bresnan, 1982] or Head-driven Phrase Structure Grammar [Pollard and Sag, 1987] N is infinite and the languages such grammars generate need not be context-free or even recursive.

Example 4. Let $G_4 = (T_4, N_4, \mathrm{S}, R_4)$ where $T_4 = \{\mathrm{a}, \mathrm{b}\}$, $N_4 = \{\mathrm{S}\} \cup \{\mathrm{A}, \mathrm{B}\}^+$ (i.e., N_4 consists of S and nonempty strings over the alphabet A, B) and $R_4 = \{\mathrm{S} \to \alpha\,\alpha : \alpha \in \{\mathrm{A}, \mathrm{B}\}^+\} \cup \{\mathrm{A} \to \mathrm{a}, \mathrm{B} \to \mathrm{b}\} \cup \{\mathrm{A}\alpha \to \mathrm{a}\,\alpha, \mathrm{B}\alpha \to \mathrm{b}\,\alpha : \alpha \in \{\mathrm{A}, \mathrm{B}\}^+\}$. G_4 generates the language $\{ww : w \in \{\mathrm{a}, \mathrm{b}\}^+\}$, which is not a context-free language. A parse tree for aabaab is shown below.

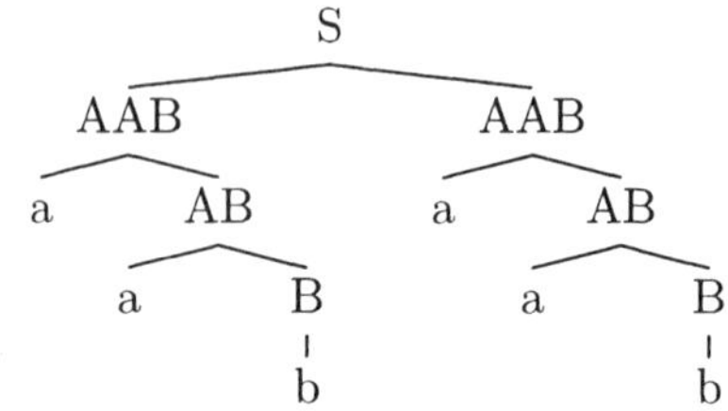

3. Probability and statistics. Obviously broad coverage is desirable—natural language is rich and diverse, and not easily held to a small set of rules. But it is hard to achieve broad coverage without massive ambiguity (a sentence may have tens of thousands of parses), and this of course complicates applications like language interpretation, language translation, and speech recognition. This is the dilemma of coverage that we referred to earlier, and it sets up a compelling role for probabilistic and statistical methods.

We will review the main probabilistic grammars and their associated theories of inference. We begin in §3.1 with probabilistic regular grammars, also known as hidden Markov models (HMM), which are the foundation of modern speech recognition systems. In §3.2 we discuss probabilistic context-free grammars, which turn out to be essentially the same thing as branching processes. We review the estimation problem, the computation problem, and the role of criticality. Finally, in §3.3, we take a more general approach to placing probabilities on grammars, which leads to Gibbs distributions, a role for Besag's *pseudolikelihood method* [Besag, 1974, Besag, 1975], various computational issues, and, all in all, an active area of research in computational linguistics.

3.1. Hidden Markov models and regular grammars. Recall that a right-linear grammar $G = (T, N, S, R)$ corresponding to a finite-state machine is characterized by rewrite rules of the form $A \to b\,B$ or $A \to \$$, where $A, B \in N$, $b \in T$, and $\$ \in T$ is a special terminal that we call an endmarker. The connection with Hidden Markov Models (HMM's) is transparent: N defines the states, R defines the allowed transitions (A can go to B if there exists a production of the form $A \to b\,B$), and the string of terminals defines the "observation." The process is "hidden" since, in general, the observations do not uniquely define the sequence of states.

In general, it is convenient to work with a "normal form" for right-linear grammars: all rules are either of the form $A \to b\,B$ or $A \to b$, where $A, B \in N$ and $b \in T$. It is easy to show that every right-linear grammar has an equivalent normal form in the sense that the two grammars produce the same language. Essentially nothing is lost, and we will usually work with a normal form.

3.1.1. Probabilities. Assume that R has no useless symbols or productions. Then the grammar G can be made into a probabilistic grammar by assigning to each nonterminal $A \in N$ a probability distribution p over productions of the form $A \to \alpha \in R$: for every $A \in N$

$$\sum_{\substack{\alpha \in (N \cup T)^+ \\ \text{s.t. } (A \to \alpha) \in R}} p(A \to \alpha) = 1 \; . \tag{1}$$

Recall that Ψ_G is the set of parse trees generated by G (see §2). If G is linear, then $\psi \in \Psi_G$ is characterized by a sequence of productions, starting from S. It is, then, straightforward to use p to define a probability P on Ψ_G: just take $P(\psi)$ (for $\psi \in \Psi_G$) to be the product of the associated production probabilities.

Example 5. Consider the right-linear grammar $G_5 = (T_5, N_5, \mathrm{S}, R_5)$, with $T_5 = \{\mathrm{a}, \mathrm{b}\}$, $N_5 = \{\mathrm{S}, \mathrm{A}\}$ and the productions (R_5) and production probabilities (p):

$$\begin{array}{lll} \mathrm{S} \rightarrow \mathrm{a\,S} & & p = .80 \\ \mathrm{S} \rightarrow \mathrm{b\,S} & & p = .01 \\ \mathrm{S} \rightarrow \mathrm{b\,A} & & p = .19 \\ \mathrm{A} \rightarrow \mathrm{b\,A} & & p = .90 \\ \mathrm{A} \rightarrow \mathrm{b} & & p = .10\ . \end{array}$$

The language is the set of strings ending with a sequence of at least two b's. The grammar is ambiguous: in general, a sequence of terminal states does not uniquely identify a sequence of productions. The sentence $aabbbb$ has three parses (determined by the placement of the production $\mathrm{S} \rightarrow \mathrm{b\,A}$), but the most likely parse, by far, is $\mathrm{S} \rightarrow \mathrm{a\,S}$, $\mathrm{S} \rightarrow \mathrm{a\,S}$, $\mathrm{S} \rightarrow \mathrm{b\,A}$, $\mathrm{A} \rightarrow \mathrm{b\,A}$, $\mathrm{A} \rightarrow \mathrm{b\,A}$, $\mathrm{A} \rightarrow \mathrm{b}$ ($P = .8 \cdot .8 \cdot .19 \cdot .9 \cdot .1$), which has a *posterior* probability of nearly .99. The corresponding parse tree is shown below.

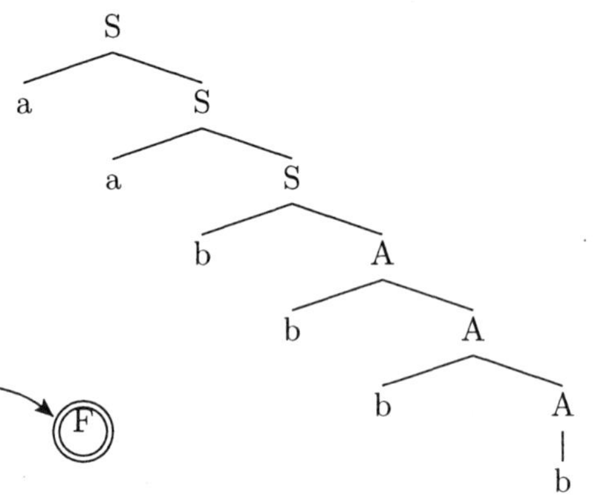

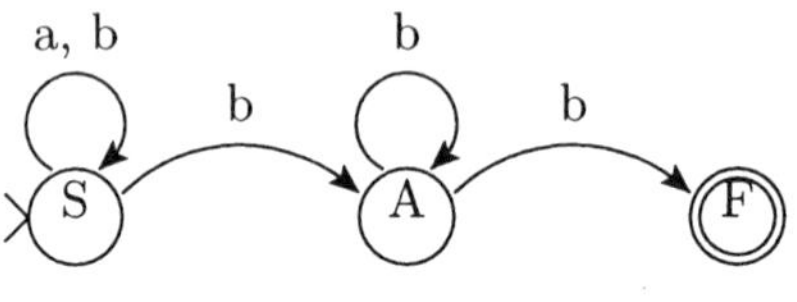

An equivalent formulation is through the associated three-state (S, A, and F) two-output (a and b) HMM also shown above: the transition probability matrix is

$$\begin{pmatrix} .81 & .19 & .00 \\ .00 & .90 & .10 \\ .00 & .00 & 1.00 \end{pmatrix}$$

where the first row and column represent S, the next represent A, and the last represent F; and the output probabilities are based on state-to-state pairs,

$$(\mathrm{S},\mathrm{S}) \to \begin{cases} a & \text{prob} = 80/81 \\ b & \text{prob} = 1/81 \end{cases}$$

$$(\mathrm{S},\mathrm{A}) \to \begin{cases} a & \text{prob} = 0 \\ b & \text{prob} = 1 \end{cases}$$

$$(\mathrm{A},\mathrm{F}) \to \begin{cases} a & \text{prob} = 0 \\ b & \text{prob} = 1 \; . \end{cases}$$

3.1.2. Inference. The problem is to estimate the transition probabilities, $p(\cdot)$, either from parsed data (examples from Ψ_G) or just from sentences (examples from L_G). Consider first the case of parsed data ("supervised learning"), and let $\psi_1, \psi_2, \ldots, \psi_n \in \Psi$ be a sequence taken iid according to P. If $f(A \to \alpha; \psi)$ is the counting function, counting the number of times the transition $A \to \alpha \in R$ occurs in ψ, then the likelihood function is

$$L = L(p; \psi_1, \ldots, \psi_n) = \prod_{i=1}^{n} \prod_{A \to \alpha \in R} p(A \to \alpha)^{f(A \to \alpha; \psi_i)} \; . \tag{2}$$

The maximum likelihood estimate is, sensibly, the relative frequency estimator:

$$\hat{p}(A \to \alpha) = \frac{\sum_{i=1}^{n} f(A \to \alpha; \psi_i)}{\sum_{i=1}^{n} \sum_{\beta \text{ s.t. } A \to \beta \in R} f(A \to \beta; \psi_i)} \; . \tag{3}$$

If a nonterminal A does not appear in the sample, then the numerator and denominator are zero, and $\hat{p}(A \to \alpha)$, $\alpha \in (N \cup T)^+$, can be assigned arbitrarily, provided it is consistent with (1).

The problem of estimating p from sentences ("unsupervised learning") is more interesting, and more important for applications. Recall that $\mathcal{Y}(\psi)$ is the yield of ψ, i.e. the sequence of terminals in ψ. Given a sentence $w \in T^+$, let Ψ_w be the set of parses which yield w: $\Psi_w = \{\psi \in \Psi : \mathcal{Y}(\psi) = w\}$. The likelihood of a sentence $w \in T^+$ is the sum of the likelihoods of its possible parses:

$$L(p; w) = \sum_{\psi \in \Psi_w} P(\psi) = \sum_{\psi \in \Psi_w} \prod_{A \to \alpha \in R} p(A \to \alpha)^{f(A \to \alpha; \psi)} \; .$$

Imagine now a sequence $\psi_1, \ldots, \psi_n$, iid according to P, for which only the corresponding yields, $w_i = \mathcal{Y}(\psi_i)$, $1 \leq i \leq n$, are observed. The likelihood function is

$$L = L(p; w_1, \ldots, w_n) = \prod_{i=1}^{n} \sum_{\psi \in \Psi_{w_i}} \prod_{A \to \alpha \in R} p(A \to \alpha)^{f(A \to \alpha; \psi_i)} \; . \tag{4}$$

To get the maximum likelihood equation, take logarithms, introduce Lagrange multipliers to enforce (1), and set the derivative with respect to $p(A \to \alpha)$ to zero:

$$(5) \quad \lambda_A + \frac{1}{\hat{p}(A\to\alpha)} \sum_{i=1}^{n} \frac{\sum_{\psi\in\Psi_{w_i}} f(A\to\alpha;\psi) \prod_{B\to\beta\in R} \hat{p}(B\to\beta)^{f(B\to\beta;\psi)}}{\sum_{\psi\in\Psi_{w_i}} \prod_{B\to\beta\in R} \hat{p}(B\to\beta)^{f(B\to\beta;\psi)}} = 0 .$$

Introduce $E_p[\cdot]$, meaning expectation under the probability on P induced by p, and solve for $\hat{p}(A \to \alpha)$:

$$(6) \qquad \hat{p}(A \to \alpha) = \frac{\sum_{i=1}^{n} E_{\hat{p}}[f(A \to \alpha;\psi)|\psi \in \Psi_{w_i}]}{\sum_{A\to\beta\in R}\sum_{i=1}^{n} E_{\hat{p}}[f(A \to \beta;\psi)|\psi \in \Psi_{w_i}]} .$$

We can't solve, directly, for $\hat{p}$, but (6) suggests an iterative approach [Baum, 1972]: start with an arbitrary $\hat{p}_0$ (but positive on R). Given $\hat{p}_t$, $t = 1, 2, \ldots$, define $\hat{p}_{t+1}$ by using $\hat{p}_t$ in the right hand side of (6):

$$(7) \qquad \hat{p}_{t+1}(A \to \alpha) = \frac{\sum_{i=1}^{n} E_{\hat{p}_t}[f(A \to \alpha;\psi)|\psi \in \Psi_{w_i}]}{\sum_{A\to\beta\in R}\sum_{i=1}^{n} E_{\hat{p}_t}[f(A \to \beta;\psi)|\psi \in \Psi_{w_i}]} .$$

Evidently, $\hat{p}_t = \hat{p}_{t+1}$ if and only if we have found a solution to the likelihood equation, $\frac{\partial}{\partial \hat{p}(A\to\alpha)} L = 0$, $\forall A \to \alpha \in R$. What's more, as shown by [Baum, 1972], it turns out that $L(\hat{p}_{t+1}; w_1, \ldots, w_n) \geq L(\hat{p}_t; w_1, \ldots, w_n)$, and the procedure finds a local maximum of the likelihood. It turns out, as well, that (7) is just an instance of the EM algorithm, which of course is more general and was discovered later by [Dempster et al., 1977].

Needless to say, nothing can be done with this unless we can actually evaluate, in a computationally feasible way, expressions like $E_{\hat{p}}[f(A \to \alpha;\psi)|\psi \in \Psi_w]$. This is one of several closely related computational problems that are part of the mechanics of working with grammars.

3.1.3. Computation. A sentence $w \in T^+$ is *parsed* by finding a sequence of productions $A \to bB \in R$ which yield w. Depending on the grammar, this corresponds more or less to an *interpretation* of w. Often, there are many parses and we say that w is ambiguous. In such cases, if there is a probability p on R then there is a probability P on Ψ, and a reasonably compelling choice of parse is the most likely parse:

$$(8) \qquad \arg\max_{\psi\in\Psi_w} P(\psi) .$$

This is the *maximum a posteriori* (MAP) estimate of ψ—obviously it minimizes the probability of error under the distribution P.

What is the probability of w? How are its parses computed? How is the most likely parse computed? These computational issues turn out to be more-or-less the same as the issue of computing $E_{\hat{p}}[f(A \to \alpha; \psi)|\psi \in \Psi_w]$ that came up in our discussion of inference. The basic structure and cost of the computational algorithm is the same for each of the four problems—compute the probability of w, compute the set of parses, compute the best parse, compute $E_{\hat{p}}$. For regular grammars, there is a simple dynamic programming solution to each of these problems, and in each case the complexity is of the order $n \cdot |R|$, where n is the length of w, and $|R|$ is the number of productions in G.

Consider the representative problem of producing the most likely parse, (8). Let $w = (b_1, \ldots, b_n) \in T^n$. There are $n - 1$ productions of the form $A_k \to b_{k+1}\, A_{k+1}$ for $k = 0, \ldots, n-2$, with $A_0 = S$, followed by a single terminating production $A_{n-1} \to b_n$. The most likely sequence of productions can be computed by a dynamic-programming type iteration: for every $A \in N$ initialize with

$$\begin{aligned} \mathcal{A}_1(A) &= S \\ \mathcal{V}_1(A) &= p(S \to b_1 A)\ . \end{aligned}$$

Then, given $\mathcal{A}_k(A)$ and $\mathcal{V}_k(A)$, for $A \in N$ and $k = 1, 2, \ldots, n-2$, compute $\mathcal{A}_{k+1}(A)$ and $\mathcal{V}_{k+1}(A)$ from

$$\begin{aligned} \mathcal{A}_{k+1}(A) &= \arg\max_{B \in N} p(B \to b_{k+1} A)\mathcal{V}_k(B) \\ \mathcal{V}_{k+1}(A) &= p(\mathcal{A}_{k+1}(A) \to b_{k+1} A)\mathcal{V}_k(\mathcal{A}_{k+1}(A))\ . \end{aligned}$$

Finally, let

$$\mathcal{A}_n = \arg\max_{B \in N} p(B \to b_n)\mathcal{V}_{n-1}(B).$$

Consider the most likely sequence of productions from S at "time 0" to A at "time k," given $b_1, \ldots, b_k$, $k = 1, \ldots, n-1$. $\mathcal{A}_k(A)$ is the state at time $k-1$ along this sequence, and $\mathcal{V}_k(A)$ is the likelihood of this sequence. Therefore, $\hat{A}_{n-1} \stackrel{def}{=} \mathcal{A}_n$ is the state at time $n-1$ associated with the most likely parse, and working backwards, the best state sequence overall is $\hat{A}_0, \hat{A}_1, \ldots, \hat{A}_{n-1}$, where

$$\hat{A}_{k-1} = \mathcal{A}_k(\hat{A}_k), \quad k = n-1, n-2, \ldots, 1\ .$$

There can be ties when more than one sequence achieves the optimum. In fact, the procedure generalizes easily to produce the best l parses, for any $l > 1$. Another modification produces all parses, while still another computes expectations E_p of the kind that appear in the EM iteration (7) or probabilities such as $P\{\mathcal{Y}(\psi) = w\}$ (these last two are, essentially, just a matter of replacing argmax by summation).

3.1.4. Speech recognition. An outstanding application of probabilistic regular grammars is to speech recognition. The approach was first proposed in the 1970's (see [Jelinek, 1997] for a survey), and has since become the dominant technology. Modern systems achieve high accuracy in multi-user continuous-speech applications. Many tricks of representation and computation are behind the successful systems, but the basic technology is nevertheless that of probabilistic regular grammars trained via EM and equipped with a dynamic programming computational engine. We will say something here, briefly and informally, about how these systems are crafted.

So far in our examples T has generally represented a vocabulary of words, but it is not words themselves that are observable in a speech recognition task. Instead, the acoustic signal is observable, and a time-localized discrete representation of this signal makes up the vocabulary T. A typical approach is to start with a spectral representation of progressive, overlapping windows, and to summarize this representation in terms of a relatively small number, perhaps 200, of possible values for each window. One way to do this is with a clustering method such as vector quantization. This ensemble of values then constitutes the terminal set T.

The state space, N, and the transition rules, R, are built from a hierarchy of models, for phonemes (which correspond to letters in speech), words, and grammars. A phoneme model might have, for example, three states representing the beginning, middle, and end of a phoneme's pronunciation, and transitions that allow, for example, remaining in the middle state as a way of modeling variable duration. The state space, N, is small—maybe three states for each of thirty or forty phonemes—making a hundred or so states. This becomes a regular grammar by associating the transitions with elements of T, representing the quantized acoustic features. Of course a realistic system must accommodate an enormous variability in the acoustic signal, even for a single speaker, and this is why probabilities are so important.

Words are modeled similarly, as a set of phonemes with a variety of allowed transition sequences representing a variety of pronunciations choices. These representations can now be expanded into basic units of phoneme pronunciation, by substituting phoneme models for phonemes. Although the transition matrix is conveniently organized by this hierarchical structure, the state space is now quite large: the number of words in the system's vocabulary (say 5,000) *times* the number of states in the phoneme models (say 150). In fact many systems model the effects of context on articulation (e.g. co-articulation), often by introducing states that represent *triplets* of phonemes ("triphones"), which can further increase the size of N, possibly dramatically.

The sequence of words uttered in continuous speech is highly constrained by syntactic and semantic conventions. These further constraints, which amount to a grammar on words, constitute a final level in the hierarchy. An obvious candidate model would be a regular grammar, with N made up of syntactically meaningful parts of speech (verb, noun, noun phrase, article, and so on). But implementations generally rely on the much simpler and less structured *trigram*. The set of states is the set of ordered word pairs, and the transitions are *a priori* only limited by noting that the second word at one unit of time must be the same as the first word at the next. Obviously, the trigram model is of no utility by itself; once again probabilities play an essential role in meaningfully restricting the coverage.

Trigrams have an enormous effective state space, which is made all the larger by expanding the words themselves in terms of word models. Of course the actual number of possible, or at least reasonable, transitions out of a state in the resulting (expanded) grammar is not so large. This fact, together with a host of computational and representational tricks and compromises, renders the dynamic programming computation feasible, so that training can be carried out in a matter of minutes or hours, and recognition can be performed at real time, all on a single user's PC.

3.2. Branching processes and context-free grammars. Despite the successes of regular grammars in speech recognition, the problems of language *understanding* and *translation* are generally better addressed with the more structured and more powerful context-free grammars. Following our development of probabilistic regular grammars in the previous section, we will address here the inter-related issues of fitting context-free grammars with probability distributions, estimating the parameters of these distributions, and computing various functionals of these distributions.

The context-free grammars $G = (T, N, S, R)$ have rules of the form $A \to \alpha, \alpha \in (N \cup T)^+$, as discussed previously in §2. There is again a normal form, known as the Chomsky normal form, which is particularly convenient when developing probabilistic versions. Specifically, one can always find a context-free grammar G', with all productions of the form $A \to BC$ or $A \to a$, $A, B, C, \in N$, $a \in T$, which produces the same language as G: $L_{G'} = L_G$. Henceforth, we will assume that context-free grammars are in the Chomsky normal form.

3.2.1. Probabilities. The goal is to put a probability distribution on the set of parse trees generated by a context-free grammar in Chomsky normal form. Ideally, the distribution will have a convenient parametric form, that allows for efficient inference and computation.

Recall from §2 that context-free grammars generate labeled, ordered trees. Given sets of nonterminals N and terminals T, let Ψ be the set of finite trees with:

(a) root node labeled S;
(b) leaf nodes labeled with elements of T;
(c) interior nodes labeled with elements of N;
(d) every nonterminal (interior) node having either two children labeled with nonterminals or one child labeled with a terminal.

Every $\psi \in \Psi$ defines a sentence $w \in T^+$: read the labels off of the terminal nodes of ψ from left to right. Consistent with the notation of §3.1, we will write $\mathcal{Y}(\psi) = w$. Conversely, every sentence $w \in T^+$ defines a subset of Ψ, which we denote by Ψ_w, consisting of all ψ with yield w ($\mathcal{Y}(\psi) = w$). A context-free grammar G defines a subset of Ψ, Ψ_G, whose collection of yields is the language, L_G, of G. We seek a probability distribution P on Ψ which concentrates on Ψ_G.

The time-honored approach to probabilistic context-free grammars is through the production probabilities $p : R \to [0,1]$, with

$$\sum_{\substack{\alpha \in N^2 \cup T \\ \text{s.t. } (A \to \alpha) \in R}} p(A \to \alpha) = 1 \ . \tag{9}$$

Following the development in §3.1, we introduce a counting function $f(A \to \alpha; \psi)$, which counts the number of instances of the rule $A \to \alpha$ in the tree ψ, i.e. the number of nonterminal nodes A whose daughter nodes define, left-to-right, the string α. Through f, p induces a probability P on Ψ:

$$P(\psi) = \prod_{(A \to \alpha) \in R} p(A \to \alpha)^{f(A \to \alpha; \psi)} \ . \tag{10}$$

It is clear enough that P concentrates on Ψ_G, and we shall see shortly that this parameterization, in terms of products of probabilities p, is particularly workable and convenient. The pair, G and P, is known as a probabilistic context-free grammar, or PCFG for short.

Branching Processes and Criticality. Notice the connection to branching processes [Harris, 1963]: Starting at S, use R, and the associated probabilities $p(\cdot)$, to expand nodes into daughter nodes until all leaf nodes are labeled with terminals (elements of T). Since branching processes display critical behavior, whereby they may or may not terminate with probability one, we should ask ourselves whether p truly defines a probability on Ψ_G—bearing in mind that Ψ includes only finite trees. Evidently, for p to induce a probability on Ψ_G ($P(\Psi_G) = 1$), the associated branching process must terminate with probability one. This may not happen, as is most simply illustrated by a bare-boned example:

Example 6. $G_6 = (T_6, N_6, \mathrm{S}, R_6)$, $T_6 = \{\mathrm{a}\}$, $N_6 = \{\mathrm{S}\}$, and R_6 includes only

$$\begin{array}{l} \mathrm{S} \to \mathrm{S\,S} \\ \mathrm{S} \to \mathrm{a}\ . \end{array}$$

Let $p(\mathrm{S} \to \mathrm{S\,S}) = q$ and $p(\mathrm{S} \to \mathrm{a}) = 1 - q$, and let S_h be the total probability of all trees with depth less than or equal to h. Then $S_2 = 1 - q$ (corresponding to $\mathrm{S} \to \mathrm{a}$) and $S_3 = (1-q) + q(1-q)^2$ (corresponding to $\mathrm{S} \to \mathrm{a}$ or $\mathrm{S} \to \mathrm{S\,S}$ followed by $\mathrm{S} \to \mathrm{a}$, $\mathrm{S} \to \mathrm{a}$). In general, $S_{h+1} = 1 - q + qS_h^2$, which is nonincreasing in q and converges to $\min(1, \frac{1-q}{q})$ as $q \uparrow 1$. Hence $P(\Psi_G) = 1$ if and only if $q \leq .5$.

More generally, it is not difficult to characterize production probabilities that put full mass on finite trees (so that $P(\Psi_G) = 1$), see for example [Grenander, 1976] or [Harris, 1963]. But the issue is largely irrelevant, since maximum likelihood estimated probabilities *always* have this property, as we shall see shortly.

3.2.2. Inference. As with probabilistic regular grammars, the production probabilities of a context-free grammar, which amount to a parameterization of the distribution P on Ψ_G, can be estimated from examples. In one scenario, we have access to a sequence $\psi_1, \ldots, \psi_n$ from Ψ_G under P. This is "supervised learning," in the sense that sentences come equipped with parses. More interesting is the problem of "unsupervised learning," wherein we observe only the yields, $\mathcal{Y}(\psi_1), \ldots, \mathcal{Y}(\psi_n)$.

In either case, the treatment of maximum likelihood estimation is essentially identical to the treatment for regular grammars. In particular, the likelihood for fully observed data is again (2), and the maximum likelihood estimator is again the relative frequency estimator (3). And, in the unsupervised case, the likelihood is again (4) and this leads to the same EM-type iteration given in (7).

Criticality. We remarked earlier that the issue is largely irrelevant. This is because estimated probabilities $\hat{p}$ are *always* proper probabilities: $\hat{P}(\Psi) = 1$ whenever $\hat{P}$ is induced by $\hat{p}$ computed from (3) or *any* iteration of (7) [Chi and Geman, 1998].

3.2.3. Computation. There are four basic computations: find the probability of a sentence $w \in T^+$, find a $\psi \in \Psi$ (or find *all* $\psi \in \Psi$) satisfying $\mathcal{Y}(\psi) = w$ ("parsing"); find

$$\arg \max_{\substack{\psi \in \Psi \text{ s.t.} \\ \mathcal{Y}(\psi) = w}} P(\psi)$$

("maximum *a posteriori*" or "optimal" parsing); compute expectations of the form $E_{\hat{p}_t}[f(A \to \alpha; \psi) | \psi \in \Psi_w]$ that arise in iterative estimation schemes like (7). The four computations turn out to be more-or-less the same, as was the case for regular grammars (§3.1.3), and there is a common dynamic-programming-like solution [Lari and Young, 1990, Lari and Young, 1991].

We illustrate with the problem of finding the probability of a string (sentence) w, under a grammar G, and under a probability distribution P concentrating on Ψ_G. For PCFGs, the dynamic-programming algorithm involves a recursion over *substrings* of the string w to be parsed. If $w = w_1 \ldots w_m$ is the string to be parsed, then let $w_{i,j} = w_i \ldots w_j$ be the substring consisting of terminals i through j, with the convention that $w_{i,i} = w_i$. The dynamic-programming algorithm works from smaller to larger substrings $w_{i,j}$, calculating the probability that $A \Rightarrow^\star w_{i,j}$ for each nonterminal $A \in N$. Because a substring of length 1 can only be generated by a unary production of the form $A \to x$, for each $i = 1, \ldots, m$, $P(A \Rightarrow^\star w_{i,i}) = p(A \to w_i)$. Now consider a substring $w_{i,j}$ of length 2 or greater. Consider any derivation $A \Rightarrow^\star w_{i,j}$. The first production used must be a binary production of the form $A \to B\,C$, with $A, B, C \in N$. That is, there must be a k between i and j such that $B \Rightarrow^* w_{i,k}$ and $C \Rightarrow^* w_{k+1,j}$. Thus the dynamic programming step involves iterating from smaller to larger substrings $w_{i,j}, 1 \leq i, j \leq m$, calculating:

$$P(A \Rightarrow^\star w_{i,j}) = \sum_{\substack{B,C \in N \\ s.t.\ A \to B\,C \in R}} p(A \to B\,C) \sum_{k=i,j-1} P(B \Rightarrow^\star w_{i,k}) P(C \Rightarrow^\star w_{k+1,j}) \,.$$

At the end of this iteration, $P(w) = P(S \Rightarrow^\star w_{1,m})$. This calculation involves applying each production once for each triple of "string positions" $0 < i \leq k < j \leq m$, so the calculation takes $O(|R|m^3)$ time.

3.3. Gibbs distributions. There are many ways to generalize. The coverage of a context-free grammar may be inadequate, and we may hope, therefore, to find a workable scheme for placing probabilities on context-sensitive grammars, or perhaps even more general grammars. Or, it may be preferable to maintain the structure of a context-free grammar, especially because of its dynamic programming principle, and instead generalize the class of probability distributions away from those induced (parameterized) by production probabilities. But nothing comes for free. Most efforts to generalize run into nearly intractable computational problems when it comes time to parse or to estimate parameters.

Many computational linguists have experimented with using Gibbs distributions, popular in statistical physics, to go beyond production-based probabilities, while nevertheless preserving the basic context-free structure. We shall take a brief look at this particular formulation, in order to illustrate the various challenges that accompany efforts to generalize the more standard probabilistic grammars.

3.3.1. Probabilities. The sample space is the same: Ψ is the set of finite trees, rooted at S, with leaf nodes labeled from elements of T and interior nodes labeled from elements of N. For convenience we will stick to Chomsky normal form, and we can therefore assume that every nonterminal node has either two children labeled from N or a single child labeled from T. Given a particular context-free grammar G, we will be interested in measures concentrating on the subset Ψ_G of Ψ. The sample space, then, is effectively Ψ_G rather than Ψ.

Gibbs measures are built from sums of more-or-less simple functions, known as "potentials" in statistical physics, defined on the sample space. In linguistics, it is more natural to call these *features* rather than potentials. Let us suppose, then, that we have identified M linguistically salient features $f_1, \ldots, f_M$, where $f_k : \Psi_G \to R$, through which we will characterize the fitness or appropriateness of a structure $\psi \in \Psi_G$. More specifically, we will construct a class of probabilities on Ψ_G which depend on $\psi \in \Psi_G$ only through $f_1(\psi), \ldots, f_M(\psi)$. Examples of features are the number of times a particular production occurs, the number of words in the yield, various measures of subject-verb agreement, and the number of embedded or independent clauses.

Gibbs distributions have the form

$$P_\theta(\psi) = \frac{1}{Z} \exp\left\{\sum_{i=1}^{M} \theta_i f_i(\psi)\right\} \tag{11}$$

where $\theta_1 \ldots, \theta_M$ are parameters, to be adjusted "by hand" or inferred from data, $\theta = (\theta_1 \ldots, \theta_M)$, and where $Z = Z(\theta)$ (known as the "partition function") normalizes so that $P_\theta(\Psi) = 1$. Evidently, we need to assume or ensure that $\sum_{\psi \in \Psi_G} \exp\{\sum_1^M \theta_i f_i(\psi)\} < \infty$. For instance, we had better require that $\theta_1 < 0$ if $M = 1$ and $f_1(\psi) = |\mathcal{Y}(\psi)|$ (the number of words in a sentence), unless of course $|\Psi_G| < \infty$.

Relation to Probabilistic Context-Free Grammars. The feature set $\{f(A \to \alpha; \psi)\}_{A \to \alpha \in R}$ represents a particularly important special case: The Gibbs distribution (11) takes on the form

$$P_\theta(\psi) = \frac{1}{Z} \exp\left\{\sum_{A \to \alpha \in R} \theta_{A \to \alpha} f(A \to \alpha; \psi)\right\}. \tag{12}$$

Evidently, we recover probabilistic context-free grammars by taking $\theta_{A \to \alpha} = \log_e p(A \to \alpha)$, where p is a system of production probabilities consistent with (9), in which case $Z = 1$. But is (12) *more general*? Are there probabilities on Ψ_G of this form that are not PCFGs? The answer turns out to be no, as was shown by [Chi, 1999] and [Abney et al., 1999]: Given a probability distribution P on Ψ_G of the form of (12), there always exists a system of production probabilities p under which P is a PCFG.

One interesting consequence relates to the issue of criticality raised in §3.2.1. Recall that a system of production probabilities p may define (through 10) an improper probability P on Ψ_G: $P(\Psi_G) < 1$. In these cases it is tempting to simply renormalize, $\bar{P}(\psi) = p(\psi)/P(\Psi_G)$, but then what kind of distribution is $\bar{P}$? It is clear enough that $\bar{P}$ is Gibbs with feature set $\{f(A \to \alpha; \psi)\}_{A \to \alpha \in R}$, so it must also be a PCFG, by the result of Chi and Abney et al. What are the new production probabilities, $\bar{p}(\cdot)$?

For each $A \in N$, consider the grammar G_A which "starts at A," i.e. replace S, the start symbol, by A. If Ψ_A is the resulting set of tree structures (rooted at A), then (12) defines a measure P_A on Ψ_A, which will have a new normalization Z_A. Consider now the production $A \to BC$,

$A, B, C \in N$. Chi's proof of the equivalence between PCFGs and Gibbs distributions of the form (12) is constructive:

$$p(A \to B\,C) = \frac{Z_B Z_C}{Z_A} e^{\theta_{A \to B\,C}}$$

is, explicitly, the production probability under which P is a PCFG. For a terminal production, $A \to a$,

$$p(A \to a) = \frac{1}{Z_A} e^{\theta_{A \to a}} .$$

Consider again example 6, in which $S \to S\,S$ with probability q and $S \to a$ with probability $1 - q$. We calculated $P(\Psi_G) = \min(1, \frac{1-q}{q})$, so renormalize and define

$$\bar{P}(\psi) = \frac{P(\psi)}{\min(1, \frac{1-q}{q})} \qquad \psi \in \Psi_G .$$

Then $\bar{P} = P$ when $q \leq .5$. In any case, $\bar{P}$ is Gibbs of the form (12), with $\theta_{S \to S\,S} = \log_e q$, $\theta_{S \to a} = \log_e(1 - q)$, and $Z_S = \min(1, \frac{1-q}{q})$. Accordingly, $\bar{P}$ is also a PCFG with production probabilities

$$\bar{p}(S \to S\,S) = q \frac{\min(1, \frac{1-q}{q}) \min(1, \frac{1-q}{q})}{\min(1, \frac{1-q}{q})} = q \min(1, \frac{1-q}{q})$$

and

$$\bar{p}(S \to a) = (1 - q) \frac{1}{\min(1, \frac{1-q}{q})} .$$

In particular, $\bar{p} = p$ when $q \leq .5$, but $\bar{p}(S \to S\,S) = 1 - q$ and $\bar{p}(S \to a) = q$ when $q > .5$.

3.3.2. Inference. The feature set $\{f_i\}_{i=1,\ldots,M}$ can accommodate arbitrary linguistic attributes and constraints, and the Gibbs model (11), therefore, has great promise as an accurate measure of linguistic fitness. But the model depends critically on the parameters $\{\theta_i\}_{i=1,\ldots,M}$, and the associated estimation problem is, unfortunately, very hard. Indeed, the problem of unsupervised learning appears to be all but intractable.

Of course if the features are simply production frequencies, then (11) is just a PCFG, with $Z = 1$, and the parameters are just log production probabilities ($\theta_{A \to \alpha} = \log_e p(A \to \alpha)$) and easy to estimate (see §3.2.2). More generally, let $\theta = (\theta_1, \ldots, \theta_M)$ and suppose that we observe a sample $\psi_1, \ldots \psi_n \in \Psi_G$ ("supervised learning"). Writing Z as $Z(\theta)$, to emphasize the dependency of the normalizing constant on θ, the likelihood function is

$$L = L(\theta; \psi_1 \ldots, \psi_n) = \prod_{i=1}^{n} \frac{1}{Z(\theta)} \exp\left\{ \sum_{j=1}^{M} \theta_j f_j(\psi_i) \right\}$$

which leads to the likelihood equations (by setting $\frac{\partial}{\partial \theta_j} \log L$ to 0):

$$E_\theta[f_j(\psi)] = \frac{1}{n} \sum_{i=1}^{n} f_j(\psi_i) \qquad 1 \leq j \leq N \tag{13}$$

where E_θ is expectation under (11). In general, Ψ_G is infinite and depending on the features $\{f_i\}_{i=1,\ldots,M}$ and the choice of θ, various sums (like $Z(\theta)$ and E_θ) could diverge and be infinite. But if these summations converge, then the likelihood function is concave. Furthermore, unless there is a linear dependence among $\{f_i\}_{i=1,\ldots,M}$ on $\{\psi_i\}_{i=1,\ldots,n}$, then the likelihood is in fact strictly concave, and there is a unique solution to (13). (If there is a linear dependence, then there are infinitely many θ values with the same likelihood.)

The favorable shape of $L(\theta; \psi_1 \ldots, \psi_n)$ suggests gradient ascent, and in fact the θ_j component of the gradient is proportional to $\frac{1}{n}\sum_{i=1}^{n} f_j(\psi_i) - E_\theta[f_j(\psi)]$. But $E_\theta[f]$ is difficult to compute (to say the least), except in some very special and largely uninteresting cases. Various efforts to use Monte Carlo methods to approximate $E_\theta[f]$, or related quantities that arise in other approaches to estimation, have been made [Abney, 1997]. But realistic grammars involve hundreds or thousands of features and complex feature structures, and under such circumstances Monte Carlo methods are notoriously slow to converge. Needless to say, the important problem of *unsupervised* learning, wherein only yields are seen, is even more daunting.

This state of affairs has prompted a number of suggestions in way of compromise and approximation. One example is the method of pseudolikelihood, which we will now discuss.

Pseudolikelihood. If the primary goal is to select good parses, then perhaps the likelihood function

$$\prod_{i=1}^{n} P_\theta(\psi_i) \tag{14}$$

asks for too much, or even the wrong thing. It might be more relevant to maximize the likelihood of the observed parses, *given the yields* $\mathcal{Y}(\psi_1), \ldots, \mathcal{Y}(\psi_n)$ [Johnson et al., 1999]:

$$\prod_{i=1}^{n} P_\theta(\psi_i | \mathcal{Y}(\psi_i)) \ . \tag{15}$$

One way to compare these criteria is to do some (loose) asymptotics. Let $P(\psi)$ denote the "true" distribution on Ψ_G (from which $\psi_1, \ldots, \psi_n$

are presumably drawn, iid), and in each case (14 and 15) compute the large-sample-size average of the log likelihood:

$$\begin{aligned}
\frac{1}{n}\log\prod_{i=1}^{n} P_\theta(\psi_i) &= \frac{1}{n}\sum_{i=1}^{n}\log P_\theta(\psi_i) \\
&\approx \sum_{\psi\in\Psi_G} P(\psi)\log P_\theta(\psi) \\
&= \sum_{\psi\in\Psi_G} P(\psi)\log P(\psi) - \sum_{\psi\in\Psi_G} P(\psi)\log\frac{P(\psi)}{P_\theta(\psi)}
\end{aligned}$$

$$\begin{aligned}
\frac{1}{n}\log\prod_{i=1}^{n} P_\theta(\psi_i|\mathcal{Y}(\psi_i)) &= \frac{1}{n}\sum_{i=1}^{n}\log P_\theta(\psi_i|\mathcal{Y}(\psi_i)) \\
&\approx \sum_{w\in T^+} P(w) \sum_{\substack{\psi\in\Psi_G \\ \text{s.t. } \mathcal{Y}(\psi)=w}} P(\psi|\mathcal{Y}(\psi))\log P_\theta(\psi|\mathcal{Y}(\psi)) \\
&= \sum_{w\in T^+} P(w) \sum_{\substack{\psi\in\Psi_G \\ \text{s.t. } \mathcal{Y}(\psi)=w}} P(\psi|\mathcal{Y}(\psi))\log P(\psi|\mathcal{Y}(\psi)) \\
&\quad - \sum_{w\in T^+} P(w) \sum_{\substack{\psi\in\Psi_G \\ \text{s.t. } \mathcal{Y}(\psi)=w}} P(\psi|\mathcal{Y}(\psi))\log\frac{P(\psi|\mathcal{Y}(\psi))}{P_\theta(\psi|\mathcal{Y}(\psi))}\,.
\end{aligned}$$

Therefore, maximizing the likelihood (14) is more or less equivalent to minimizing the Kullback-Leibler divergence between $P(\psi)$ and $P_\theta(\psi)$, whereas maximizing the "pseudolikelihood" (15) is more or less equivalent to minimizing the Kullback-Leibler divergence between $P(\psi|\mathcal{Y}(\psi))$ and $P_\theta(\psi|\mathcal{Y}(\psi))$—averaged over yields. Perhaps this latter minimization makes more sense, given the goal of producing good parses.

Maximization of (15) is an instance of Besag's remarkably effective *pseudolikelihood method* [Besag, 1974, Besag, 1975], which is commonly used for estimating parameters of Gibbs distributions. The computations involved are generally much easier than what is involved in maximizing the ordinary likelihood function (14). Take a look at the gradient of the logarithm of (15): the θ_j component is proportional to

$$\frac{1}{n}\sum_{i=1}^{n} f_j(\psi_i) - \frac{1}{n}\sum_{i=1}^{n} E_\theta[f_j(\psi)|\mathcal{Y}(\psi)=\mathcal{Y}(\psi_i)]\,. \tag{16}$$

Compare this to the gradient of the likelihood function, which involves $E_\theta[f_j(\psi)]$ instead of $\frac{1}{n}\sum_{i=1}^{n} E_\theta[f_j(\psi)|\mathcal{Y}(\psi)=\mathcal{Y}(\psi_i)]$. $E_\theta[f_j(\psi)]$ is essentially intractable, whereas $E_\theta[f_j(\psi)|\mathcal{Y}(\psi)]$ can be computed directly from the set of parses of the sentence $\mathcal{Y}(\psi)$. (In practice there is often massive ambiguity, and the number of parses may be too large to feasibly consider. Such cases require some form of pruning or approximation.)

Thus gradient ascent of the pseudolikelihood function is (at least approximately) computationally feasible. This is particularly useful since the Hessian of the logarithm of the pseudolikelihood function is non-positive, and therefore there are no local maxima. What's more, under mild conditions pseudolikelihood estimators (i.e. maximizers of (15)) are consistent [Chi, 1998].

4. Generalizations and other directions. There are a large number of extensions and applications of the grammatical tools just outlined. Treebank corpora, which consist of the hand-constructed parses of tens of thousands of sentences, are an extremely important resource for developing stochastic grammars [Marcus et al., 1993]. For example, the parses in a treebank can be used to generate, more or less automatically, a PCFG. Productions can be simply "read off" of the parse trees, and production probabilities can be estimated from relative frequencies, as explained in §3.1.2. Such PCFGs typically have on the order of 50 nonterminals and 15,000 productions. While the average number of parses per sentence is astronomical (we estimate greater than 10^{60}), the dynamic programming methods described in §3.2.3 are quite tractable, involving perhaps only hundreds of thousands of operations.

PCFGs derived from treebanks are moderately effective in parsing natural language [Charniak, 1996]. But the actual probabilities generated by these models (e.g. the probability of a given sentence) are considerably worse than those generated by other much simpler kinds of models, such as trigram models. This is presumably because these PCFGs ignore lexical dependencies between pairs or triples of words. For example, a typical treebank PCFG might contain the productions $\mathrm{VP} \to \mathrm{V\,NP}, \mathrm{V} \to \mathrm{eat}$ and $\mathrm{NP} \to \mathrm{pizza}$, in order to generate the string *eat pizza*. But since noun phrases such as *airplanes* are presumably also generated by productions such as $\mathrm{NP} \to \mathrm{airplanes}$, this grammar also generates unlikely strings such as *eat airplanes*.

One way of avoiding this difficulty is to *lexicalize* the grammar, i.e., to "split" the nonterminals so that they encode the "head" word of the phrase that they rewrite to. In the previous example, the corresponding lexicalized productions are $\mathrm{VP_{eat}} \to \mathrm{V_{eat}\,NP_{pizza}}, \mathrm{V_{eat}} \to \mathrm{eat}$ and $\mathrm{NP_{pizza}} \to \mathrm{pizza}$. This permits the grammar to capture some of the lexical selectional preferences of verbs and other heads of phrases for specific head words. This technique of splitting the nonterminals is very general, and can be used to encode other kinds of nonlocal dependencies as well [Gazdar et al., 1985]. In fact, the state of the art probabilistic parsers can be regarded as PCFG parsers operating with very large, highly structured, nonterminals. Of course, this nonterminal splitting dramatically increases the number of nonterminals N and the number of productions R in the grammar, and this complicates both the computational problem [Eisner and Satta, 1999] and, more seriously, inference. While it is straight-forward to lexicalize the productions of a context-free grammar, many or even most productions in the resulting grammar will not actually appear even in a large treebank. Developing methods for accurately estimating the probability of such productions by somehow exploiting the structure of the lexically split nonterminals is a central theme of much of the research in statistical parsing [Collins, 1996, Charniak, 1997].

While most current statistical parsers are elaborations of the PCFG approach just specified, there are a number of alternative approaches that are attracting interest. Because some natural languages are not context-free languages (as mentioned earlier), most linguistic theories of syntax incorporate context-sensitivity in some form or other. That is, according to these theories the set of trees corresponding to the sentences of a human language is not necessarily generated by a context-free grammar, and therefore the PCFG methods described above cannot be used to define a probability distribution over such sets of trees. One alternative is to employ the more general Gibbs models, discussed above in §3.3 (see for example [Abney, 1997]). Currently, approaches that apply Gibbs models build on previously existing "unification grammars" [Johnson et al., 1999], but this may not be optimal, as these grammars were initially designed to be used non-stochastically.

REFERENCES

[Abney et al., 1999] ABNEY, STEVEN, DAVID MCALLESTER, AND FERNANDO PEREIRA. 1999. Relating probabilistic grammars and automata. In *Proceedings of the 37th Annual Meeting of the Association for Computational Linguistics*, pages 542–549, San Francisco. Morgan Kaufmann.

[Abney, 1997] ABNEY, STEVEN P. 1997. Stochastic Attribute-Value Grammars. *Computational Linguistics*, **23**(4):597–617.

[Baum, 1972] BAUM, L.E. 1972. An inequality and associated maximization techniques in statistical estimation of probabilistic functions of Markov processes. *Inequalities*, **3**:1–8.

[Besag, 1974] BESAG, J. 1974. Spatial interaction and the statistical analysis of lattice systems (with discussion). *Journal of the Royal Statistical Society, Series D*, **36**:192–236.

[Besag, 1975] BESAG, J. 1975. Statistical analysis of non-lattice data. *The Statistician*, **24**:179–195.

[Charniak, 1996] CHARNIAK, EUGENE. 1996. Tree-bank grammars. In *Proceedings of the Thirteenth National Conference on Artificial Intelligence*, pages 1031–1036, Menlo Park. AAAI Press/MIT Press.

[Charniak, 1997] CHARNIAK, EUGENE. 1997. Statistical parsing with a context-free grammar and word statistics. In *Proceedings of the Fourteenth National Conference on Artificial Intelligence*, Menlo Park. AAAI Press/MIT Press.

[Chi, 1998] CHI, ZHIYI. 1998. *Probability Models for Complex Systems*. PhD thesis, Brown University.

[Chi, 1999] CHI, ZHIYI. 1999. Statistical properties of probabilistic context-free grammars. *Computational Linguistics*, **25**(1):131–160.

[Chi and Geman, 1998] CHI, ZHIYI AND STUART GEMAN. 1998. Estimation of probabilistic context-free grammars. *Computational Linguistics*, **24**(2):299–305.

[Chomsky, 1957] CHOMSKY, NOAM. 1957. *Syntactic Structures*. Mouton, The Hague.

[Collins, 1996] COLLINS, M.J. 1996. A new statistical parser based on bigram lexical dependencies. In *The Proceedings of the 34th Annual Meeting of the Association for Computational Linguistics*, pages 184–191, San Francisco. The Association for Computational Linguistics, Morgan Kaufmann.

[Culy, 1985] CULY, CHRISTOPHER. 1985. The complexity of the vocabulary of Bambara. *Linguistics and Philosophy*, **8**(3):345–352.

[Dempster et al., 1977] DEMPSTER, A., N. LAIRD, AND D. RUBIN. 1977. Maximum likelihood from incomplete data via the EM algorithm. *Journal of the Royal Statistical Society, Series B*, **39**:1–38.

[Eisner and Satta, 1999] EISNER, JASON AND GIORGIO SATTA. 1999. Efficient parsing for bilexical context-free grammars and head automaton grammars. In *Proceedings of the 37th Annual Meeting of the Association for Computational Linguistics*, pages 457–464, 1999.

[Foo, 1974] FOO, K.S. 1974. *Syntactic Methods in Pattern Recognition*. Academic Press.

[Foo, 1982] FOO, K.S. 1982. *Syntactic Pattern Recognition and Applications*. Prentice-Hall.

[Garey and Johnson, 1979] GAREY, MICHAEL R. AND DAVID S. JOHNSON. 1979. *Computers and Intractability: A Guide to the Theory of NP-Completeness*. W.H. Freeman and Company, New York.

[Gazdar et al., 1985] GAZDAR, GERALD, EWAN KLEIN, GEOFFREY PULLUM, AND IVAN SAG. 1985. *Generalized Phrase Structure Grammar*. Basil Blackwell, Oxford.

[Grenander, 1976] GRENANDER, ULF. 1976. *Lectures in Pattern Theory. Volume 1: Pattern Synthesis*. Springer, Berlin.

[Harris, 1963] HARRIS, T.E. 1963. *The Theory of Branching Processes*. Springer, Berlin.

[Hopcroft and Ullman, 1979] HOPCROFT, JOHN E. AND JEFFREY D. ULLMAN. 1979. *Introduction to Automata Theory, Languages and Computation*. Addison-Wesley.

[Jelinek, 1997] JELINEK, FREDERICK. 1997. *Statistical Methods for Speech Recognition*. The MIT Press, Cambridge, Massachusetts.

[Johnson et al., 1999] JOHNSON, MARK, STUART GEMAN, STEPHEN CANON, ZHIYI CHI, AND STEFAN RIEZLER. 1999. Estimators for stochastic "unification-based" grammars. In *The Proceedings of the 37th Annual Conference of the Association for Computational Linguistics*, pages 535–541, San Francisco. Morgan Kaufmann.

[Kaplan and Bresnan, 1982] KAPLAN, RONALD M. AND JOAN BRESNAN. 1982. Lexical-Functional Grammar: A formal system for grammatical representation. In Joan Bresnan, editor, *The Mental Representation of Grammatical Relations*, Chapter 4, pages 173–281. The MIT Press.

[Kay et al., 1994] KAY, MARTIN, JEAN MARK GAVRON, AND PETER NORVIG. 1994. *Verbmobil: a translation system for face-to-face dialog*. CSLI Press, Stanford, California.

[Lari and Young, 1990] LARI, K. AND S.J. YOUNG. 1990. The estimation of Stochastic Context-Free Grammars using the Inside-Outside algorithm. *Computer Speech and Language*, **4**(35–56).

[Lari and Young, 1991] LARI, K. AND S.J. YOUNG. 1991. Applications of Stochastic Context-Free Grammars using the Inside-Outside algorithm. *Computer Speech and Language*, **5**:237–257.

[Marcus et al., 1993] MARCUS, MICHELL P., BEATRICE SANTORINI, AND MARY ANN MARCINKIEWICZ. 1993. Building a large annotated corpus of English: The Penn Treebank. *Computational Linguistics*, **19**(2):313–330.

[Pollard and Sag, 1987] POLLARD, CARL AND IVAN A. SAG. 1987. *Information-based Syntax and Semantics*. Number 13 in CSLI Lecture Notes Series. Chicago University Press, Chicago.

[Shieber, 1985] SHIEBER, STUART M. 1985. Evidence against the Context-Freeness of natural language. *Linguistics and Philosophy*, **8**(3):333–344.

[Shieber, 1986] SHIEBER, STUART M. 1986. *An Introduction to Unification-based Approaches to Grammar*. CSLI Lecture Notes Series. Chicago University Press, Chicago.

THREE ISSUES IN MODERN LANGUAGE MODELING

DIETRICH KLAKOW*

Abstract. In this paper we discuss three issues in modern language modeling. The first one is the question of a quality measure for language models, the second is language model smoothing and the third is the question of how to build good long-range language models. In all three cases some results are given indicating possible directions of further research.

Key words. Language models, quality measures, perplexity, smoothing, long-range correlations.

1. Introduction. Language models (LM) are very often a component of speech and natural language processing systems. They assign a probability to any sentence of a language. The use of language models in speech recognition systems is well known for a long time and any modern commercial or academic speech recognizer uses them in one form or other [1]. Closely related is the use in machine translation systems [2] where a language model of the target language is used. Relatively new is the language model approach to information retrieval [3] where the query is the language model history and the documents are to be predicted. It may even be applied to question answering. When asking an open question like "The name of the capital of Nepal is X?" filling the open slot X is just the language modeling task given the previous words of the question as the history.

This paper extends the issues raised by the author being a panelist at the language modeling workshop of the Institute of Mathematics and its Applications. The three sections will discuss the three issues raised in the abstract.

2. Correlation of word error rate and perplexity. How can the value of a language model for speech recognition be evaluated with little effort? Perplexity (PP, it measures the predictive power of a language model) is simple to calculate. It is defined by:

$$\log \mathrm{PP} = -\sum_{w,h} \frac{N_{\mathrm{test}}(w,h)}{N_{\mathrm{test}}(h)} \cdot \log p_{\mathrm{LM}}(w|h) \tag{1}$$

where $N_{\mathrm{test}}(w,h)$ and $N_{\mathrm{test}}(h)$ are frequencies on a test corpus for word w following a history of words h and $p_{\mathrm{LM}}(w|h)$ is the probability assigned to that sequence by the language model.

The possible correlation of word-error-rate (WER, the fraction of words of a text, miss-recognized by a speech recognizer) words) and perplexity has been an issue in the literature for quite some time now, but

*Philips GmbH Forschungslaboratorien, Weisshausstr.2, D-52066 Aachen, Germany (dietrich.klakow@philips.com).

the value of perplexity as a quality measure for language models in speech recognition has been questioned by many people. This triggered the development of new evaluation metrics [4–8]. The topic of a good quality measure for language models deserves much attention for two reasons:

- **Target function of LM development:** a clear and mathematically tractable target function for LM construction allows for mathematically well defined procedures. Such a target function completely defines the problem. Then the LM task is just to find better language model structures and to optimize their free parameters.
- **Fast development of speech recognition systems:** An indirect quality measure (as compared to WER) of LMs allows LM development mostly decoupled from acoustic training and optimizing the recognizer as such. This is essential to speed up the process of setting up new speech recognition systems.

Perplexity is one of the quality measures that used to be very popular but has been questioned during the last years. For the reasons given above perplexity has clear advantages. We only have to know how well it correlates with word-error-rate.

An important new aspect from our investigations [9] is the observation that both perplexity and WER are subject to measurement errors. This is mostly due to the finite size of our test samples. It is straightforward to develop uncertainty measures for WER and PP and details can be found in [9] or any basic mathematics book [10]. For all results shown here we picked a 95% confidence level.

Based on this we can derive uncertainty measures for the correlation coefficient of WER and PP. On top of the uncertainty coming from the measurement for one individual LM we now also have to take into account the number of LMs used to measure the correlation. The more LMs built the better.

Motivated by Fig. 1 our hypothesis is that there is a power-law relation of the form

$$\text{WER} = b\,\text{PP}^a \tag{2}$$

where a and b are free parameters. We found that they depend on the data set (e.g. WSJ or Broadcast News) under investigation.

To actually test the correlation we built 450 different language models using the whole variety of available techniques: backing-off models, linear interpolation, class-models, cache-models and FMA-adapted models [11]. Note, that the language models with the lowest perplexities shown are highly tuned state-of-the-art trigrams. All those models were trained on different portions of Wall Street Journal Corpus (WSJ) which contains about 80000 articles (40 million running words). We use a vocabulary of 64000 words. For testing the adaptation spoke of the 1994 DARPA evaluation was used. There, articles (development-set and test-set with about

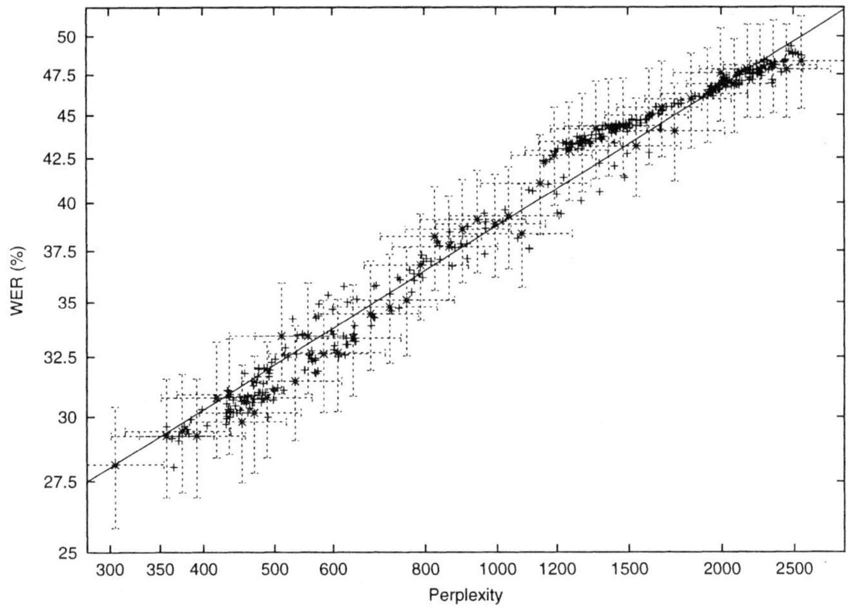

FIG. 1. *Correlation of WER and perplexity tested on data from the DARPA 1994 evaluation (spoke 4). Only a small fraction of the error bars are shown, to keep the power law fit visible.*

2000 words) with special topics (like "Jackie Kennedy") were provided. The task was to optimizes and adapt language models to the particular topics.

The results of our experiments given are in Fig. 1. Each point corresponds to one of the 450 language models. Only a small fraction of the error bars are shown to keep the power law fit visible. We observe that the power law fit nicely runs through the error-bars. The optimal fit parameters in (2) are $a = 0.270 \pm 0.002$ and $b = 6.0 \pm 0.01$. Those are not universal values but they depend on the corpus!

The correlation coefficient is given in Tab. 1. In addition we now also show results for the 1996 and 1997 DARPA Hub4-evaluation data. For a perfect correlation, the correlation coefficient r should be one. We observe that $r = 1$ is always within the error-bars and hence we have no indication that the power-law relation (2) is not valid. Please note that the fact that both values (and all correlation-coefficient values given in the literature) are smaller than one is not the result of a systematic deviation but a fact coming from the definition of the correlation coefficient.

In summary: we have observed no indication, that WER and perplexity are not perfectly correlated. However, these are only two data-sets investigated. We will perform the same analysis on all future speech recog-

TABLE 1
Measured correlation coefficients r and their error bars.

Data	Correlation r
Hub4: 96 + 97	0.978 ± 0.073
DARPA Eval 1994: "Kennedy"	0.993 ± 0.048

nition tasks we are going to work on, but collecting data for a huge number of really different language models is a time-consuming endeavor. We would like to invite others to join the undertaking.

3. Smoothing of language models. Smoothing of language models attracted much attention for a very long time. However for backing-off language models the discussion calmed down during the last few years as most people started to think that there is very little room for further improvement.

A well established method is absolute discounting with marginal backing-off [12]. It is defined by a very simple structure:

$$p(w|h_N)=\begin{cases} \dfrac{\text{Count}(h_N,w)-d}{\text{Count}(h_N)}+\alpha(h_N)\cdot\beta(w|h_{N-1}) & \text{if Count}(h_N,w)>0,\\ \alpha(h_N)\cdot\beta(w|h_{N-1}) & \text{if Count}(h_N,w)=0,\end{cases}$$

with the discounting parameter d ($0 \leq d \leq 1$) and the dedicated backing-off distribution $\beta(w|h)$ which is normalized $\sum_w \beta(w|h) = 1$.

Absolute discounting refers to the fact that d is subtracted from the observed counts. Marginal backing-off means that special backing-off distributions are used rather than smoothed relative frequencies. How to calculate the optimal backing-off distributions was described by Kneser and Ney [12].

Can we do better smoothing? To answer this question we want to first turn to a very basic observation: Zipfs law [13, 14]. To observe this behavior on a corpus, the frequency of the words in the corpus is counted, this list is sorted by frequency and then for each word the position in the list (the rank) is plotted versus the frequency on a doubly logarithmic scale. The result is shown in Fig. 2. We looked at two different "texts". One is the novel "Crime and Punishment" by the Russian poet Dostoevsky. The other is the Philips research speech recognizer, a typical piece of C-code. The observation is that in both cases the statistics is nearly a power law even though the exponents differ.

This was observed for the first time by Zipf and since then several models have been developed, the most well known by Mandelbrot [15]. He used the model of a "string of letters" chopped randomly into pieces. This behavior is very general and can be observed very often in nature. Some examples for systems with power law distributions:

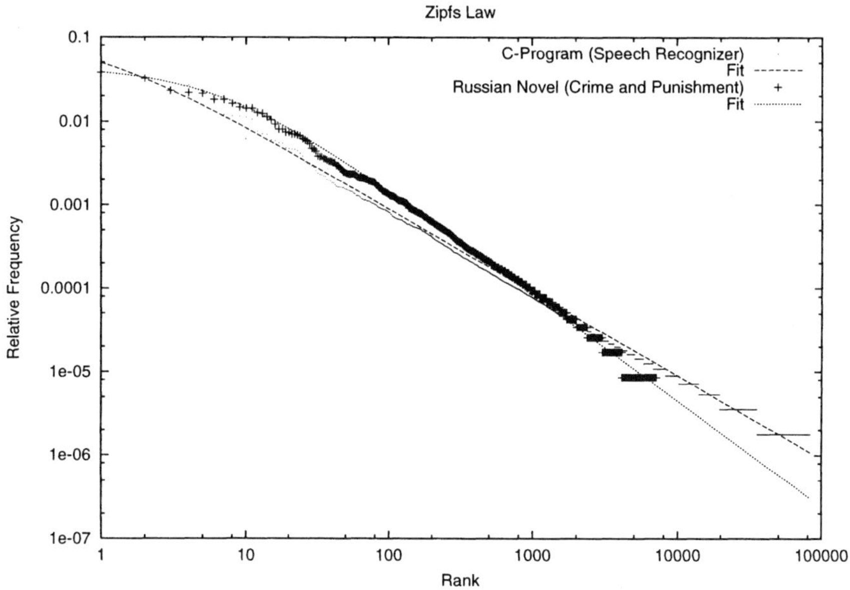

FIG. 2. *Zipf law demonstrated for a natural text and a C-program.*

- Smash a piece of glass like a window pane and do statistics of the size of the fragments
- Measure the file sizes of your hard disc and plot the results accordingly
- Consider on the strength of earth quakes

Always a nearly power law behavior can be observed. The function describing the relation is:

$$f(w) = \frac{\text{Count}(w)}{\text{TotalCount}} \approx \frac{\mu}{(c + r(w))^B} \tag{3}$$

where $r(w)$ is the rank of the word in the sorted list and $f(w)$ the corresponding frequency. This function has two parameters: B and c. Here μ serves as a scaling but can also be used to normalize the distribution. For all the experiments described above only these parameters vary.

We can and should use this observation to create a new smoothed LM type. There is no need to actually estimate probabilities or frequencies. All we have to do is to estimate the three parameters and the rank of the words. This has to be compared with the task of estimating probabilities for about a hundred thousand words. Hence we have very much simplified the estimation problem.

To actually do language modeling we proceed as follows:

- **Estimate the rank of each word:** All words are sorted according to their frequency in the training corpus. In case of equal frequency, a huge background corpus is used to decide which word to rank higher.

- **Obtain Probabilities:** The actual probabilities can either be estimated from relative frequency in a huge background corpus or from (3) where the parameters are estimated on the training corpus.

The experiments were again performed on the DARPA Spoke 4 data from the 1994 evaluation. In addition to the topic "Kennedy" we also used stories about "Korea". The background corpus are the 40 million words from Wall Street journal. The results are given in Tab. 2. Training on the background corpus WSJ gives very high perplexities (first line). The domain specific training material supplied for the evaluation gave much better results (next two lines). But the models are undertrained as most words from the 64000-word vocabulary are not observed in this small 2000-running-words training data. When we now perform the sorting procedure as described above and use the probabilities from the full WSJ-corpus we get a very significant improvement (last but one line) and using the Mandelbrot function (3) with parameters tuned on the domain specific adaptation corpus gives an additional boost (last line).

We can conclude that we have demonstrated a new smoothing method which doesn't have any zero-frequency problem because it uses ranks of words and no probabilities have to be estimated. We did this to show that there is still room for improved smoothing.

TABLE 2
Improved smoothing of unigrams (BO: Backing-off smoothing).

Model	$PP_{Kennedy}$	PP_{Korea}
BO Unigram WSJ	2318	1583
BO Unigram "Kennedy"	1539	-
BO Unigram "Korea"	-	1277
Zipf Unigram (WSJ)	1205	892
Zipf Unigram (Fit)	1176	886

4. Modeling long range dependencies. So far trigrams are very popular in speech recognition systems. They have a very limited context but they seem to work quite well. Still, speech recognition systems and statistical translation systems tend to produce output that locally looks reasonable but globally is inconsistent and humans can easily spot this. One approach to cure this is the combination of grammars with trigrams [16, 17]. But this is not the only way to approach long-range dependencies in language.

To motivate our approach we want to start with simple observations from the Wall Street Journal Corpus, which we also hold true on other corpora (British National Corpus, Broadcast News, Verbmobil,...).

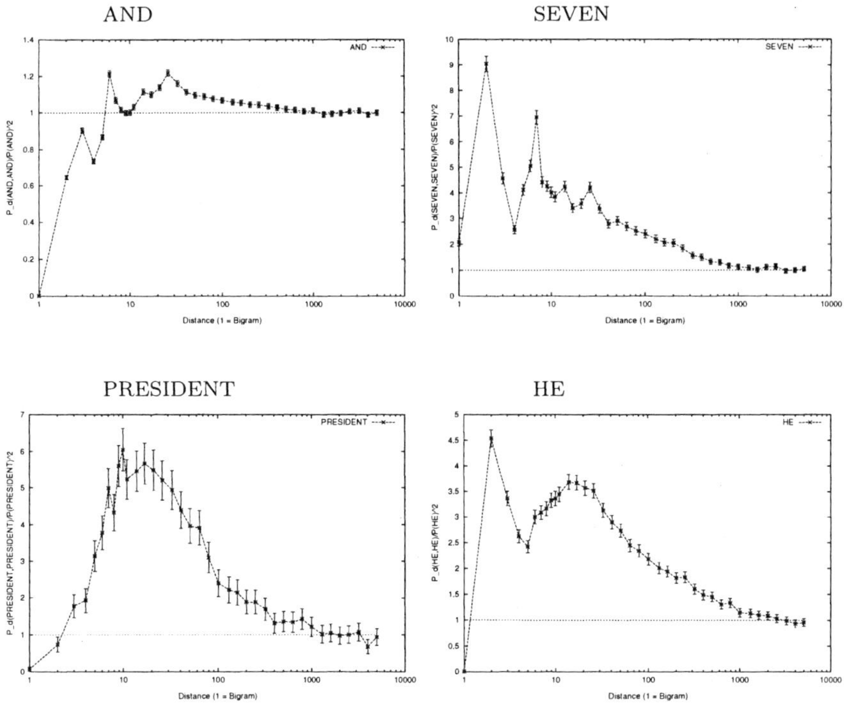

FIG. 3. *Pair auto-correlation functions.*

In Fig. 3 the pair auto-correlation function

$$c_d(w) = \frac{p_d(w,w)}{p(w)^2} \tag{4}$$

is given for four example words. Here, $p_d(w,w)$ is the probability that the word w occurs now and is observed again after skipping d words and $p(w)$ is the unigram distribution. We have the obvious property

$$\lim_{d\to\infty} c_d(w) = 1 \ . \tag{5}$$

For four different words Fig. 3 shows that it is indeed the case that after skipping about a thousand words in between the value 1 is reached. However each word has its individual pattern as to how it approaches this limit. A short-function word like "and" shows at short distances a strong anti-correlation and then approaches this limit rapidly. "President" shows a broad bump stemming from purely semantic relations within one newspaper article. The other two examples show mixed behavior. The very short range positive correlation for "seven" comes from patterns like 7x7 where x is another digit, which relates to references to Boeing airplanes. In general, we observed a very individual pair-correlation function for every word and also each pair-correlation of any pair of words has its own characteristics.

We developed a method to combine generalized pair correlation functions which we called log-linear interpolation[18]. It is a generalization of an adaptation technique we proposed in [11]. Related work on maximum-entropy models can be found in [19] and [20]. Log-linear interpolation can be viewed as a simplified version of maximum-entropy models and is defined by

$$p(w|h) = \frac{1}{Z_\lambda(h)} \prod_i p_i(w|h)^{\lambda_i} \tag{6}$$

where p_i are the different component language models to be combined and $Z_\lambda(h)$ is the normalization. The free parameters to be optimized are the λ_i. The component language models may be usual trigram or distance bigrams where one word at a certain distance in the history predicts w. This would model the same information as measured by the pair-correlation function. Of course the component models could also be distance trigrams or even higher order models of any pattern of skipped words.

To build a general long-range language model using log-linear interpolation we propose the following scheme:

- Start with a base model like a trigram.
- Add all distance bigram models up to a defined range and optimize the parameters (i.e. exponents) using a maximum likelihood criterion.
- For all possible skip patterns look at the distance trigrams that add information on top of the already existing model built in the previous step. This algorithm should work like the feature selection algorithms proposed for maximum entropy language models [21].
- Add all selected distance trigrams to the model built so far and optimize the exponents.
- ...repeat the previous two steps for all possible distance 4-grams, 5-grams...

To get an impression about the potential of the above described scheme we built a language model with an effective 10-gram context. As the base model we used a backing-off 5-gram and combined it with distance bigrams and distance trigrams. We have used the specialized structure

$$p(w|h) = \frac{1}{Z_\lambda(h)} p_5(w|h)^{\lambda_5} \prod_{i=0}^{8} \left(\frac{p_2^i(w|h_{-i})}{p(w)} \right)^{\lambda_2^i} \prod_{j=3}^{7} \left(\frac{p_3^j(w|h_0 h_{-j-1})}{p(w|h_0)} \right)^{\lambda_3^j} \tag{7}$$

where h_0 is the word immediately preceding w and all older words of the history have negative indices. Also, λ_2^i and λ_3^j are the exponents of the i-th distance-bigram and the j-th distance-trigram respectively.

In Tab. 3 we give the results for increasing context length. The results are produced on the WSJ-corpus. The training corpus is the same as described in the previous section but the test data is a closed vocabulary task again from Wall Street Journal with a vocabulary of 5000 words. We observe a steady improvement when increasing the context. The bigram and trigram are traditional backing-off models. The other models follow the formula (7) only the upper bounds of the products and the range of the base 5-gram may be adjusted to the effective language model range. The 10-gram has a perplexity 30% lower than the trigram. The corresponding speech recognition experiments are described in [22].

TABLE 3
Perplexities for an increasing language model range.

LM-Range	2	3	4	6	10
PP	112.7	60.4	50.4	45.4	43.3

Given the scheme described above several questions arise

- How does this scheme perform in general?
- Is this scheme simpler than using a grammar?
- Is it more powerful than using a grammar?
- What happens if such a model and a grammar are combined?

5. Conclusion. In conclusion we can observe that language modeling as a research topic has still important unresolved issues. Three of them have been illustrated:

- Perplexity as a quality measure may be better than often thought. In particular a careful correlation analysis shows that there is no indication that word-error-rate and perplexity are not correlated. However investigations on other corpora, on other languages and other groups should be done along the lines outlined.
- The smoothing of language models is not a closed topic. We demonstrated by a simple toy language model that even unigrams can be dramatically improved by better smoothing.
- Long-range language models will lead to more consistent output of our systems. We suggested a method of modeling but its relation to grammars needs further investigation.

Acknowledgment. The author would like to thank Jochen Peters, Roni Rosenfeld and Harry Printz for many stimulating discussions.

REFERENCES

[1] F. JELINEK, *Statsitical Method's for Speech Recognition*, The MIT Press, 1997.
[2] H. SAWAF, K. SCHUTZ, AND H. NEY, *On the Use of Grammar Based Language Models for Statistical Machine Translation*, Proc. 6th Intl. Workshop on Parsing Technologies, 2000, pp. 231.
[3] J. PONTE AND W. CROFT, *A Language Modeling Approach to Information Retrieval*, Research and Development in Information Retrieval, 1998, pp. 275.

[4] P. Clarkson and T. Robinson, *Towards Improved Language Model Evaluation Measures*, Proc. Eurospeech, 1999, pp. 2707.

[5] A. Ito, M. Kohda, and M. Ostendorf, *A new Metric for stochastic Language Model Evaluation*, Proc. Eurospeech, 1999, pp. 1591.

[6] R. Iyer, M. Ostendorf, and M. Meteer *Analyzing and Predicting Language Model Improvements*, Proc. ASRU, 1997, pp. 254.

[7] S. Chen, D. Beeferman, and R. Rosenfeld, *Evaluation metrics for language models*, Proc. DARPA Broadcast news transcription and understanding workshop, 1998.

[8] H. Printz and P. Olsen, *Theory and Practice of Acoustic Confusability*, Proc. ASR, 2000, pp. 77.

[9] D. Klakow and J. Peters, *Testing the Correlation of Word Error Rate and Perplexity*, accepted for publication in Speech Communications.

[10] I.N. Bronshtein and K.A. Semendyayev, *Handbook of Mathematics*, Springer, 1997.

[11] R. Kneser, J. Peters, and D. Klakow, *Language Model Adaptation using Dynamic Marginals*, Proc Eurospeech, 1997, pp. 1971.

[12] R. Kneser and H. Ney, *Improved backing-off for m-gram language modeling*, Proc. ICASSP, 1995, pp. 181.

[13] G. Zipf, *The Psycho-biology of Language*, Houghton Mifflin, Boston, 1935.

[14] C. Manning and H. Schütze, *Foundation of Statistical Natural Language Processing*, The MIT Press, 1999.

[15] B. Mandelbrot, *The Fractal Geometry of Nature*, W.H. Freeman and Company, 1977.

[16] C. Chelba and F. Jelinek, *Exploiting syntactic structure for language modeling*, Proc. of COLING-ACL, 1998.

[17] E. Charniak, *Immediate-Head Parsing for Language Models*, Proc. ACL, 2001, pp. 124.

[18] D. Klakow, *Log-Linear Interpolation of Language Models*, Proc. ICSLP, 1998, pp. 1695.

[19] R. Rosenfeld, *A Maximum Entropy Approach to Adaptive Statistical Language Modeling*, Computer, Speech, and Language, 1996, pp. 10.

[20] S. Chen, K. Seymore, and R. Rosenfeld, *Topic Adaptation for Language Modeling Using Unnormalized Exponential Models*, Proc. ICASSP, 1998, pp. 681.

[21] H. Printz, *Fast Computation of Maximum Entropy/Minimum Divergence Feature Gain*, Proc. ICSLP, 1998, pp. 2083.

[22] C. Neukirchen, D. Klakow, and X. Aubert *Generation and Expansion of Word Graphs using Long Span Context Information*, Proc. ICASSP, 2001, pp. 41.

STOCHASTIC ANALYSIS OF STRUCTURED LANGUAGE MODELING

FREDERICK JELINEK*

Abstract. As previously introduced, the Structured Language Model (SLM) operated with the help of a stack from which less probable sub-parse entries were purged before further words were generated. In this article we generalize the CKY algorithm to obtain a chart which allows the direct computation of language model probabilities thus rendering the stacks unnecessary. An analysis of the behavior of the SLM leads to a generalization of the Inside – Outside algorithm and thus to rigorous EM type re-estimation of the SLM parameters. The derived algorithms are computationally expensive but their demands can be mitigated by use of appropriate thresholding.

1. Introduction. The structured language model (SLM) was developed to allow a speech recognizer to assign a priori probabilities to words and do so based on a wider past context than is available to the state-of-the-art trigram language model. It is then not surprising that the use of the SLM results in lower perplexities and lower error probabilities [1, 2].[1]

The SLM generates a string of words $w_0, w_1, w_2, ..., w_n, w_{n+1}$, where w_i, $i = 1, ..., n$ are elements of a vocabulary $\mathcal{V}$, $w_0 = < s >$ (the beginning of sentence marker) and $w_{n+1} = < /s >$ (the end of sentence marker). During its operation, the SLM also generates a *parse* consisting of a binary tree whose nodes are marked by *headwords*. The headword at the apex of the final tree is $< s >$. The tree structure and its headwords arise from the operation of a SLM component called the *constructor* (see below).

When its performance was previously evaluated [1], the SLM's operation involved a set of stacks containing partial parses, the less probable of the latter were being purged. The statistical parameters of that version of the SLM were trained by a re-estimation procedure based on $N-$best final parses.

This article deals with the stochastic properties of the SLM that, for the sake of exposition, is at first somewhat simplified as compared with the full-blown SLM presented previously [1]. This simplified SLM (SSLM) is fully lexical – no non-terminals or part-of-speech tags are used. As a direct consequence, the resulting parses contain no unary branches. All simplifications are removed in Section 6 and all algorithms are extended to the complete SLM version introduced in earlier publications [1, 2].

Among the stochastic properties with which we will be concerned are the following ones:

- The probability of the generated sentence based on a generalization of the CKY algorithm [4–6].

*Center for Language and Speech Processing, Johns Hopkins University, 3400 N. Charles St., Baltimore, MD 21218.

[1]Additional results related to the original SLM formulation can be found in the following references: [11–19].

- The probability of the next word given the sentence prefix.
- The probability of the most probable parse.
- A full EM style re-estimation algorithm for the statistical parameters underlying the SLM – a generalization of the Inside – Outside algorithm [7].
- Various subsidiary probabilities necessary for the computation of the above quantities of interest.

The algorithms of this article allow the running of the SLM without the use of any stacks thus, unfortunately, increasing the required computational load. Thus we pay with complexity for rigor.

2. A simplified structured language model. The simplified structured language model (SSLM) generates a string of words $w_0, w_1, w_2, ..., w_n$, w_{n+1}, where w_i, $i = 1, ..., n$ are elements of a vocabulary $\mathcal{V}$, $w_0 =< s >$ (the beginning of sentence marker) and $w_{n+1} =< /s >$ (the end of sentence marker). During its operation, the SSLM also generates a *parse* consisting of a binary tree whose nodes are marked by *headwords*. The headword at the apex of the final tree will be $< s >$. Headwords arise from the operation of the *constructor* (see below).

A node of a parse tree dominates a phrase consisting of the sequence of words associated with the leaves of the sub-tree emanating from the node. Intuitively and ideally, the headword associated with the node should be that word of the phrase (dominated by the node) which best represents the entire phrase and could function as that phrase. Immediately after a word is generated it belongs only to its own phrase whose headword is the word itself.

The SSLM operates left–to–right, building up the parse structure in a bottom–up manner. At any given stage of the word generation by the SSLM, the *exposed* headwords are those headwords of the current partial parse which are not (*yet*) part of a higher phrase with a head of its own (i.e., are not the progeny of another headword). Thus, in Figure 1, at the time just after the word AS is generated, the exposed headwords are $< s >$, SHOW, HAS, AS. We now specify precisely the operation of the SSLM. It is based on constructor moves and *predictor* moves. These are specifically arranged so that

- A full sentence is parsed by a complete binary tree
- The trigram language model is a special case of the SSLM. It is the result of a degenerate choice of the constructor statistics: $Q(\textbf{null}|h_{-2} = v, h_{-1} = v^{'}) = 1$ for all $v, v^{'} \in \mathcal{V}$.[2]

The SSLM operation is then as follows:

1. **Constructor moves**: The constructor looks at the pair of right-most (last) *exposed* headwords, h_{-2}, h_{-1} and performs an *action* a with probability $Q(a|\ h_{-2}, h_{-1})$ where $a \in$ {adjoin **right**,

[2]The meaning of constructor statistics $Q(a|h_{-2}, h_{-1})$ will be made clear shortly.

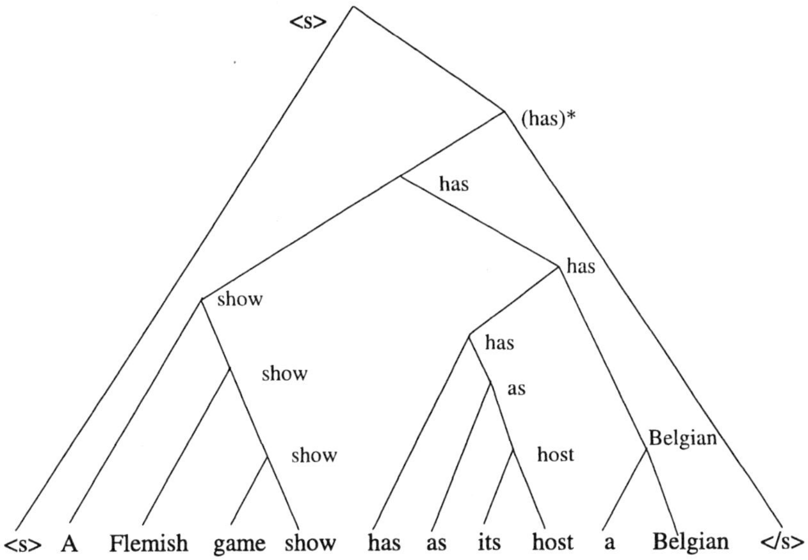

FIG. 1. *Parse by the simplified structured language model.*

adjoin **right***, adjoin **left**, adjoin **left***, **null**}. The specifications of the five possible actions are:[3]

- adjoin **right**: create an apex, mark it by the identity of h_{-1} and connect it by a leftward branch[4] to the (formerly) exposed headword h_{-2} and by a rightward branch to the exposed headword h_{-1} (i.e., the headword h_{-1} is percolated up by one tree level). Increase the indices of the current exposed headwords $h_{-3}, h_{-4}, \dots$ by 1. These headwords together with h_{-1} become the new exposed headwords $h'_{-1}, h'_{-2}, h'_{-3}, \dots$ I.e., $h'_{-1} = h_{-1}$, and $h'_{-i} = h_{-i-1}$ for $i = 2, 3, \dots$
- adjoin **right***: create an apex, mark it by the identity of the word corresponding to h_{-1}, attach to it the marker $*$,[5] and connect it by a leftward branch to the (formerly) exposed headword h_{-2} and by a rightward branch to the exposed headword h_{-1} (i.e., the headword h_{-1} is percolated up by one tree level). Increase the indices of the current exposed

[3]The actions adjoin **right*** and adjoin **left*** are necessary to assure that the trigram language model be a special case of the SSLM. This case will result from a degenerate choice of the constructor statistics: $Q(\mathbf{null}|h_{-2} = v, h_{-1} = v') = 1$ for all $v, v' \in \mathcal{V}$.

[4]Aiming down from the apex.

[5]I.e., if either $h_{-1} = v$, or $h_{-1} = v^*$, then the newly created apex will be marked by $h'_{-1} = v^*$.

headwords $h_{-3}, h_{-4}, ...$ by 1. These headwords together with $(h_{-1})^*$ become the new exposed headwords $h'_{-1}, h'_{-2}, h'_{-3}, ...$ I.e., $h'_{-1} = (h_{-1})^*$, and $h'_{-i} = h_{-i-1}$ for $i = 2, 3, ...$

- adjoin **left**: create an apex, mark it by the identity of h_{-2} and connect it by a rightward branch to the (formerly) exposed headword h_{-2} and by a leftward branch to the exposed headword h_{-1} (i.e., the headword h_{-2} is percolated one tree level up). Increase the indices of the new apex, as well as those of the current exposed headwords $h_{-3}, h_{-4}, ...$ by 1. These headwords together with h_{-1} become the new exposed headwords $h'_{-1}, h'_{-2}, h'_{-3}, ...$ I.e., $h'_{-1} = h_{-2}$, and $h'_{-i} = h_{-i-1}$ for $i = 2, 3, ...$
- adjoin **left***: create an apex, mark it by the identity of the word corresponding to h_{-2}, attach to it the marker * , and connect it by a rightward branch to the (formerly) exposed headword h_{-2} and by a leftward branch to the exposed headword h_{-1} (i.e., the headword h_{-2} is percolated one tree level up). Increase the indices of the new apex, as well as those of the current exposed headwords $h_{-3}, h_{-4}, ...$ by 1. These headwords thus become the new exposed headwords $h'_{-1}, h'_{-2}, h'_{-3}, ...$ I.e., $h'_{-1} = (h_{-2})^*, h'_{-i} = h_{-i-1}$ for $i = 2, 3, ...$
- **null**: leave headword indexing and current parse structure as they are and pass control to the predictor.

If $a \neq$**null**, then the constructor stays in control and chooses the next action $a' \in$ {adjoin **right**, adjoin **right***, adjoin **left**, adjoin **left***, **null**} with probability $Q(a' \mid h'_{-2}, h'_{-1})$ where the latest (possibly newly created) headword indexation is used. If $a =$ **null**, the constructor suspends operation and the control is passed to the *predictor*.

Note that a **null** move means that the right-most exposed headword will eventually be connected to the right. An **adjoin** move connects the right-most exposed headword to the left.

2. **Predictor moves**: The predictor generates the next word w_j with probability $P(w_j = v \mid h_{-2}, h_{-1})$, $v \in \mathcal{V} \cup </s>$. The indexing of the current headwords $h_{-1}, h_{-2}, h_{-3}, ...$ is decreased by 1 and the newly generated word becomes the right-most exposed headword so that $h'_{-1} = w_j, h'_{-i} = h_{-i+1}$ for $i = 2, 3, ...$ Control is then passed to the constructor.

The operation of the SSLM ends when the parser completes the tree, marking its apex by the headword $<s>$.[6]

[6]It will turn out that the operation and statistical parameter values of the SSLM are such that the only possible headword of a complete tree whose leaves are $<s>$ $, w_1, w_2, ..., w_n, </s>$ is $<s>$.

To complete the description of the operation of the parser, we have to prescribe particular values to certain statistical parameters:

Start of operation: The predictor generates the first word w_1 with probability $P_1(w_1 = v) = P(w_1 = v| < s >)$, $v \in \mathcal{V}$. The initial headwords (both exposed) become $h_{-2} =< s >$, $h_{-1} = w_1$. Control is passed to the constructor.

Special constructor probabilities:

- If $h_{-1} = v,\ v \in \mathcal{V}$ then

$$Q(a|h_{-2} =< s >, h_{-1}) = \begin{cases} 1 & \text{if } a = \textbf{null} \\ 0 & \text{otherwise} \end{cases} \tag{1}$$

- If $h_{-1} = v^*$, $v \in \mathcal{V}$ then

$$Q(a|h_{-2} =< s >, h_{-1}) = \begin{cases} 1 & \text{if } a = \text{adjoin } \textbf{left} \\ 0 & \text{otherwise} \end{cases} \tag{2}$$

- If $h_{-2} = v,\ v \in \mathcal{V}$ then

$$Q(a|h_{-2}, h_{-1} =< /s >) = \begin{cases} 1 & \text{if } a = \text{adjoin } \textbf{left}^* \\ 0 & \text{otherwise} \end{cases} \tag{3}$$

- If $h_{-1} = v^*$, $v \in \mathcal{V}$ and $h_{-2} \neq < s >$ then

$$Q(a|h_{-2}, h_{-1}) = 0 \quad \text{for } a \in \{\text{adjoin } \textbf{right}, \text{ adjoin } \textbf{left}, \textbf{ null}\} \tag{4}$$

The special constructor probabilities have the following consequences:

1. Formula (1) assures that the $< s >$ marking the beginning of the sentence will not be a part of the parse tree until the tree is completed.
2. Formula (2) assures that the completed parse tree will have $< s >$ attached to its apex.
3. Formulas (3) and (4) assure that once the end of sentence marker $< /s >$ is generated by the predictor, the parse tree will be completed.
4. Formula (3) highlights the last exposed headword h_{-2} of the sentence and by attaching an asterisk to it marks it for forced attachment to the previous exposed headword.
5. Formula (4) forces attachment of the two last exposed headwords h_{-2} and h_{-1}into a phrase with either h_{-2} or h_{-1} being percolated up as the headword of the new phrase.

The special constructor and predictor probabilities assure that the final parse of any word string has the appearance of Figure 2.

Experienced readers will note that the SSLM is a generalization of a shift – reduce parser with *adjoin* corresponding to *reduce* and *predict* to *shift.* The particular non-context free nature of the SSLM is interesting because its word generation depends on exposed headwords and therefore potentially on the entire word string already generated.

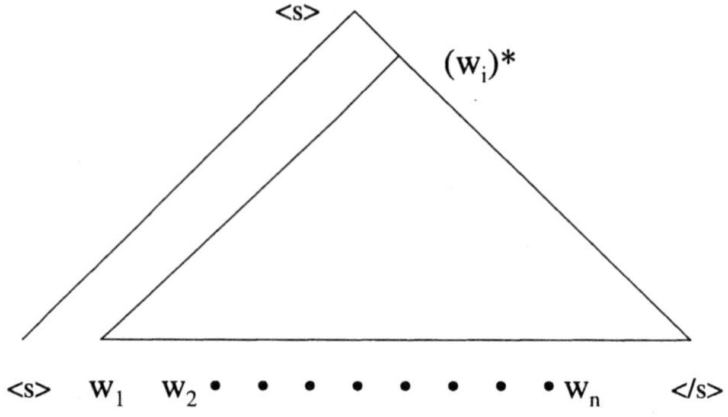

FIG. 2. *Form of a complete SLM parse.*

3. Some notation. Let the generated sequence be denoted by $\mathbf{W} = < s >, w_1, w_2, ..., w_n, < /s >$, and let $\mathbf{T}$ denote a complete (binary) parse of $\mathbf{W}$, that is, one whose sole exposed headword is $< s >$. Further, let $\mathbf{W}^i = < s >, w_1, ..., w_i$ denote a prefix of the complete sentence $\mathbf{W}$, and let $\mathbf{T}^i$ denote a partial parse structure built by the SSLM constructor on top of $\mathbf{W}^i$. Clearly, $\mathbf{W} = \mathbf{W}^{n+1}$ and $\mathbf{W}^{i+1} = \mathbf{W}^i, w_{i+1}$. Finally, let $h_{-j}(\mathbf{T}^i)$ denote the jth exposed headword of the structure $\mathbf{T}^i$, $j = 1, 2, ..., k$ where $h_{-k}(\mathbf{T}^i) = < s >$.

We will be interested in such quantities as $P(\mathbf{W})$, $P(\mathbf{T}, \mathbf{W})$, $P(\mathbf{W}^i)$, $P(\mathbf{T}^i, \mathbf{W}^i)$, $P(w_{i+1}|\mathbf{W}^i)$, etc.

Because of the nature of the SSLM operation specified in Section 2, the computation of $P(\mathbf{T}, \mathbf{W})$ or $P(\mathbf{T}^i, \mathbf{W}^i)$ is straight-forward. In fact, given a "legal" parse structure $\mathbf{T}^i$ of a prefix $\mathbf{W}^i$, there is a unique sequence of constructor and predictor moves that results in the pair $\mathbf{T}^i, \mathbf{W}^i$. For instance, for the parse of Figure 1, the sub-parse $\mathbf{T}^6$ corresponding to the prefix $< s >$ A FLEMISH GAME SHOW HAS AS results from the following sequence of SSLM moves: pred(A), **null**, pred(FLEMISH), **null**, pred(GAME), **null**, pred(SHOW), adjoin **right**, adjoin **right**, adjoin **right**, **null**, pred(HAS), **null**, pred(AS), **null**. $P(\mathbf{T}^i, \mathbf{W}^i)$ is simply equal to the product of the probabilities of the SSLM moves that result in $\mathbf{T}^i, \mathbf{W}^i$. More formally,

$$P(\mathbf{T}^i, \mathbf{W}^i) = \prod_{j=1}^{i} P\big(w_j | h_{-2}(\mathbf{T}^{j-1}), h_{-2}(\mathbf{T}^{j-1})\big) \times \prod_{l=1}^{m(j)} Q(a_{j,l} | \mathbf{T}^{j-1}, w_j, a_{j,1}, ..., a_{j,l-1}) \tag{5}$$

where $a_{j,m(j)}$ = **null**, $a_{j,l} \in$ {adjoin **left**, adjoin **right**}, $l = 1, 2, ..., m(j) - 1$ are the actions taken by the constructor after w_j (and before w_{j+1}) has been generated by the predictor. $\mathbf{T}^j$ results from $\mathbf{T}^{j-1}$ after actions $a_{j,1}, ..., a_{j,m(j)}$ have been performed by the constructor, and $\mathbf{T}^{j-1}$ is the built-up structure just before control passes to the predictor which will generate w_j. Furthermore, in (5) the actions $a_{j,1}, ..., a_{j,l-1}$ performed in succession on the structure $[\mathbf{T}^{j-1}, w_j]$ result in a structure having some particular pair of last two exposed headwords $h(l-1)_{-2}, h(l-1)_{-1}$, and we define

$$Q(a_{j,l}|\mathbf{T}^{j-1}, w_j, a_{j,1}, ..., a_{j,l-1}) \doteq Q(a_{j,l}|h(l-1)_{-2}, h(l-1)_{-1}). \tag{6}$$

Strictly speaking, (5) applies to $i < n + 1$ only. It may not reflect the "end moves" that complete the parse.

4. A chart parsing algorithm. We will now develop a chart parsing algorithm [4–6] that will enable us to calculate $P(\mathbf{W})$ when the word string was generated by the SSLM. The results of this algorithm will facilitate the calculation of $P(\mathbf{W}^i)$ and thus allow an implementation of the SSLM language model that is alternative to the one in [1]. Furthermore, it will lead to a Viterbi-like determination of the most probable parse

$$\widehat{\mathbf{T}} = \arg\max_{\mathbf{T}} P(\mathbf{T}, \mathbf{W}) \tag{7}$$

and form the basis of a parameter re-estimation procedure that is a generalization of the Inside – Outside algorithm of Baker [7].

4.1. Calculating $P(\mathbf{W})$. As before, $\mathbf{W}$ denotes a string of words $w_0, w_1, w_2, ..., w_n, w_{n+1}$ that form the complete sentence, where w_i, $i = 1, ..., n$ are elements of a vocabulary $\mathcal{V}$, $w_0 =< s >$ (the beginning of sentence marker, generated with probability 1) and $w_{n+1} =< /s >$ (the end of sentence marker). The first word, w_1 is generated with probability $P_1(w_1) = P(w_1| < s >)$, the rest with probabilities $P(w_i|h_{-2}, h_{-1})$ where h_{-2}, h_{-1} are the most recent exposed headwords valid at the time of generation of w_i. The algorithm we will develop will be computationally complex (see below) exactly because with different probability different headword pairs h_{-2}, h_{-1} determine the parser's moves, and h_{-2}, h_{-1} can in principle be any pair of succeeding words belonging to the sentence prefix $\mathbf{W}^{i-1}$.

Our algorithm will proceed left-to-right.[7] The probabilities of phrases covering word position spans $< i, j >$, $i \in \{0, 1, ..., j\}$ [8] will be calculated

[7]So can the famous CKY algorithm [4–6] that will turn out to be similar to ours. As a matter of fact, it will be obvious from formula (9) below that the presented algorithm can also be run from bottom up, just as the CKY algorithm usually is, but such a direction would be computationally wasteful because is could not take full advantage of the thresholding suggested in Section 8.

[8]The span $< i, j >$ refers to the words $w_i, w_{i+1}, ..., w_j$.

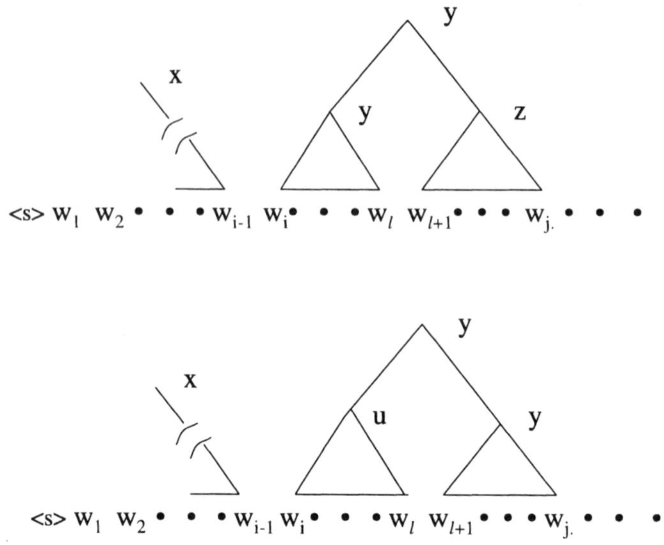

FIG. 3. *Diagram illustrating inside probability recursions.*

after the corresponding information concerning spans $< k, j-1 >, k = 0, 1, ..., j-1$ and $< l, j >$, $l = i+1, ..., j$ had been determined.

We will be interested in the *inside* probabilities $P(\mathbf{w}_{i+1}^{j}, y[i,j]|w_i, x)$ that, given that x is the last exposed headword preceding time i and that w_i is generated, the following words $\mathbf{w}_{i+1}^{j} = w_{i+1}..., w_j$ are generated and y becomes the headword of the phrase $w_i, w_{i+1}..., w_j$.[9] Figure 3 illustrates two ways in which the described situation may arise.

In fact, the first way to generate $w_{i+1}, ..., w_j$ and create a phrase spanning $< i, j >$ whose headword is y, given that the headword of the preceding phrase is x and the word w_i was generated, is:

- a string $w_{i+1}, ..., w_l$ is generated,
- a phrase spanning $< i, l >$ is formed whose headword is y (and preceding that phrase is another phrase whose headword is x),
- the word w_{l+1} is generated "from" its two preceding headwords (i.e., x, y)

[9] I.e.,

$$P(\mathbf{w}_{i+1}^{j}, y[i,j]|w_i, x) \doteq P(w_{i+1}, ..., w_j, h(w_i, w_{i+1}, ..., w_j) = y|w_i, h_{-1}(\mathbf{T}^{i-1}) = x)$$

where $h(w_i, w_{i+1}, ..., w_j) = \varepsilon$ (empty symbol) if $w_i, w_{i+1}, ..., w_j$ do not form a phrase, and

$$P(\mathbf{w}_{i+1}^{i}, y[i,i]|w_i, x) \doteq P(h(w_i) = y|w_i, h_{-1}(\mathbf{T}^{i-1}) = x) = \begin{cases} 1 & \text{if } y = w_i \\ 0 & \text{otherwise} \end{cases}$$

- the string $w_{l+2}, ..., w_j$ is generated and the span $< l+1, j >$ forms a following phrase whose headword is, say, z (and the headword of its preceding phrase must be y!)
- and finally, the two phrases are joined as one phrase whose headword is y.

The just described process can be embodied in the formula

$$\sum_{l=i}^{j-1} \sum_{z} P^*(w_{l+1}|x, y)\, P(\mathbf{w}_{i+1}^{l}, y[i,l]|w_i, x) \\ \times P(\mathbf{w}_{l+2}^{j}, z[l+1, j]|w_{l+1}, y)\, Q(\mathbf{left}|y, z)$$

where

(8) $$P^*(w|h_{-2}, h_{-1}) = P(w|h_{-2}, h_{-1})\, Q(\mathbf{null}|h_{-2}, h_{-1})$$

The second way[10] to create a phrase whose headword is y and to generate $w_{i+1}, ..., w_j$, given that the headword of the preceding phrase is x and the word w_i was generated, is almost the same as the one described above, except that the first of the two phrases is headed by some headword u and the second phrase by headword y, and when these two phrases are joined it is the second headword, y, which is percolated upward to head the overall phrase. Of course, in this case w_{l+1} is generated "from" its preceding two headwords, x and u. This second process is embodied in the formula

$$\sum_{l=i}^{j-1} \sum_{u} P^*(w_{l+1}|x, u)\, P(\mathbf{w}_{i+1}^{l}, u[i,l]|w_i, x) \\ \times P(\mathbf{w}_{l+2}^{j}, y[l+1, j]|w_{l+1}, u)\, Q(\mathbf{right}|u, y)$$

We may thus conclude that for $j > i, i \in \{0, 1, 2, ..., n\}$

(9) $$\begin{aligned} P(\mathbf{w}_{i+1}^{j}, y[i,j]|w_i, x) = \\ &\sum_{l=i}^{j-1} \sum_{z} P^*(w_{l+1}|x, y)\, P(\mathbf{w}_{i+1}^{l}, y[i,l]|w_i, x) \\ &\quad \times P(\mathbf{w}_{l+2}^{j}, z[l+1, j]|w_{l+1}, y)\, Q(\mathbf{left}|y, z) \\ &+ \sum_{l=i}^{j-1} \sum_{u} P^*(w_{l+1}|x, u)\, P(\mathbf{w}_{i+1}^{l}, u[i,l]|w_i, x) \\ &\quad \times P(\mathbf{w}_{l+2}^{j}, y[l+1, j]|w_{l+1}, u)\, Q(\mathbf{right}|u, y) \end{aligned}$$

where

(10) $$P(\mathbf{w}_{i+1}^{j}, y[i,j]|w_i, x) = 0 \\ \text{if } x \notin \{w_0, ..., w_{i-1}\} \text{ or } y \notin \{w_i, ..., w_j\} \text{ or } i > j.$$

[10] See the second diagram of Figure 3.

The boundary conditions for the recursion (9) are

$$
\begin{aligned}
P(\mathbf{w}_{i+1}^{i}, y[i,i] | w_i, x) = P(h(w_i) = y | w_i, h_{-1}(\mathbf{T}^{i-1}) = x) = 1 \\
\text{for } x \in \{w_0, ..., w_{i-1}\}, y = w_i
\end{aligned} \tag{11}
$$

and the probability we are interested in is given by

$$
P(\mathbf{W}) = P(\mathbf{w}_2^{n+1}, </s> [1, n+1] | w_1, <s>)\, P_1(w_1). \tag{12}
$$

It may be useful to illustrate the carrying out of the above recursion by a simple example. We will create the corresponding *chart* (also referred to as the *parse triangle*) for the sentence fragment $<s>$ FRESHMAN BASKETBALL PLAYER $</s>$. To simplify the presentation we will abbreviate the preceding by $<s>$ F B P $</s>$. To fit into into Table 1 we will further simplify the probabilities $P(\mathbf{w}_{i+1}^{l}, u[i,l] | w_i, x)$ by omitting the redundant $\mathbf{w}_{i+1}^{l}$, thus obtaining $P(u[i,l]|w_i, x)$, one of the entries in the ith row and l^{th} column of the parse triangle. The parse triangle of Table 1 contains all the cases that the recursion (9) generates. As a further illustration, the probability $P(\text{P}[2,3]|\text{B,F})$ would be computed as follows:

$$
\begin{aligned}
P(\text{P}[2,3]|\text{B,F}) &= Q(\mathbf{null}|\text{F,B}) \times P(\text{P}|\text{F,B}) \times P(\text{B}[2,2]|\text{B,F}) \\
&\times P(\text{P}[3,3]|\text{P,B}) \times Q(\mathbf{right}|\ \text{B,P}).
\end{aligned}
$$

TABLE 1
Parse triangle of the SSLM.

<s>	freshman (F)	basketball (B)	player (P)	</s>
$P(\langle s\rangle[0,0]\mid\langle s\rangle,\langle/s\rangle)$				$P(\langle/s\rangle[0,4]\mid\langle s\rangle,\langle/s\rangle)$
	$P(\text{F}[1,1]\mid\text{F},\langle s\rangle)$	$P(\text{B}[1,2]\mid\text{F},\langle s\rangle)$ $P(\text{F}[1,2]\mid\text{F},\langle s\rangle)$	$P(\text{P}[1,3]\mid\text{F},\langle s\rangle)$ $P(\text{B}[1,3]\mid\text{F},\langle s\rangle)$ $P(\text{F}[1,3]\mid\text{F},\langle s\rangle)$	$P(\langle/s\rangle[1,4]\mid\text{F},\langle s\rangle)$
		$P(\text{B}[2,2]\mid\text{B,F})$	$P(\text{P}[2,3]\mid\text{B,F})$ $P(\text{B}[2,3]\mid\text{B,F})$	
			$P(\text{P}[3,3]\mid\text{P,B})$ $P(\text{P}[3,3]\mid\text{P,F})$	
				$P(\langle/s\rangle[4,4]\mid\langle/s\rangle,\text{P})$

4.2. Computing $P(w_{i+1}|\mathbf{W}^i)$. We can now use the concepts and notation developed in Section 4.1 to compute left-to-right probabilities of word generation by the SSLM. Let $P(\mathbf{W}^{i+1}, x)$ denote the probability that the sequence $w_0, w_1, w_2, ..., w_i, w_{i+1}$ is generated in a manner resulting in

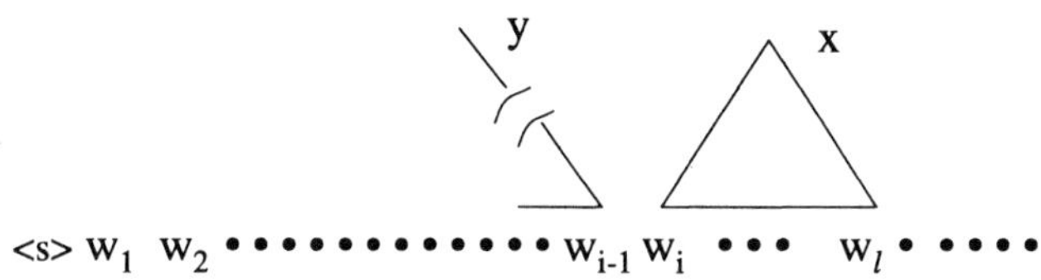

FIG. 4. *Diagram illustrating the basis of the recursion (13).*

any *construction*[11] $\mathbf{T}^i$ whose last exposed headword is x.[12] Further, define the set of words $\mathcal{W}^i = \{w_0, w_1, w_2, ..., w_i\}$. Then we have the following recursion:

$$P(\mathbf{W}^{l+1}, x) = \sum_{i=1}^{l} \sum_{y \in \mathcal{W}^{i-1}} P(\mathbf{W}^i, y)\, P(\mathbf{w}_{i+1}^l, x[i,l] | w_i, y)\, P^*(w_{l+1} | y, x) \quad \text{for } x \in \mathcal{W}^l \tag{13}$$

with the initial condition

$$P(\mathbf{W}^1, x) = \begin{cases} P_1(w_1) & x = < s > \\ 0 & x \neq < s > . \end{cases}$$

The situation corresponding to the general term in the sum (13) is depicted in Figure 4.

For the example sentence $< s >$ FRESHMAN BASKETBALL PLAYER $< /s >$ the probability $P(\mathbf{W}^3, \text{B})$ is given by the following formula:

$$\begin{aligned} P(\mathbf{W}^3, \text{B}) &= P(\mathbf{W}^2, \text{F})\, P(\text{B}[2,2] | \text{B}, \text{F})\, Q(\mathbf{null} | \text{F}, \text{B})\, P(\text{P} | \text{F}, \text{B}) \\ &\quad + P(\mathbf{W}^1, < s >)\, P(\text{B}[1,2] | \text{F}, < s >) \\ &\quad \times Q(P(\text{P} | < s >, \text{B}). \end{aligned}$$

It follows directly from the definition of $P(\mathbf{W}^{i+1}, x)$ that

$$P(w_0, w_1, w_2, ..., w_i, w_{i+1}) = \sum_{x \in \mathcal{W}^i} P(\mathbf{W}^{i+1}, x)$$

---

[11] By construction $\mathbf{T}^i$ we mean a sub-tree covering $\mathbf{W}^i$ that can be generated in the process of operating the SSLM.

[12] That is, $\mathbf{T}^i$ is a construction "covering" $\mathbf{W}^i = w_0, w_1, w_2, ..., w_i$, the constructor passes control to the predictor which then generates the next word w_{i+1}. Thus

$$P(\mathbf{W}^{i+1}, x) = \sum_{\mathbf{T}^i} P(\mathbf{W}^i, w_{i+1}, h_{-1}(\mathbf{T}^i) = x).$$

and therefore

$$(14) \qquad P(w_{i+1}|w_0, w_1, w_2, ..., w_i) = \frac{\sum_{x \in \mathcal{W}^i} P(\mathbf{W}^{i+1}, x)}{\sum_{y \in \mathcal{W}^{i-1}} P(\mathbf{W}^i, y)}.$$

4.3. Finding the most probable parse $\widehat{\mathbf{T}}$. We are next interested in finding the most probable parse $\widehat{\mathbf{T}}$ specified by (7). Consider any parse $\mathbf{T}$ of the full sentence $\mathbf{W}$. Since the pair $\mathbf{T}, \mathbf{W}$ is generated in a unique sequence of parser actions, we may label every node $< i, j >$[13] of the parse tree by the appropriate headword pair x, y, where y is the headword of the phrase $< i, j >$ corresponding to the node, and x is the preceding exposed headword at the time the constructor created the node. Let us denote the subtree corresponding to the node in question by $V(x, y, i, j)$. Now consider any other possible subtree spanning the same interval $< i, j >$ and having y as its headword, and denote this subtree by $V^{'}(x, y, i, j)$.[14] Clearly, if the probability of generating $w_{i+1}, ..., w_j$ (given w_i and the structure of $\mathbf{T}$ pertaining to $< 1, i-1 >$) and creating the subtree $V^{'}(x, y, i, j)$ exceeds the corresponding probability of generating $w_{i+1}, ..., w_j$ and creating $V(x, y, i, j)$, then $P(\mathbf{T}^{'}, \mathbf{W}) > P(\mathbf{T}, \mathbf{W})$ where $\mathbf{T}^{'}$ arises from $\mathbf{T}$ by replacing in the latter the subtree $V(x, y, i, j)$ by the subtree $V^{'}(x, y, i, j)$.

From the above observation it is now clear how to find the most probable parse $\widehat{\mathbf{T}}$. Namely, given that x is the last exposed headword corresponding to $\mathbf{W}^{i-1}$ and w_i is generated, let $\widehat{R}(\mathbf{w}_{i+1}^j, y[i,j]|w_i, x)$ denote the probability of the most probable sequence of moves that generate the following words $w_{i+1} ..., w_j$ with y becoming the headword of the phrase $w_i, w_{i+1} ..., w_j$. Then we have for $j > i, i \in \{0, 1, 2, ..., n\}$ that

$$\begin{aligned}
&\widehat{R}(\mathbf{w}_{i+1}^j, y[i,j]|w_i, x) \; = \\
&\max \Bigg\{ \max_{l \in \{i, j-1\}, z} \Big[P^*(w_{l+1}|x, y) \widehat{R}(\mathbf{w}_{i+1}^l, y[i,l]|w_i, x) \\
(15) \qquad &\qquad \times \widehat{R}(\mathbf{w}_{l+2}^j, z[l+1, j]|w_{l+1}, y)\, Q(\mathbf{left}|y, z) \Big], \\
&\max_{l \in \{i, j-1\}, u} \Big[P^*(w_{l+1}|x, u) \widehat{R}(\mathbf{w}_{i+1}^l, u[i,j]|w_i, x) \\
&\qquad \times \widehat{R}(\mathbf{w}_{l+2}^j, y[l+1, j]|w_i, u)\, Q(\mathbf{right}|u, y) \Big] \Bigg\}
\end{aligned}$$

where

$$\widehat{R}(\mathbf{w}_{i+1}^j, y[i,j]|w_i, x) = 0$$
$$\text{if } x \notin \{w_0, ..., w_{i-1}\} \;\text{ or }\; y \notin \{w_i, ..., w_j\} \;\text{ or }\; i > j.$$

[13]In a given parse tree $\mathbf{T}$, every node corresponds to some particular phrase span $< i, j >$ and is therefore uniquely identified by it.

[14]The preceding exposed headword does not change, so it must still be x.

The boundary conditions for the recursion (15) are

$$\widehat{R}(\mathbf{w}_{i+1}^{i}, w_i[i,i]|w_i, x) = P(h(w_i) = w_i|w_i, h_{-1}(\mathbf{T}^{i-1}) = x) = 1 \quad \text{for } x \in \{w_0, ..., w_{i-1}\}$$

and the probability we are interested in will be given by

$$P(\widehat{\mathbf{T}}, \mathbf{W}) = \widehat{R}(\mathbf{w}_2^{n+1}, </s>[1, n+1]|w_1, <s>)\, P_1(w_1).$$

Obviously, the tree $\widehat{\mathbf{T}}$ itself can be obtained by a back-trace of relations (15) starting from the tree apex $\widehat{\mathbf{T}}(<s>, </s>, 0, n+1)$.

5. An EM-type re-estimation algorithm for the SSLM. We need to derive re-estimation formulas for the basic statistical parameters of the SSLM: $P(w|h_{-2}, h_{-1})$ and $Q(a|h_{-2}, h_{-1})$. We will be inspired by the Inside – Outside algorithm of Baker [7]. We will generalize that approach to the SSLM whose structure is considerably more complex than that of a context free grammar.

5.1. Computing the "outside" probabilities. We will first derive formulas for $P(\mathbf{W}, x, y[i,j])$, the probability that $\mathbf{W}$ was produced by some tree $\mathbf{T}$ that has a phrase spanning $< i, j >$ whose headword is y (not necessarily exposed) and the immediately preceding exposed headword is x. More formally,

(16)
$$\begin{aligned} &P(\mathbf{W}, x, y[i,j]) \\ &\quad \doteq P\big(w_0, w_1, ..., w_{n+1}, h_{-1}(w_0, ..., w_{i-1}) = x, h(w_i, ..., w_j) = y\big) \\ &\quad = P\big(w_0, w_1, ..., w_i, h_{-1}(w_0, ..., w_{i-1}) = x\big) \\ &\qquad \times\ P\big(w_{i+1}, ..., w_j, h(w_i, ..., w_j) = y|w_i, h_{-1}(w_0, ..., w_{i-1}) = x,\big) \\ &\qquad \times\ P\big(w_{j+1}, ..., w_{n+1}|h_{-1}(w_0, ..., w_{i-1}) = x, h(w_i, ..., w_j) = y\big). \end{aligned}$$

Now the middle term on the right-hand side of (16) was designated by $P(\mathbf{w}_{i+1}^{j}, y[i,j]|w_i, x)$ (an "inner" probability) and can be computed by the recursion (9). We need a way to compute the product of the outer terms (outer probabilities)[15]

$$\begin{aligned} &P(\mathbf{w}_0^i, \mathbf{w}_{j+1}^{n+1}, x[i-1]\,; y[i,j]) \\ &\quad \doteq P\big(w_0, w_1, ..., w_i, h_{-1}(w_0, ..., w_{i-1}) = x\big) \\ &\qquad \times P\big(w_{j+1}, ..., w_{n+1}|h_{-1}(w_0, ..., w_{i-1}) = x, h(w_i, ..., w_j) = y\big). \end{aligned} \tag{17}$$

We thus have

$$P(\mathbf{W}, x, y[i,j]) = P(\mathbf{w}_0^i, \mathbf{w}_{j+1}^{n+1}, x[i-1];\ y[i,j])\, P(\mathbf{w}_{i+1}^{j}, y[i,j]|w_i, x). \tag{18}$$

[15]In $P(\mathbf{w}_0^i, \mathbf{w}_{j+1}^{n+1}, x[i-1]\,; y[i,j])$ we use a semicolon rather that a vertical slash to indicate that this is a product of probabilities and not a probability itself. The semicolon avoids a possible problem in equation (18).

We will obtain a recursion for $P(\mathbf{w}_0^i, \mathbf{w}_{j+1}^{n+1}, x[i-1]\,; y[i,j])$ based on the four cases presented in Figure 6 that illustrate what the situation of Figure 5 (that pertains to $P(\mathbf{w}_0^i, \mathbf{w}_{j+1}^{n+1}, x[i-1]\,; y[i,j])$) may lead to. The four cases correspond to the four double sums on the right-hand side of the following equation valid for $x \in \mathcal{W}^{i-1}, y \in \{w_i, ..., w_j\}$:[16]

$$
\begin{aligned}
&P(\mathbf{w}_0^i, \mathbf{w}_{j+1}^{n+1}, x[i-1]\,; y[i,j]) \\
&= \sum_{l=1}^{i-1} \sum_{z \in \mathcal{W}^{i-l-1}} \Big[P(\mathbf{w}_0^{i-l}, \mathbf{w}_{j+1}^{n+1}, z[i-l-1]; x[i-l,j]) \\
&\qquad \times P(\mathbf{w}_{i-l+1}^{i-1}, x[i-l,i-1] | w_{i-l}, z) P^*(w_i | z, x)\, Q(\mathbf{left} | x, y)\Big] \\
&\quad + \sum_{l=1}^{i-1} \sum_{z \in \mathcal{W}^{i-l-1}} \Big[P(\mathbf{w}_0^{i-l}, \mathbf{w}_{j+1}^{n+1}, z[i-l-1]; y[i-l,j]) \\
&\qquad \times P(\mathbf{w}_{i-l+1}^{i-1}, x[i-l,i-1] | w_{i-l}, z) P^*(w_i | z, x) Q(\mathbf{right} | x, y)\Big] \\
&\quad + \sum_{m=1}^{n-j+1} \sum_{u \in \mathcal{W}_{j+1}^{j+m}} \Big[P(\mathbf{w}_0^i, \mathbf{w}_{j+m+1}^{n+1}, x[i-1]; y[i,j+m]) \\
&\qquad \times P(\mathbf{w}_{j+2}^{j+m}, u[j+1,j+m] | w_{j+1}, y) P^*(w_{j+1} | x, y) Q(\mathbf{left} | y, u)\Big] \\
&\quad + \sum_{m=1}^{n-j+1} \sum_{u \in \mathcal{W}_{j+1}^{j+m}} \Big[P(\mathbf{w}_0^i, \mathbf{w}_{j+m+1}^{n+1}, x[i-1]; u[i,j+m]) \\
&\qquad \times P(\mathbf{w}_{j+2}^{j+m}, u[j+1,j+m] | w_{j+1}, y) P^*(w_{j+1} | x, y) Q(\mathbf{right} | y, u)\Big]
\end{aligned}
\tag{19}
$$

where $P^*(w_{j+1}|x, y)$ is defined by (8). Of course,

$$
\begin{gathered}
P(\mathbf{w}_0^i, \mathbf{w}_{j+1}^{n+1}, x[i-1]\,; y[i,j]) \doteq 0 \\
\text{if either } x \notin \mathcal{W}^{i-1} \text{ or } y \notin \{w_i, ..., w_j\}.
\end{gathered}
\tag{20}
$$

The above recursion allows for the computation of $P(\mathbf{w}_0^i, \mathbf{w}_{j+1}^{n+1}, x[i-1];\ y[i,j])$ provided the values of $P(\mathbf{w}_{i+1}^j, y[i,j] | w_i, z)$ are known beforehand (they were presumably obtained using (9)), as are the values $P(\mathbf{w}_0^{i-l}, \mathbf{w}_{j+1}^{n+1}, z[i-l-1]; x[i-l,j])$, $l = 1, 2, ..., i-1$ and $P(\mathbf{w}_0^i, \mathbf{w}_{j+m+1}^{n+1}, x[i-1]\,; y[i, j+m])$, $m = 1, 2, ..., n-j+1$.

In order to start the recursion process (19) we need the boundary condition

$$
P(\mathbf{w}_0^1, \mathbf{w}_{n+2}^{n+1}, x[0]; y[1, n+1]) = \begin{cases} P_1(w_1) & \text{if } x = <s>,\ y = </s> \\ 0 & \text{otherwise} \end{cases}
\tag{21}
$$

which reflects the requirement, pointed out in Section 2, that the final parse has the appearance of Figure 2.

[16]Below we use the set notation $\mathcal{W}_i^j \doteq \{w_i, w_{i+1}, ..., w_j\}$.

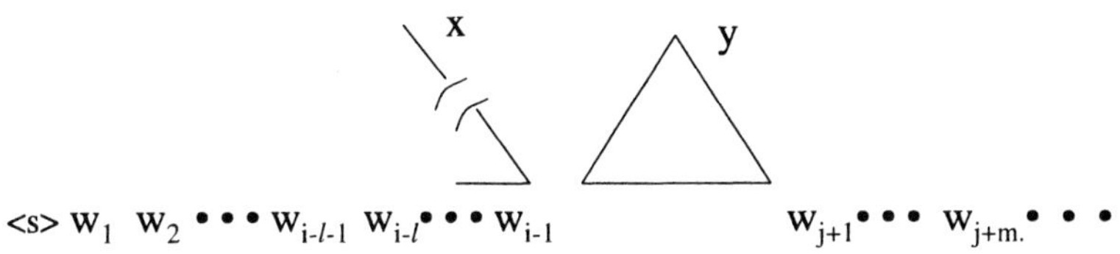

FIG. 5. *Diagram illustrating the parse situation* $P(\mathbf{w}_0^i, \mathbf{w}_{j+1}^{n+1}, x[i-1]\ ; y[i,j])$.

We can make a partial check on the correctness of these boundary conditions by substituting into (18) for values $i = 1$, $j = n+1$, $x = < s >$, $y = < /s >$:

$$\begin{aligned} &P(\mathbf{W}, < s >, < /s > [1, n+1]) \\ &\quad = P(\mathbf{w}_0^1, \mathbf{w}_{n+2}^{n+1}, < s > [i-1]; < /s > [1, n+1]) \\ &\qquad \times P(\mathbf{w}_2^{n+1}, < /s > [i,j] | w_1, < s >) \\ &\quad = P_1(w_1) P(\mathbf{w}_2^{n+1}, < /s > [i,j] | w_1, < s >) \end{aligned}$$

which agrees with (12).

For the example sentence $< s >$ FRESHMAN BASKETBALL PLAYER $< /s >$ the probability $P(\text{F}; \text{B}[2,2])$ is given by the following formula(in the formula, we use a simplified notation, replacing $P(\mathbf{w}_0^i, \mathbf{w}_{j+1}^{n+1}, \mathbf{x}[i-1]; \mathbf{y}[i,j])$ by $P(\mathbf{x}; \mathbf{y}[i,j])$ which unambiguously specifies the former):

$$\begin{aligned} &P(\text{F}; \text{B}[2,2]) \\ &\quad = P(<\!\text{S}\!>; \text{B}[1,2]) P(\text{F}[1,1] | \text{F}, <\!\text{S}\!>) Q(\mathbf{null} | <\!\text{S}\!>, \text{F}) \\ &\qquad \times P(\text{B} | <\!\text{S}\!>, \text{F}) Q(\mathbf{right} | \text{F}, \text{B}) \\ &\quad + P(<\!\text{S}\!>; \text{F}[1,2]) P(\text{F}[1,1] | \text{F}, <\!\text{S}\!>) Q(\mathbf{null} | <\!\text{S}\!>, \text{F}) \\ &\qquad \times P(\text{B} | <\!\text{S}\!>, \text{F}) Q(\mathbf{left} | \text{F}, \text{B}) \\ &\quad + P(\text{F}; \text{B}[2,3]) P(\text{P}[3,3] | \text{P}, \text{B}) Q(\mathbf{null} | \text{F}, \text{B}) P(\text{P} | \text{F}, \text{B}) Q(\mathbf{left} | \text{B}, \text{P}) \\ &\quad + P(\text{F}; \text{P}[2,3]) P(\text{P}[3,3] | \text{P}, \text{B}) Q(\mathbf{null} | \text{F}, \text{B}) P(\text{P} | \text{F}, \text{B}) Q(\mathbf{right} | \text{B}, \text{P}). \end{aligned} \tag{22}$$

5.2. The re-estimation formulas. We now need to use the inside and outside probabilities derived in Sections 4.1 and 5.1 to obtain formulas for re-estimating $P(v|h_{-2}, h_{-1})$ and $Q(a|h_{-2}, h_{-1})$. We will do so with the help of the following quantities found on the right-hand side of (19):

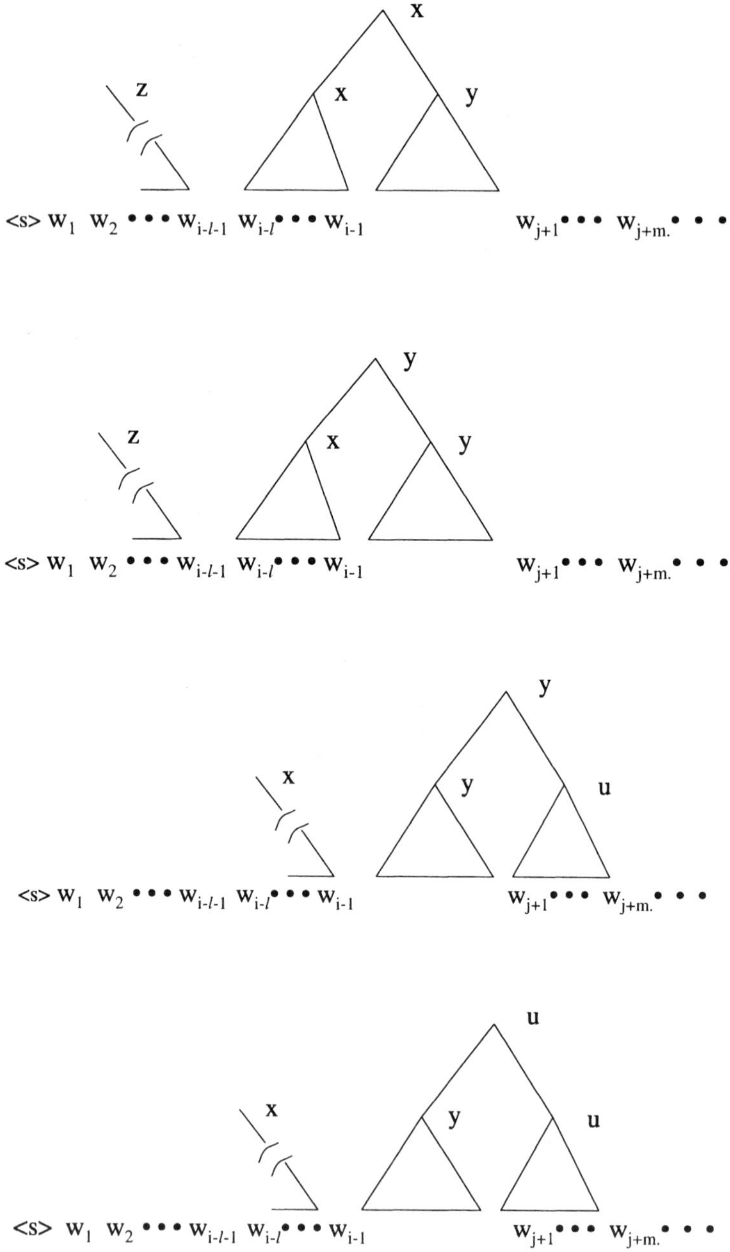

FIG. 6. *Diagrams illustrating outside probability recursions.*

$$C_K(x,y,i,j,\mathbf{left}) \doteq \frac{1}{P(\mathbf{W}_K)} P(\mathbf{w}_{i+1}^{j}, y[i,j]|w_i,x) \\ \times \sum_z \sum_{l=1}^{i-1} \Big[P(\mathbf{w}_0^{i-l}, \mathbf{w}_{j+1}^{n+1}, z[i-l-1]\,; x[i-l,j]) \\ \times P(\mathbf{w}_{i-l+1}^{i-1}, x[i-l,i-1]|w_{i-l},z)\, P^*(w_i|z,x)\, Q(\mathbf{left}|x,y) \Big] \tag{23}$$

$$C_K(x,y,i,j,\mathbf{right}) \doteq \frac{1}{P(\mathbf{W}_K)} P(\mathbf{w}_{i+1}^{j}, y[i,j]|w_i,x) \\ \times \sum_z \sum_{l=1}^{i-1} \Big[P(\mathbf{w}_0^{i-l}, \mathbf{w}_{j+1}^{n+1}, z[i-l-1]\,; y[i-l,j]) \\ \times P(\mathbf{w}_{i-l+1}^{i-1}, x[i-l,i-1]|w_{i-l},z) P^*(w_i|z,x) Q(\mathbf{right}|x,y) \Big] \tag{24}$$

$$C_K(x,y,i,j,\mathbf{null}) \doteq \frac{1}{P(\mathbf{W}_K)} P(\mathbf{w}_{i+1}^{j}, y[i,j]|w_i,x) \\ \times \Big\{ \sum_u \sum_{m=1}^{n-j+1} \Big[P(\mathbf{w}_0^{i}, \mathbf{w}_{j+m+1}^{n+1}, x[i-1]\,; y[i,j+m]) \\ \times P(\mathbf{w}_{j+2}^{j+m}, u[j+1,j+m]|w_{j+1},y) P^*(w_{j+1}|x,y) Q(\mathbf{left}|y,u) \Big] \\ + \sum_u \sum_{m=1}^{n-j+1} \Big[P(\mathbf{w}_0^{i}, \mathbf{w}_{j+m+1}^{n+1}, x[i-1]; u[i,j+m]) \\ \times P(\mathbf{w}_{j+2}^{j+m}, u[j+1,j+m]|w_{j+1},y) P^*(w_{j+1}|x,y) Q(\mathbf{right}|y,u) \Big] \Big\} \tag{25}$$

where the index K refers to the Kth of the M sentences constituting the training data.[17]

It is clear that $C_K(x,y,i,j,\mathbf{left})$ corresponds to the first case depicted in Figure 6, $C_K(x,y,i,j,\mathbf{right})$ to the second case, and $C_K(x,y,i,j,\mathbf{null})$ to the last two cases of Figure 6. Thus, defining counter "contents" (n_K is the length of the Kth sentence)

$$CC(x,y,\mathbf{left}) \doteq \sum_{K=1}^{M} \sum_{i=1}^{n_K+1} \sum_{j=i}^{n_K+1} C_K(x,y,i,j,\mathbf{left})$$

$$CC(x,y,\mathbf{right}) \doteq \sum_{K=1}^{M} \sum_{i=1}^{n_K+1} \sum_{j=i}^{n_K+1} C_K(x,y,i,j,\mathbf{right})$$

$$CC(x,y,\mathbf{null}) \doteq \sum_{K=1}^{M} \sum_{i=1}^{n_K} \sum_{j=i}^{n_K} C_K(x,y,i,j,\mathbf{null})$$

[17] Not to complicate the notation, we did not bother to associate the index K with the words w_j and subsequences $\mathbf{w}_j^l$ of the Kth sentence $\mathbf{W}_K$. However, the meaning is implied.

we get the re-estimates

$$Q'(a|h_{-2}=x,h_{-1}=y)=\frac{CC(x,y,a)}{\sum_{a'}CC(x,y,a')}. \tag{26}$$

We can similarly use the quantities (25) for re-estimating $P(v|h_{-2},h_{-1})$. In fact, let

$$CC(x,y,v)\doteq\sum_{K=1}^{M}\sum_{i=1}^{n_K}\sum_{j=i}^{n_K}C_K(x,y,i,j,\mathbf{null})\,\delta(v,w_{j+1}) \tag{27}$$

then

$$P'(v|h_{-2}=x,h_{-1}=y)=\frac{CC(x,y,v)}{\sum_{v'}CC(x,y,v')}. \tag{28}$$

Of course, $P_1(v)$ need not be re-estimated. It is equal to the relative frequency in the M training sentences of the initial words $w_1(K)$ being equal to v:

$$P_1(v)=\frac{1}{M}\sum_{K=1}^{M}\delta(v,w_1(K)). \tag{29}$$

6. Extension of training to full structured language models. We will now extend our results to the complete structured language model (SLM) that has both binary and unary constructor actions [1]. It has a more complex constructor than does the SSLM and an additional module, the *tagger*. Headwords h will be replaced by *heads* $\mathbf{h}=(h^1,h^2)$ where h^1 is a headword and h^2 is a tag or a non-terminal. Let us describe briefly the operation of the SLM:[18]

- Depending on the last two exposed heads, the *predictor* generates the next word w_i with probability $P(w_i|\mathbf{h}_{-2},\mathbf{h}_{-1})$.
- Depending on the last exposed head and on w_i, the *tagger* tags w_i by a part of speech $g\in\mathcal{G}$ with probability $P(g|w_i,\mathbf{h}_{-1})$.
  - Heads shift: $\mathbf{h}'_{-i-1}=\mathbf{h}_{-i},\ i=1,2,...$
  - A new last exposed head is created: $\mathbf{h}'_{-1}=(h^1_{-1},h^2_{-1})=(w_i,g)$
- The constructor operates essentially as described in Section 2 according to a probability $Q(\mathbf{a}|\mathbf{h}_{-2},\mathbf{h}_{-1})$, but with an enlarged action alphabet. That is, $\mathbf{a}\in\{(\mathbf{right}||\gamma),(\mathbf{right}^*||\gamma),(\mathbf{left}||\gamma),(\mathbf{left}^*||\gamma),(\mathbf{up}||\gamma),\mathbf{null}\}$ where $\gamma\in\Gamma$, the set of non-terminal symbols.

[18] We will be brief, basing our exposition on the assumption that the reader is by now familiar with the operation of the SSLM as described in Section 2.

– (**right**$||\gamma$) means create an apex with downward connections to $\mathbf{h}_{-2}$ and $\mathbf{h}_{-1}$. Label the apex by $\mathbf{h}'_{-1} = (h^1_{-1}, \gamma)$. Let $\mathbf{h}'_{-i} = \mathbf{h}_{-i-1}, i = 2, 3, \ldots$
– (**right***$||\gamma$) means create an apex with downward connections to $\mathbf{h}_{-2}$ and $\mathbf{h}_{-1}$. Label the apex by $\mathbf{h}'_{-1} = (h^1_{-1}, \gamma)^*$. Let $\mathbf{h}'_{-i} = \mathbf{h}_{-i-1}, i = 2, 3, \ldots$
– (**left**$||\gamma$) means create an apex with downward connections to $\mathbf{h}_{-2}$ and $\mathbf{h}_{-1}$. Label the apex by $\mathbf{h}'_{-1} = (h^1_{-2}, \gamma)$. Let $\mathbf{h}'_{-i} = \mathbf{h}_{-i-1}, i = 2, 3, \ldots$
– (**left***$||\gamma$) means create an apex with downward connections to $\mathbf{h}_{-2}$ and $\mathbf{h}_{-1}$. Label the apex by $\mathbf{h}'_{-1} = (h^1_{-2}, \gamma)^*$. Let $\mathbf{h}'_{-i} = \mathbf{h}_{-i-1}, i = 2, 3, \ldots$
– (**up**$||\gamma$) means create an apex with a downward connection to $\mathbf{h}_{-1}$ only. Label the apex by $\mathbf{h}'_{-1} = (h^1_{-1}, \gamma)$. Let $\mathbf{h}'_{-i} = \mathbf{h}_{-i}, i = 2, 3, \ldots$
– **null** means pass control to the predictor.

The operation of the SLM ends when the parser marks its apex by the head $< s >$.[19]

Start of operation: The predictor generates the first word w_1 with probability $P_1(w_1 = v) = P(w_1 = v| < s >), v \in \mathcal{V}$. The tagger then tags w_1 by the part of speech g with probability $P(g|w_1, < s >), g \in \mathcal{G}$. The initial heads (both exposed) become $\mathbf{h}_{-2} = (< s >, < s >)$, $\mathbf{h}_{-1} = (w_1, g)$. Control is passed to the constructor.

Restriction: For all $\mathbf{h}_{-2}, h^1_{-1}, \gamma_0$ and $j = 0, 1, \ldots$[20]

$$Q((\mathbf{up}||\gamma_0)|\mathbf{h}_{-2}, (h^1_{-1}, \gamma_j)) \prod_{i=1}^{j} Q((\mathbf{up}||\gamma_i)|\mathbf{h}_{-2}, (h^1_{-1}, \gamma_{i-1})) = 0. \tag{30}$$

Special constructor probabilities:

- If $\mathbf{h}_{-1} = (v, \gamma)$, $v \in \mathcal{V}$ then

$$Q(\mathbf{a}|\mathbf{h}_{-2} = (< s >, < s >), \mathbf{h}_{-1}) = \begin{cases} 1 & \text{if } \mathbf{a} = \mathbf{null} \\ 0 & \text{otherwise.} \end{cases} \tag{31}$$

- If $\mathbf{h}_{-1} \in \{(v, \gamma)^*, (< /s >, < /s >)\}$, $v \in \mathcal{V}$ then

$$Q(\mathbf{a}|\mathbf{h}_{-2} = (< s >, < s >), \mathbf{h}_{-1}) = \begin{cases} 1 & \text{if } \mathbf{a} = (\mathbf{left}|| < s >) \\ 0 & \text{otherwise.} \end{cases} \tag{32}$$

[19] Formally, this head is $\mathbf{h} = (< s > , < s >)$, but we may sometimes omit writing the second component. Similarly, the head corresponding to the end of sentence symbol is $\mathbf{h} = (< /s > , < /s >)$. This, of course, means that when the tagger is called upon to tag $< /s >$, it tags it with probability one by the part of speech $< /s >$.

[20] I.e., **up** actions cannot cycle.

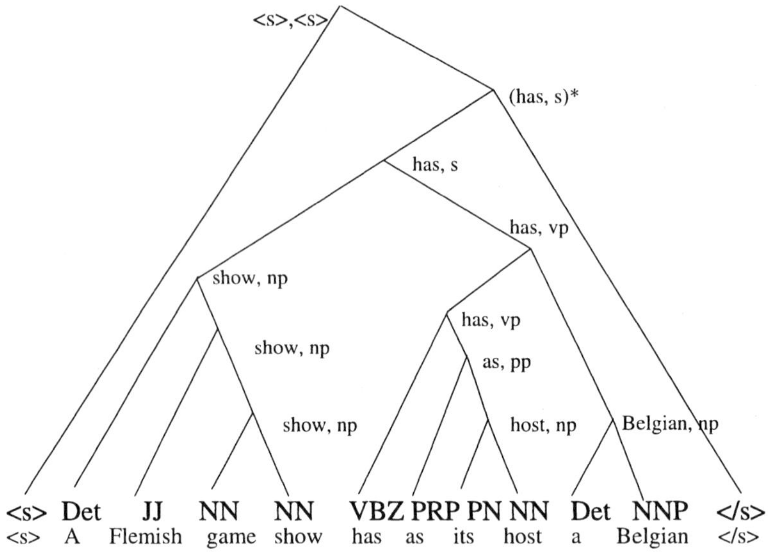

FIG. 7. *Parse by the complete structured language model.*

- If $\mathbf{h}_{-2} = (v, \gamma)$, $v \in \mathcal{V}$ then

$$Q(\mathbf{a}|\mathbf{h}_{-2}, \mathbf{h}_{-1} = </s>) = \begin{cases} 1 & \text{if } \mathbf{a} = (\mathbf{left}^*||\gamma) \\ 0 & \text{otherwise.} \end{cases} \tag{33}$$

- If $\mathbf{h}_{-1} = (v, \gamma)^*$, $v \in \mathcal{V}$ and $\mathbf{h}_{-2} \neq (<s>, <s>)$ then

$$Q(\mathbf{a}|\mathbf{h}_{-2}, \mathbf{h}_{-1}) = 0 \text{ for } \mathbf{a} \in \{\mathbf{right}, \mathbf{left}, \mathbf{null}\}. \tag{34}$$

Special predictor probabilities:

- If $\mathbf{h}_{-2} \neq <s>$ then

$$P(</s> \mid \mathbf{h}_{-2}, \mathbf{h}_{-1}) = 0. \tag{35}$$

Figure 7 illustrates one possible parse (the "correct" one) resulting from the operation of the complete SLM on the sentence of Figure 1.

A good way to regard the increase in complexity of the full SLM (compared to the original, simplified version) is to view it as an enlargement of the headword vocabulary. We will now adjust the recursions of Sections 4.1, 4.2 and 5.1 to reflect the new situation.[21] We will find that certain scalar arguments in the preceding formulas will be replaced by their appropriate vector counterparts denoted in boldface.

[21] Adjustment of Section 4.3 is left to the reader since it is very similar to that of Section 4.1. In fact, the only difference between formula (9) and (15) is that sums in the former are replaced by maxima in the latter.

We will first adjust the equations (9) through (11). We get:
For $j > i, i \in \{0, 1, 2, ..., n\}$

$$
\begin{aligned}
& P(\mathbf{w}_{i+1}^{j}, \mathbf{y}[i,j] | w_i, \mathbf{x}) \\
& \quad = \sum_{\gamma \in \Gamma(\mathbf{x},\mathbf{y})} P(\mathbf{w}_{i+1}^{j}, (y^1, \gamma)[i,j] | w_i, \mathbf{x})\, Q((\mathbf{up} || y^2)\, | \mathbf{x}, (y^1, \gamma)) \\
& \quad + \sum_{l=i}^{j-1} \sum_{\mathbf{z}} \sum_{\gamma} \Big[P^*(w_{l+1} | \mathbf{x}, (y^1, \gamma))\, P(\mathbf{w}_{i+1}^{l}, (y^1, \gamma)[i,l] | w_i, \mathbf{x}) \\
& \quad \times P(\mathbf{w}_{l+2}^{j}, \mathbf{z}[l+1, j] | w_{l+1}, (y^1, \gamma))\, Q((\mathbf{left} || y^2)\, | (y^1, \gamma), \mathbf{z}) \Big] \\
& \quad + \sum_{l=i}^{j-1} \sum_{\mathbf{u}} \sum_{\gamma} \Big[P^*(w_{l+1} | \mathbf{x}, \mathbf{u})\, P(\mathbf{w}_{i+1}^{l}, \mathbf{u}[i,l] | w_i, \mathbf{x}) \\
& \quad \times P(\mathbf{w}_{l+2}^{j}, (y^1, \gamma)[l+1, j] | w_{l+1}, \mathbf{u})\, Q((\mathbf{right} || y^2)\, | \mathbf{u}, (y^1, \gamma)) \Big]
\end{aligned} \tag{36}
$$

where

$$P^*(w | \mathbf{h}_{-2}, \mathbf{h}_{-1}) = P(w | \mathbf{h}_{-2}, \mathbf{h}_{-1})\, Q(\mathbf{null} | \mathbf{h}_{-2}, \mathbf{h}_{-1}) \tag{37}$$

and

$$
\begin{gathered}
P(\mathbf{w}_{i+1}^{j}, (y^1, \gamma)[i,j] | w_i, \mathbf{x}) = 0 \\
\text{if } x^1 \notin \{w_0, ..., w_{i-1}\} \text{ or } y^1 \notin \{w_i, ..., w_j\} \text{ or } i > j
\end{gathered} \tag{38}
$$

and $\Gamma(\mathbf{x}, \mathbf{y})$ is an appropriate subset of the non-terminal set Γ as discussed below.

The boundary conditions for the recursion (36) are

$$
\begin{aligned}
P(\mathbf{w}_{i+1}^{i}, (w_i, \gamma)[i,i] | w_i, \mathbf{x}) &= P(\mathbf{h} = (w_i, \gamma) | w_i, h_{-1}(\mathbf{T}^{i-1}) = \mathbf{x}) \\
&= P(\gamma | w_i, \mathbf{x}) \qquad \text{for } x^1 \in \{w_0, ..., w_{i-1}\}
\end{aligned} \tag{39}
$$

and the final probability we are interested in remains

$$P(\mathbf{W}) = P(\mathbf{w}_2^{n+1}, </s> [1, n+1] | w_1, <s>)\, P_1(w_1).$$

A certain subtlety must be observed in evaluating (36): For every pair $(\mathbf{x}, y^1)$ the probabilities $P(\mathbf{w}_{i+1}^{j}, (y^1, \gamma)[i,j] | w_i, \mathbf{x})$ must be evaluated for different non-terminals γ in a particular order. Because of the restriction (30) such an order exists assuring that the values of $P(\mathbf{w}_{i+1}^{j}, (y^1, \gamma)[i,j] | w_i, \mathbf{x})$ for $\gamma \in \Gamma(\mathbf{x}, \mathbf{y})$ are fully known before $P(\mathbf{w}_{i+1}^{j}, \mathbf{y}[i,j] | w_i, \mathbf{x})$ is computed.

This completes the adjustment of the formulas of Section 4.1.

Before proceeding further we can return to the example fragment $< s >$ FRESHMAN BASKETBALL PLAYER $< /s >$ and perform for it the operations (36) when the basic production probabilities are as given in Table 2 (these are realistic quantities obtained after training on

TABLE 2
Probabilities needed to compute inside-outside probabilities.

Probability	Value
$P(\text{F}\mid(</s>,</s>),(<s>,<s>))$	7.05E-5
$P(\text{B}\mid(<s>,<s>),(\text{F},\text{NN}))$	1.26E-1
$P(\text{P}\mid(<s>,<s>),(\text{B},\text{NP}))$	3.22E-6
$P(\text{P}\mid(\text{F},\text{NN}),(\text{B},\text{NN}))$	2.37E-1
$P(\text{P}\mid(\text{F},\text{NN}),(\text{B},\text{NNP}))$	3.18E-5
$P(</s>\mid(<s>,<s>),(\text{P},\text{NP}))$	3.56E-1
$P(\text{NN}\mid\text{F},<s>)$	9.95E-1
$P(\text{NN}\mid\text{B},\text{NN})$	9.94E-1
$P(\text{NNP}\mid\text{B},\text{NN})$	5.03E-3
$P(\text{NN}\mid\text{P},\text{NP})$	9.34E-1
$P(\text{NNP}\mid\text{P},\text{NP})$	5.62E-2
$P(\text{NN}\mid\text{P},\text{NN})$	9.65E-1
$P(\text{NNP}\mid\text{P},\text{NN})$	3.46E-2
$P(\text{NN}\mid\text{P},\text{NNP})$	6.65E-1
$P(\text{NNP}\mid\text{P},\text{NNP})$	3.31E-1
$P(\text{NULL\_}\mid(<s>,<s>),(\text{F},\text{NN}))$	8.59E-1
$P(\text{NULL\_}\mid(<s>,<s>),(\text{B},\text{NP}))$	1.00
$P(\text{NULL\_}\mid(\text{F},\text{NN}),(\text{B},\text{NN}))$	6.64E-1
$P(\text{NULL\_}\mid(\text{F},\text{NN}),(\text{B},\text{NNP}))$	7.85E-1
$P(\text{AR\_NP}\mid(\text{F},\text{NN}),(\text{B},\text{NN}))$	1.23E-1
$P(\text{AR\_NP}\mid(\text{F},\text{NN}),(\text{B},\text{NNP}))$	3.11E-2
$P(\text{AR\_NP'}\mid(\text{B},\text{NN}),(\text{P},\text{NN}))$	8.98E-1
$P(\text{AR\_NP'}\mid(\text{B},\text{NNP}),(\text{P},\text{NN}))$	7.42E-1
$P(\text{AR\_NP'}\mid(\text{B},\text{NN}),(\text{P},\text{NNP}))$	1.24E-1
$P(\text{AR\_NP'}\mid(\text{B},\text{NNP}),(\text{P},\text{NNP}))$	2.13E-1
$P(\text{AR\_NP}\mid(\text{B},\text{NN}),(\text{P},\text{NN}))$	4.93E-2
$P(\text{AR\_NP}\mid(\text{B},\text{NNP}),(\text{P},\text{NN}))$	5.38E-2
$P(\text{AR\_NP}\mid(\text{B},\text{NN}),(\text{P},\text{NNP}))$	2.82E-1
$P(\text{AR\_NP}\mid(\text{B},\text{NNP}),(\text{P},\text{NNP}))$	2.47E-1
$P(\text{AR\_NP}\mid(\text{B},\text{NP}),(\text{P},\text{NN}))$	4.55E-1
$P(\text{AR\_NP}\mid(\text{B},\text{NP}),(\text{P},\text{NNP}))$	2.82E-2
$P(\text{AR\_NP}\mid(\text{F},\text{NN}),(\text{P},\text{NP}))$	1.66E-2
$P(\text{AR\_NP}\mid(\text{F},\text{NN}),(\text{P},\text{NP'}))$	8.45E-1

the U Penn Treebank [3]). Table 3 then represents the parse triangle chart containing the inside probabilities (36) for all the relevant spans. Finally, the following is the detailed calculation for the inside probability $P(\text{BP},(\text{P},NP)[1,3]|\text{F},(<s>,<s>))$ which we abbreviate as $P((\text{P},NP)[1,3]|\text{F},(<s>,<s>))$:

TABLE 3
Inside probability table.

<s>	freshman (F)	basketball (B)	player (P)	</s>
P((<s>. <s>)| <s>.(</s>. </s>)) = 1				P((</s>. </s>)| <s>.(</s>. </s>)) = 3.11E−7
	P((F.NN)|F..(<s>. <s>)) = 9.95E−1	P((B.NP)|F.(<s>. <s>)) = 1.32E−2	P((P.NP)|F.(<s>. <s>)) = 1.24E−2	P((</s>. </s>)|F.(<s>. <s>)) = 4.41E−3
		P((B.NN)|B.(F.NN)) = 9.94E−1 P((B.NNP)|B.(F.NN)) = 5.03E−3	P((P.NP)|B.(F.NN)) = 9.0E−3 P((P.NP')|B.(F.NN)) = 1.36E−1	
			P((P.NN)|P.(B.NP)) = 9.34E−1 P((P.NNP)|P.(B.NP)) = 5.62E−2 P((P.NN)|P.(B.NN)) = 9.65E−1 P((P.NNP)|P.(B.NN)) = 3.46E−2 P((P.NN)|P.(B.NNP)) = 6.65E−1 P((P.NNP)|P.(B.NNP)) = 3.31E−1	
				P((</s>. </s>)| </s>.(P.NP)) = 1

$$
\begin{aligned}
&P((\textsc{p},\mathrm{NP})[1,3]|\textsc{f},(<\textsc{s}>,<\textsc{s}>))\\
&\quad= P((\textsc{f},\mathrm{NN})[1,1]|\textsc{f},(<\textsc{s}>,<\textsc{s}>)) \times P((\textsc{p},\mathrm{NP})[2,3]|\textsc{b},(\textsc{f},\mathrm{NN}))\\
&\qquad\times P(\mathrm{NULL\_}|(<\textsc{s}>,<\textsc{s}>),(\textsc{f},\mathrm{NN}))\\
&\qquad\times P(\textsc{b}|(<\textsc{s}>,<\textsc{s}>),(\textsc{f},\mathrm{NN})) \times P(\mathrm{AR\_NP}|(\textsc{f},\mathrm{NN}),(\textsc{p},\mathrm{NP}))\\
&\quad+ P((\textsc{f},\mathrm{NN})[1,1]|\textsc{f},(<\textsc{s}>,<\textsc{s}>)) \times P((\textsc{p},\mathrm{NP'})[2,3]|\textsc{b},(\textsc{f},\mathrm{NN}))\\
&\qquad\times P(\mathrm{NULL\_}|(<\textsc{s}>,<\textsc{s}>),(\textsc{f},\mathrm{NN}))\\
&\qquad\times P(\textsc{b}|(<\textsc{s}>,<\textsc{s}>),(\textsc{f},\mathrm{NN})) \times P(\mathrm{AR\_NP}|(\textsc{f},\mathrm{NN}),(\textsc{p},\mathrm{NP'}))\\
&\quad+ P((\textsc{b},\mathrm{NP})[1,2]|\textsc{f},(<\textsc{s}>,<\textsc{s}>)) \times P((\textsc{p},\mathrm{NN})[3,3]|\textsc{p},(\textsc{b},\mathrm{NP}))\\
&\qquad\times P(\mathrm{NULL\_}|(<\textsc{s}>,<\textsc{s}>),(\textsc{b},\mathrm{NP}))\\
(40)\quad&\qquad\times P(\textsc{p}|(<\textsc{s}>,<\textsc{s}>),(\textsc{b},\mathrm{NP})) \times P(\mathrm{AR\_NP}|(\textsc{b},\mathrm{NP}),(\textsc{p},\mathrm{NN}))\\
&\quad+ P((\textsc{b},\mathrm{NP})[1,2]|\textsc{f},(<\textsc{s}>,<\textsc{s}>)) \times P((\textsc{p},\mathrm{NNP})[3,3]|\textsc{p},(\textsc{b},\mathrm{NP}))\\
&\qquad\times P(\mathrm{NULL\_}|(<\textsc{s}>,<\textsc{s}>),(\textsc{b},\mathrm{NP}))\\
&\qquad\times P(\textsc{p}|(<\textsc{s}>,<\textsc{s}>),(\textsc{b},\mathrm{NP})) \times P(\mathrm{AR\_NP}|(\textsc{b},\mathrm{NP}),(\textsc{p},\mathrm{NNP}))\\
&\quad= 0.995 \times 9.0 \times 10^{-3} \times 0.859 \times 0.126 \times 0.0166\\
&\qquad+ 0.995 \times 0.136 \times 0.859 \times 0.126 \times 0.845\\
&\qquad+ 0.0132 \times 0.934 \times 1 \times 3.22 \times 10^{-6} \times 0.455\\
&\qquad+ 0.0132 \times 0.136 \times 1 \times 3.22 \times 10^{-6} \times 0.0282\\
&\quad= 1.24 \times 10^{-2}
\end{aligned}
$$

We are now in the position to compute the probabilities $P(\mathbf{W}^i)$.[22] The required recursion replacing (13) is

$$
\begin{aligned}
P(\mathbf{W}^{l+1},\mathbf{x}) &= \sum_{i=1}^{l} \sum_{y^1 \in \mathcal{W}^{i-1}} \sum_{\gamma} \Big[P(\mathbf{W}^i,(y^1,\gamma))\\
(41)\qquad &\qquad \times P(\mathbf{w}_{i+1}^{l},\mathbf{x}[i,l]|w_i,(y^1,\gamma))\, P^*(w_{l+1}|(y^1,\gamma),\mathbf{x})\Big]\\
&\quad \text{for}\quad x^1 \in \mathcal{W}^l = \{w_1, w_2, ..., w_l\}
\end{aligned}
$$

with the initial condition

$$
P(\mathbf{W}^1,\mathbf{x}) = \begin{cases} P_1(w_1) & \mathbf{x} = < s >, < s > \\ 0 & \mathbf{x} \neq < s >, < s > . \end{cases}
$$

It then follows from (41) that

$$
P(w_0, w_1, w_2, ..., w_i, w_{i+1}) = \sum_{x^1 \in \mathcal{W}^i} \sum_{\gamma} P(\mathbf{W}^{i+1},(x^1,\gamma)).
$$

[22] Compare with the results of Section 4.2.

In refining the formulas of Section 5.1 we get new recursions for the outer probabilities $P(\mathbf{w}_0^i, \mathbf{w}_{j+1}^{n+1}, \mathbf{x}[i-1]\ ;\ \mathbf{y}[i,j])$ involved in

$$P(\mathbf{W}, \mathbf{x}, \mathbf{y}[i,j]) = P(\mathbf{w}_0^i, \mathbf{w}_{j+1}^{n+1}, \mathbf{x}[i-1]\ ;\ \mathbf{y}[i,j])\, P(\mathbf{w}_{i+1}^j, \mathbf{y}[i,j] | w_i, \mathbf{x}). \tag{42}$$

For $x^1 \in \mathcal{W}^{i-1}$ and $y^1 \in \{w_i, ..., w_j\}$, the required formulas are

$$\begin{aligned} &P(\mathbf{w}_0^i, \mathbf{w}_{j+1}^{n+1}, \mathbf{x}[i-1]\,;\mathbf{y}[i,j]) \\ &\quad = \sum_{\gamma \in \Delta(\mathbf{x},\mathbf{y})} P(\mathbf{w}_0^i, \mathbf{w}_{j+1}^{n+1}, \mathbf{x}[i-1]\,; (y^1,\gamma)[i,j]) Q((\mathbf{up}||\gamma)\,|\mathbf{x},\mathbf{y}) \\ &\quad + \sum_{l=1}^{i-1}\sum_{\mathbf{z}}\sum_{\gamma} \Big[P(\mathbf{w}_0^{i-l}, \mathbf{w}_{j+1}^{n+1}, \mathbf{z}[i-l-1]; (x^1,\gamma)[i-l,j]) \\ &\quad\quad \times P(\mathbf{w}_{i-l+1}^{i-1}, \mathbf{x}[i-l,i-1] | w_{i-l}, \mathbf{z})\, P^*(w_i|\mathbf{z},\mathbf{x})\, Q((\mathbf{left}||\gamma)\,|\mathbf{x},\mathbf{y})\Big] \\ &\quad + \sum_{l=1}^{i-1}\sum_{\mathbf{z}}\sum_{\gamma} \Big[P(\mathbf{w}_0^{i-l}, \mathbf{w}_{j+1}^{n+1}, \mathbf{z}[i-l-1]; (y^1,\gamma)[i-l,j]) \\ &\quad\quad \times P(\mathbf{w}_{i-l+1}^{i-1}, \mathbf{x}[i-l,i-1] | w_{i-l}, \mathbf{z}) P^*(w_i|\mathbf{z},\mathbf{x}) Q((\mathbf{right}||\gamma)\,|\mathbf{x},\mathbf{y})\Big] \\ &\quad + \sum_{m=1}^{n-j+1}\sum_{\mathbf{u}}\sum_{\gamma} \Big[P(\mathbf{w}_0^i, \mathbf{w}_{j+m+1}^{n+1}, \mathbf{x}[i-1]; (y^1,\gamma)[i,j+m]) \\ &\quad\quad \times P(\mathbf{w}_{j+2}^{j+m}, \mathbf{u}[j+1,j+m] | w_{j+1}, \mathbf{y}) P^*(w_{j+1}|\mathbf{x},\mathbf{y}) Q((\mathbf{left}||\gamma)\,|\mathbf{y},\mathbf{u})\Big] \\ &\quad + \sum_{m=1}^{n-j+1}\sum_{\mathbf{u}}\sum_{\gamma} \Big[P(\mathbf{w}_0^i, \mathbf{w}_{j+m+1}^{n+1}, \mathbf{x}[i-1]; (u^1,\gamma)[i,j+m]) \\ &\quad\quad \times P(\mathbf{w}_{j+2}^{j+m}, \mathbf{u}[j+1,j+m] | w_{j+1}, \mathbf{y}) P^*(w_{j+1}|\mathbf{x},\mathbf{y}) Q((\mathbf{right}||\gamma)\,|\mathbf{y},\mathbf{u})\Big] \end{aligned} \tag{43}$$

where $P^*(w_{j+1}|\mathbf{x},\mathbf{y})$ is defined by (37). If either $x^1 \notin \mathcal{W}^{i-1}$ or $y^1 \notin \{w_i, ..., w_j\}$ then we define[23]

$$P(\mathbf{w}_0^i, \mathbf{w}_{j+1}^{n+1}, \mathbf{x}[i-1]\,;\mathbf{y}[i,j]) \doteq 0. \tag{44}$$

Again, the probabilities (43) for $(\mathbf{x},y^1)$ must be evaluated in an appropriate order to assure that $P(\mathbf{w}_0^i, \mathbf{w}_{j+1}^{n+1}, \mathbf{x}[i-1]\,; (y^1,\gamma)[i,j])$ for $\gamma \in \Delta(\mathbf{x},\mathbf{y})$ is known. This order is actually the reverse order implied by $\Gamma(\mathbf{x},\mathbf{y})$ and can be determined because the restriction (30) applies.

In order to use (43) we need the boundary conditions

$$\begin{aligned} &P(\mathbf{w}_0^1, \mathbf{w}_{n+2}^{n+1}, \mathbf{x}[0]\,;\mathbf{y}[1,n+1]) \\ &\quad = \begin{cases} P_1(w_1) & \text{if } \mathbf{x} = <s>, <s> \text{ and } \mathbf{y} = </s>, </s> \\ 0 & \text{otherwise.} \end{cases} \end{aligned} \tag{45}$$

[23]This definition is made use of in the summations $\sum_{\mathbf{z}}$ and $\sum_{\mathbf{u}}$ in (43) and elsewhere, thus simplifying notation. Consequently $\sum_{\mathbf{z}}$ could also have been written $\sum_{z^1 \in \mathcal{W}^{i-l-1}} \sum_{\delta \in \Gamma}$, etc.

We now return to the example fragment $< s >$ FRESHMAN BASKETBALL PLAYER $< /s >$ for the last time and perform for it the operations (43). Table 4 then represents the parse triangle chart containing the outside probabilities (36) for all the relevant spans. The following is the detailed calculation for the outside probability $P(<\mathbf{s}>\text{FB}, \text{P}</s>, (\text{F}, NN)[1]; (\text{B}, NN)[2,2])$ which we abbreviate as $P((\text{F}, NN); (\text{B}, NN)[2,2])$:

$$
\begin{aligned}
&P((\text{F,NN});\ (\text{B,NN})[2,2]) \\
&\quad = P((\text{F,NN});\ (\text{P,NP})[2,3]) \times P((\text{P,NN})[3,3]|\text{P},(\text{B,NN})) \\
&\qquad \times P(\text{NULL\_}|(\text{F,NN}),(\text{B,NN})) \times P(\text{P}|(\text{F,NN}),(\text{B,NN})) \\
&\qquad \times P(\text{AR\_NP}|(\text{B,NN}),(\text{P,NN})) \\
&\qquad + P((\text{F,NN});\ (\text{P,NP'})[2,3]) \times P((\text{P,NN})[3,3]|\text{P},(\text{B,NN})) \\
&\qquad \times P(\text{NULL\_}|(\text{F,NN}),(\text{B,NN})) \times P(\text{P}|(\text{F,NN}),(\text{B,NN})) \\
&\qquad \times P(\text{AR\_NP'}|(\text{B,NN}),(\text{P,NN})) \\
&\qquad + P((\text{F,NN});\ (\text{P,NP})[2,3]) \times P((\text{P,NNP})[3,3]|\text{P},(\text{B,NN})) \\
&\qquad \times P(\text{NULL\_}|(\text{F,NN}),(\text{B,NN})) \times P(\text{P}|(\text{F,NN}),(\text{B,NN})) \\
&\qquad \times P(\text{AR\_NP}|(\text{B,NN}),(\text{P,NNP})) \\
&\qquad + P((\text{F,NN});\ (\text{P,NP'})[2,3]) \times P((\text{P,NNP})[3,3]|\text{P},(\text{B,NN})) \\
&\qquad \times P(\text{NULL\_}|(\text{F,NN}),(\text{B,NN})) \times P(\text{P}|(\text{F,NN}),(\text{B,NN})) \\
&\qquad \times P(\text{AR\_NP'}|(\text{B,NN}),(\text{P,NNP})) \\
&\qquad + P((<\text{s}>, <\text{s}>);\ (\text{B,NP})[1,2]) \\
&\qquad \times P((\text{F,NN})[1,1]|\text{F},(<\text{s}>, <\text{s}>)) \\
&\qquad \times P(\text{NULL\_}|(<\text{s}>, <\text{s}>),(\text{F,NN})) \\
&\qquad \times P(\text{B}|(<\text{s}>, <\text{s}>),(\text{F,NN})) \\
&\qquad \times P(\text{AR\_NP}|(\text{F,NN}),(\text{B,NN})) \\
&\quad = 4.49 \times 10^{-8} \times 0.965 \times 0.664 \times 0.237 \times 0.0493 \\
&\qquad + 2.28 \times 10^{-6} \times 0.965 \times 0.664 \times 0.237 \times 0.898 \\
&\qquad + 4.49 \times 10^{-8} \times 0.0346 \times 0.664 \times 0.237 \times 0.282 \\
&\qquad + 2.28 \times 10^{-6} \times 0.0346 \times 0.664 \times 0.237 \times 0.124 \\
&\qquad + 3.45 \times 10^{-11} \times 0.995 \times 0.859 \times 0.126 \times 0.123 \\
&\quad = 3.13 \times 10^{-7}.
\end{aligned}
\tag{46}
$$

Finally, to obtain formulas for re-estimating $P(v|\mathbf{h}_{-2}, \mathbf{h}_{-1})$ and $Q(\mathbf{a}|\mathbf{h}_{-1}, \mathbf{h}_{-2})$ we proceed as follows:

TABLE 4
Outside probability table.

<S>	freshman (F)	basketball (B)	player (P)	</S>
P((</S>.</S>):(<S>.<S>)) = 3.11E−7				P((</S>.</S>):(</S>.</S>)) = 1
	P((<S>.<S>):(F.NN)) = 3.11E−7	P((<S>.<S>):(B.NP)) = 3.45E−11	P((<S>.<S>):(P.NP)) = 2.51E−5	P((<S>.<S>):(</S>.</S>)) = 7.05E−5
		P((F.NN):(B.NN)) = 3.13E−7 P((F.NN):(B.NNP)) = 3.23E−11	P((F.NN):(P.NP)) = 4.49E−8 P((F.NN):(P.NP')) = 2.28E−6	
			P((B.NP):(P.NN)) = 4.85E−13 P((B.NP):(P.NNP)) = 3.01E−14 P((B.NN):(P.NN)) = 3.21E−7 P((B.NN):(P.NNP)) = 4.62E−8 P((B.NNP):(P.NN)) = 2.13E−13 P((B.NNP):(P.NNP)) = 6.24E−14	
				P((P.NP):(</S>.</S>)) = 3.11E−7

$$C_K(\mathbf{x},\mathbf{y},i,j,(\mathbf{left}||\gamma)) \doteq \frac{1}{P(\mathbf{W}_K)} P(\mathbf{w}_{i+1}^{j},\mathbf{y}[i,j]|w_i,\mathbf{x})$$

$$\times \sum_{\mathbf{z}} \sum_{l=1}^{i-1} \Big[P(\mathbf{w}_0^{i-l},\mathbf{w}_{j+1}^{n+1},\mathbf{z}[i-l-1]\,;(x^1,\gamma)[i-l,j]) \tag{47}$$

$$\times P(\mathbf{w}_{i-l+1}^{i-1},\mathbf{x}[i-l,i-1]|w_{i-l},\mathbf{z})\,P^*(w_i|\mathbf{z},\mathbf{x})\,Q((\mathbf{left}||\gamma)\,|\mathbf{x},\mathbf{y})\Big]$$

$$C_K(\mathbf{x},\mathbf{y},i,j,(\mathbf{right}||\gamma)) \doteq \frac{1}{P(\mathbf{W}_K)} P(\mathbf{w}_{i+1}^{j},\mathbf{y}[i,j]|w_i,\mathbf{x})$$

$$\times \sum_{\mathbf{z}} \sum_{l=1}^{i-1} \Big[P(\mathbf{w}_0^{i-l},\mathbf{w}_{j+1}^{n+1},\mathbf{z}[i-l-1]\,;(y^1,\gamma)[i-l,j]) \tag{48}$$

$$\times P(\mathbf{w}_{i-l+1}^{i-1},\mathbf{x}[i-l,i-1]|w_{i-l},\mathbf{z})P^*(w_i|\mathbf{z},\mathbf{x})Q((\mathbf{right}||\gamma)\,|\mathbf{x},\mathbf{y})\Big]$$

$$\begin{aligned} C_K(\mathbf{x},\mathbf{y},i,j,(\mathbf{up}||\gamma)) &\doteq \frac{1}{P(\mathbf{W}_K)} P(\mathbf{w}_{i+1}^{j},\mathbf{y}[i,j]|w_i,\mathbf{x}) \\ &\times \sum_{\gamma\in\Delta(\mathbf{x},\mathbf{y})} P(\mathbf{w}_0^{i},\mathbf{w}_{j+1}^{n+1},\mathbf{x}[i-1]\,;(y^1,\gamma)[i,j])Q((\mathbf{up}||\gamma)\,|\mathbf{x},\mathbf{y}) \end{aligned} \tag{49}$$

$$C_K(\mathbf{x},\mathbf{y},i,j,\mathbf{null}) \doteq \frac{1}{P(\mathbf{W}_K)} P(\mathbf{w}_{i+1}^{j},\mathbf{y}[i,j]|w_i,\mathbf{x})$$

$$\times \Bigg\{\sum_{\mathbf{u}} \sum_{m=1}^{n-j+1} \sum_{\gamma} \Big[P(\mathbf{w}_0^{i},\mathbf{w}_{j+m+1}^{n+1},\mathbf{x}[i-1]\,;(y^1,\gamma)[i,j+m])$$

$$\times P(\mathbf{w}_{j+2}^{j+m},\mathbf{u}[j+1,j+m]|w_{j+1},\mathbf{y})P^*(w_{j+1}|x,y)Q((\mathbf{left}||\gamma)\,|\mathbf{y},\mathbf{u})\Big] \tag{50}$$

$$+\sum_{\mathbf{u}} \sum_{m=1}^{n-j+1} \sum_{\gamma} \Big[P(\mathbf{w}_0^{i},\mathbf{w}_{j+m+1}^{n+1},\mathbf{x}[i-1];(u^1,\gamma)[i,j+m])$$

$$\times P(\mathbf{w}_{j+2}^{j+m},\mathbf{u}[j+1,j+m]|w_{j+1},\mathbf{y})P^*(w_{j+1}|\mathbf{x},\mathbf{y})Q((\mathbf{right}||\gamma)\,|\mathbf{y},\mathbf{u})\Big]\Bigg\}$$

Thus, defining counter "contents" (n_K is the length of the Kth sentence)

$$CC(\mathbf{x},\mathbf{y},(\mathbf{left}||\gamma)) \doteq \sum_{K=1}^{M}\sum_{i=1}^{n_K+1}\sum_{j=i}^{n_K+1} C_K(\mathbf{x},\mathbf{y},i,j,(\mathbf{left}||\gamma))$$

$$CC(\mathbf{x},\mathbf{y},(\mathbf{right}||\gamma)) \doteq \sum_{K=1}^{M}\sum_{i=1}^{n_K+1}\sum_{j=i}^{n_K+1} C_K(\mathbf{x},\mathbf{y},i,j,(\mathbf{right}||\gamma))$$

$$CC(\mathbf{x},\mathbf{y},(\mathbf{up}||\gamma)) \doteq \sum_{K=1}^{M}\sum_{i=1}^{n_K+1}\sum_{j=i}^{n_K+1} C_K(\mathbf{x},\mathbf{y},i,j,(\mathbf{up}||\gamma))$$

$$CC(\mathbf{x},\mathbf{y},\mathbf{null}) \doteq \sum_{K=1}^{M}\sum_{i=1}^{n_K}\sum_{j=i}^{n_K} C_K(\mathbf{x},\mathbf{y},i,j,\mathbf{null})$$

we get the re-estimates

$$Q'(\mathbf{a}|h_{-2}=\mathbf{x},h_{-1}=\mathbf{y}) = \frac{CC(\mathbf{x},\mathbf{y},\mathbf{a})}{\sum_{\mathbf{a}'} CC(\mathbf{x},\mathbf{y},\mathbf{a}')}\ . \tag{51}$$

We can similarly use the quantities (47) through (50) for re-estimating $P(v|\mathbf{h}_{-2},\mathbf{h}_{-1})$. In fact, let

$$CC(\mathbf{x},\mathbf{y},v) \doteq \sum_{K=1}^{M}\sum_{i=1}^{n_K}\sum_{j=i}^{n_K} C_K(\mathbf{x},\mathbf{y},i,j,\mathbf{null})\,\delta(v,w_{j+1})$$

then

$$P'(v|h_{-2}=\mathbf{x},h_{-1}=\mathbf{y}) = \frac{CC(\mathbf{x},\mathbf{y},v)}{\sum_{v'} CC(\mathbf{x},\mathbf{y},v')}\ . \tag{52}$$

As before, $P_1(v)$ need not be re-estimated:

$$P_1(v) = \frac{1}{M}\sum_{K=1}^{M}\delta(v,w_1(K)).$$

Finally, the re-estimation of tagger probabilities is given by the formula[24]

$$P(g|y,\mathbf{h}_{-1}=\mathbf{x}) = \frac{CC(\mathbf{x},(y,g))}{\sum_{g'\in\mathcal{G}} CC(\mathbf{x},(y,g'))} \qquad g\in\mathcal{G} \tag{53}$$

where

$$CC(\mathbf{x},\mathbf{y}) = CC(\mathbf{x},\mathbf{y},\mathbf{null}) + \sum_{\gamma\in\Gamma}\Big[CC(\mathbf{x},\mathbf{y},(\mathbf{left}||\gamma)) + CC(\mathbf{x},\mathbf{y},(\mathbf{right}||\gamma))\Big]\ .$$

[24] Note that a headword y can be tagged by a part of speech g only if the phrase whose head is $\mathbf{y}$ consists of the single word y.

7. The problem of complexity. The recursion formulas (9) and (19) for the SSLM are very computing intensive, and the formulas (36) and (43) even more so. Referring to (9), the $< i, j >$ element of the "inside" chart need in general contain $i \times (j - i + 1)$ entries, one for each permissible headword pair x, y. The more complex chart for (36) has for each word pair x^1, y^1 which appears in it as many as KL entries where K and L are the numbers of different non-terminals that the headwords x^1 and y^1 can represent, respectively.

The question then is: what shortcuts can we take? The following observation shows that the SSLM by itself would not produce adequate parses:

Consider the parse of Figure 7. On the third level the headword pair HAS, AS forms a phrase having the headword HAS. But we would not have wanted to join HAS with AS on the first level, that is, *prematurely*! What prevents this joining in the SLM are the parts of speech VBZ and PRP by which the tagger had tagged HAS and AS, respectively. At the same time, the joining of HAS with AS is facilitated on the third level by their respective attached non-terminals VBZ and PP.

So if we wished to simplify, we could perhaps get away with the parametrization

$$Q(\mathbf{a}|\mathbf{x}, \mathbf{y}) = Q(\mathbf{a}|x^2, y^2) \text{ and } P(w|\mathbf{x}, \mathbf{y}) = P(w|x^1, y^1)$$

while the tagger distribution would continue to be given by $P(g|w, \mathbf{x})$. Alas, such a simplification would not materially reduce the computing effort required to carry out the recursions (36) and (43).

It is worth noting that from the point of view of *sparseness* of data, we could in principle be able to estimate constructor and tagger probabilities having the parametric forms $Q(\mathbf{a}|\mathbf{h}_{-3}, \mathbf{h}_{-2}, \mathbf{h}_{-1})$ and $P(g|w, \mathbf{h}_{-2}, \mathbf{h}_{-1})$. Indeed, $P(g|w, \mathbf{h}_{-2}, \mathbf{h}_{-1})$ has the memory range involved in standard HMM tagging, and $Q(\mathbf{a}|\mathbf{h}_{-3}, \mathbf{h}_{-2}, \mathbf{h}_{-1})$ would enhance the power of the constructor by moving it decisively beyond context freedom. Unfortunately, it follows from the recursions (36) and (43) that the computational price for accommodating this adjustment would be intolerable.

8. Shortcuts in the computation of the recursion algorithms. It is the nature of the SLM that a phrase spanning $< i, j >$ can have, with positive probability, as its headword any of the words $\{w_i, ..., w_j\}$. As a result, analysis of the chart parsing algorithm reveals its complexity to be proportional to n^6. This would make all the algorithms of this article impractical, unless schemes can be devised that would purge from the charts a substantial fraction of their entries.

8.1. Thresholding in the computation of inside probabilities. Note first that the product $P(\mathbf{W}^i, \mathbf{x})\, P(\mathbf{w}_{i+1}^j, \mathbf{y}[i, j]|w_i, \mathbf{x})$ denotes the probability that $\mathbf{W}^{i+j}$ is generated, the last exposed head of $\mathbf{W}^{i-1}$ is $\mathbf{x}$, and that the span $< i, j >$ is a phrase whose head is $\mathbf{y}$.

Observe next[25] that for a fixed span $< i,j >$ the products $P(\mathbf{W}^i, \mathbf{x})$ $P(\mathbf{w}_{i+1}^j, \mathbf{y}[i,j]|w_i, \mathbf{x})$ are comparable to each other regardless of the identity of $x^1 \in \{w_0, w_1, ..., w_{i-1}\}$ and $y^1 \in \{w_i, ..., w_j\}$. They can thus be thresholded with respect to $\max_{\mathbf{v},\mathbf{z}} P(\mathbf{W}^i, \mathbf{v})\, P(\mathbf{w}_{i+1}^j, \mathbf{z}[i,j]|w_i, \mathbf{v})$. That is, for further computation of inside probabilities $P(\mathbf{w}_{i+1}^j, \mathbf{y}[i,j]|w_i, \mathbf{x})$ can be set to 0 if[26]

$$P(\mathbf{W}^i, \mathbf{x})\, P(\mathbf{w}_{i+1}^j, \mathbf{y}[i,j]|w_i, \mathbf{x}) << \max_{\mathbf{v},\mathbf{z}} P(\mathbf{W}^i, \mathbf{v})\, P(\mathbf{w}_{i+1}^j, \mathbf{z}[i,j]|w_i, \mathbf{v}). \tag{54}$$

Of course, it must be kept in mind that, as always, thresholding is only an opportunistic device: The fact that (54) holds does not mean with probability 1 that $P(\mathbf{w}_{i+1}^j, \mathbf{y}[i,j]|w_i, \mathbf{x})$ will not become useful in some highly probable parse. For instance, $P(\mathbf{w}_{j+2}^k, \mathbf{z}[j+1,k]|w_{j+1}, \mathbf{y})$ may be very large and thus compensate for the relatively small value of $P(\mathbf{W}^i, \mathbf{x})\, P(\mathbf{w}_{i+1}^j, \mathbf{y}[i,j]|w_i, \mathbf{x})$. Thus the head $\mathbf{y}$ might be "needed" to complete the parse that corresponds to the probability $P(\mathbf{w}_{i+1}^k, \mathbf{z}[i,k]|w_i, \mathbf{x})$.

Next note that if $P(\mathbf{W}^i, \mathbf{y}) \ll \max_{\mathbf{z}} P(\mathbf{W}^i, \mathbf{z})$ then it is unlikely that a high probability parse will account for the interval $< 0, i-1 >$ with a sub-parse whose last exposed head is $\mathbf{y}$. In such a case then the calculation of $P(\mathbf{w}_{i+1}^j, \mathbf{x}[i,j]|w_i, \mathbf{y})$, $j \in \{i+1, ..., n+1\}$ will probably not be needed (for any $\mathbf{x}$) because the sub-parse corresponding to $P(\mathbf{w}_{i+1}^j, \mathbf{x}[i,j]|w_i, \mathbf{y})$ is a continuation of sub-parses whose total probability is very low. Again, the fact that $P(\mathbf{W}^i, \mathbf{y})$ is small does not mean that the head $\mathbf{y}$ cannot become useful in producing the future. I.e., it is still possible (though unlikely) that for some $\mathbf{x}$ and j, $P(\mathbf{w}_{i+1}^j, \mathbf{x}[i,j]|w_i, \mathbf{y})$ will be so large that at least some parses over the interval $< 0, j >$ that have $\mathbf{y}$ as the last exposed head at time $i-1$ will have a substantial probability mass.

So, if we are willing to take the risk that thresholding involves, the probabilities (36) and (41) should be computed as follows:

1. Once $P(\mathbf{W}^{i+1}, \mathbf{x})$ and $P(\mathbf{w}_{j+1}^i, \mathbf{y}[j,i]|w_j, \mathbf{x})$, $j = 0, 1, ..., i$, $i = 0, 1, ..., l$ are known for all allowed values of $\mathbf{x}$ and $\mathbf{y}$,[27] probabilities $P(\mathbf{w}_{k+1}^{l+1}, \mathbf{z}[k, l+1]|w_k, \mathbf{v})$ are computed in the sequence $k = l+1, l, ..., 0$ for each allowed $\mathbf{z}$ and those values of $\mathbf{v}$ that have not been previously zeroed out.[28]

---

[25] Again, anything relating to the quantities $P(\mathbf{w}_{i+1}^j, \mathbf{y}[i,j]|w_i, \mathbf{x})$ applies equally to the quantities $\widehat{R}(\mathbf{w}_{i+1}^j, \mathbf{y}[i,j]|w_i, \mathbf{x})$.

[26] The author is indebted to Mark Johnson who pointed out this improvement of the thresholding regime.

[27] Allowed are $\mathbf{x} = (x^1, \gamma)$ and $\mathbf{y} = (y^1, \gamma)$ where $x^1 \in \{w_0, ..., w_{j-1}\}$ and $y^1 \in \{w_j, ..., w_i\}$.

[28] Zeroing is carried out in step 4 that follows.

2. For each span $< k, l+1 >$ just computed, set to 0 all probabilities $P(\mathbf{w}_{k+1}^{l+1}, \mathbf{y}[k, l+1] | w_k, \mathbf{x})$ satisfying

$$P(\mathbf{W}^k, \mathbf{x})\, P(\mathbf{w}_{k+1}^{l+1}, \mathbf{y}[k, l+1] | w_k, \mathbf{x}) \ll \max_{\mathbf{v},\mathbf{z}} P(\mathbf{W}^i, \mathbf{v})\, P(\mathbf{w}_{k+1}^{l+1}, \mathbf{v}[k, l+1] | w_k, \mathbf{z}).$$

3. Use equation (41) to compute $P(\mathbf{W}^{l+1}, \mathbf{x})$ for the various heads $\mathbf{x}$.
4. Zero out all probabilities $P(\mathbf{w}_{l+2}^{k}, \mathbf{z}[l+1, k] | w_{l+1}, \mathbf{v})$, $k = l + 2, l+3, ..., n+1$ for all allowed heads $\mathbf{v}$ such that $P(\mathbf{W}^{l+1}, \mathbf{v}) \ll \max_{\mathbf{x}} P(\mathbf{W}^{l+1}, \mathbf{x})$. The zeroed out probabilities will not be computed in the future when the time comes to compute the corresponding positions in the chart.

Obviously, the thresholds implied in steps 2 and 4 above must be selected experimentally. Using them will short-cut the computation process while carrying the danger that occasional desirable parses will be discarded.

8.2. Thresholding in computation of outside probabilities. Having taken care of limiting the amount of computation for (36) and (41), let us consider the recursion (43). It is clear from equation (42) that $P(\mathbf{w}_0^i, \mathbf{w}_{j+1}^{n+1}, \mathbf{x}[i-1]\,; \mathbf{y}[i, j])$ will not be needed in any re-estimation formula if $P(\mathbf{w}_{i+1}^{j}, \mathbf{y}[j, i] | w_j, \mathbf{x}) = 0$. This can also be seen from the counter contributions (47) through (50).

However, we must check whether $P(\mathbf{w}_0^i, \mathbf{w}_{j+1}^{n+1}, \mathbf{x}[i-1]\,; \mathbf{y}[i, j])$ may not have to be computed because it might be needed on the right-hand side of the recurrence (43). Fortunately such is not the case. In fact, if $P(\mathbf{W}, \mathbf{x}, \mathbf{y}[i, j]) = 0$ then the event

$\mathbf{W}$ is generated, $\mathbf{y}$ is a head of the span $< i, j >$, and $\mathbf{x}$ is the preceding exposed headword

just cannot arise, so this situation is entirely analogous to the one where either $x^1 \notin \mathcal{W}^{i-1}$ or $y^1 \notin \{w_i, ..., w_j\}$. Consequently, if $P(\mathbf{W}, \mathbf{x}, \mathbf{y}[i, j]) = 0$ then the *reduction*[29] implied by $P(\mathbf{w}_0^i, \mathbf{w}_{j+1}^{n+1}, \mathbf{x}[i-1]\,; \mathbf{y}[i, j])$ is illegitimate and cannot be used in any formula computing other probabilities.

The conclusion of the preceding paragraph is valid if $P(\mathbf{w}_{i+1}^{j}, \mathbf{y}[j, i] | w_j, \mathbf{x}) = 0$, but we want to cut down on computation by coming to the same conclusion even if all we know is that $P(\mathbf{w}_{i+1}^{j}, \mathbf{y}[j, i] | w_j, \mathbf{x}) \approx 0$. Thus we are advocating here the not always valid[30] approximation of setting $P(\mathbf{w}_0^i, \mathbf{w}_{j+1}^{n+1}, \mathbf{x}[i-1]\,; \mathbf{y}[i, j])$ to 0 whenever our previous thresholding already set $P(\mathbf{w}_{i+1}^{j}, \mathbf{y}[j, i] | w_j, \mathbf{x})$ to 0.

A final saving in computation may be obtained by tracing back the chart corresponding to the inside probabilities. That is, starting with the chart contents for the span $< 1, n >$ (corresponding to

[29]We are using here the terminology of shift – reduce parsing.

[30]For the reasons given in the preceding sub-section.

$P(\mathbf{w}_2^n, \mathbf{y}[1,n]|w_1, < s >)$ for various values of $\mathbf{y}$) find and mark the sub-parse pairs $P(\mathbf{w}_2^j, \mathbf{x}[1,j]|w_1, < s >), P(\mathbf{w}_{j+1}^n, \mathbf{z}[j+1,n]|w_{j+1}, \mathbf{x})$ that resulted in $P(\mathbf{w}_2^n, \mathbf{y}[1,n]|w_1, < s >)$. Perform this recursively. When the process is completed, eliminate from the chart (i.e., set to 0) all the sub-parses that are not marked. Computations $P(\mathbf{w}_0^i, \mathbf{w}_{j+1}^{n+1}, \mathbf{x}[i-1]\,; \mathbf{y}[i,j])$ will thus be performed only for those positions which remain in the chart.

8.3. Limiting non-terminal productions. The straight-forward way of training the Structured Language Model is to initialize the statistics from reliable parses taken either from an appropriate treebank [3], or from a corpus parsed by some automatic parser [8–10]. This is what was done in previous work on the SLM [1, 2].

If this initialization is based on a sufficiently large corpus, then it is reasonable to assume that all allowable reductions $\gamma_1, \gamma_2 \to \gamma$ and $\gamma_0 \to \gamma$ have taken place in it. This can be used to limit the effort arising from the sums $\sum_\gamma$ appearing in (36) and (43).

If we assume that the initialization corpus is identical to the training corpus then the problem of out-of-vocabulary words does not arise. Nevertheless, we need to smooth the initial statistics. This can be accomplished by letting words have all part of speech tags the dictionary allows, and by assigning a positive probability only to reductions $(x_1^1, \gamma_1), (x_2^1, \gamma_2) \to (x_i^1, \gamma)$ that correspond to reductions $\gamma_1, \gamma_2 \to \gamma$ that were actually found in the initialization corpus.

9. Smoothing of statistical parameters. Just as in a trigram language model, the parameter values extracted in training will suffer from sparseness. In order to use them on test data, they will have to be subjected to smoothing. Let us, for instance, consider the predictor in the SSLM setting. The re-estimation formulas specify its values in equation (28) which we repeat in a slightly altered form

$$f(v|h_{-2} = x, h_{-1} = y) = \frac{CC(x,y,v)}{\sum_{v'} CC(x,y,v')} \tag{55}$$

with the function CC defined in (27). The value of $f(v|h_{-2} = x, h_{-1} = y)$ of interest is the one obtained in the last iteration of the re-estimation algorithm.

Assuming linear interpolation, the probability used for test purposes would be given by

$$\begin{aligned} &\hat{P}_\lambda(v|h_{-2} = x, h_{-1} = y) \\ &= \lambda f(v|h_{-2} = x, h_{-1} = y) + (1-\lambda)\,\hat{P}(v|h_{-1} = y) \end{aligned} \tag{56}$$

where $\hat{P}(v|h_{-1} = y)$ denotes a bigram probability smoothed according to the same principles being described here. The value of λ in (56) is a function of the "bucket" that it belongs to. Buckets would normally depend on counts, and the appropriate count would be equal to $CC(x,y) = \sum_{v'} CC(x,y,v')$ obtained during training.

Unfortunately, there is a potential problem. The counts $CC(x,y)$ are an accumulation of fractional counts which have a different character from what we are used to in trigram language modeling. In the latter, counts represent the number of times a situation actually arose. Here the pair x, y may represent a combination of exposed headwords that is totally unreasonable from a parsing point of view. From every sentence in which the pair appears it will then contribute a small value to the total count $CC(x,y)$. Nevertheless, there may be many sentences in which the word pair x, y does appear. So the eventual count $CC(x,y)$ may end up to be respectable. At the same time, there may exist pairs x', y' that are appropriate heads which appear in few sentences and as a result the count $CC(x', y')$ may fall into the same bucket as does $CC(x,y)$. But we surely want to use *different* λ's for the two situations!

The appropriate solution may be obtained by noticing that the count can be thought of as made up of two factors: the number of times, M, the x, y pair could conceivably be headwords (roughly equal to the number of sentences in which they appear), and the probability that if they could be headwords, they actually are. So therefore

$$\lambda = \lambda\Big(M, \frac{CC(x,y)}{M}\Big).$$

Now λ's must be estimated by running the re-estimation algorithm on heldout data. Assuming that the headword pair x, y belongs to the kth bucket[31] and that CC_H denotes the CC value extracted from the heldout set,[32] the contribution to the new value $\lambda'(k)$ due to the triplet x, y, v will be[33]

$$CC_H(x,y,v)\,\frac{\lambda(k)\, f(v|h_{-2}=x, h_{-1}=y)}{\lambda(k)\, f(v|h_{-2}=x, h_{-1}=y) + (1-\lambda(k))\,\hat{P}(v|h_{-1}=y)}$$

where $\lambda(k)$ denotes the previous λ value for that bucket.

Generalization of this smoothing to SLM is straight forward, as is the specification of smoothing of constructor and tagger parameters.

Acknowledgement. The author wishes to thank Peng Xu, who constructed the tables presented in this paper and carried out the necessary computations. Mr. Xu took care of the formatting and held invaluable discussions with the author concerning the SLM.

[31] Even though the buckets are two-dimensional, they can be numbered in sequence.

[32] Values depend on the probabilities $\hat{P}_\lambda(v|h_{-2}=x, h_{-1}=y)$ and $\hat{Q}_\lambda(a|h_{-2}=x, h_{-1}=y)$ which change with each iteration as the values of the $\lambda-$parameters change.

[33] We assume iterative re-estimation.

REFERENCES

[1] C. Chelba and F. Jelinek, "Structured Language Modeling," *Computer Speech and Language,* Vol. **14**, No. 4, October 2000.

[2] C. Chelba and F. Jelinek, "Exploiting Syntactic Structure for Language Modeling," *Proceedings of COLING – ACL,* Vol. **1**, pp. 225 – 231, Montreal, Canada, August 10–14, 1998.

[3] M. Marcus and B. Santorini, "Building a Large Annotated Corpus of English: the Penn Treebank," *Computational Linguistics,* Vol. **19**, No. 2, pp. 313–330, June 1993.

[4] J. Cocke, unpublished notes.

[5] T. Kasami, "An efficient recognition and syntax algorithm for context-free languages," *Scientific Report AFCRL-65-758,* Air Force Cambridge Research Lab., Bedford MA, 1965.

[6] D.H. Younger, "Recognition and Parsing of Context Free Languages in Time N^3," *Information and Control,* Vol. **10**, pp. 198–208, 1967.

[7] J.K. Baker, "Trainable Grammars for Speech Recognition," *Proceedings of the Spring Conference of the Acoustical Society of America*, pp. 547–550, Boston MA, 1979.

[8] A. Ratnaparkhi, "A Linear Observed Time Statistical Parser Based on Maximum Entropy Models," *Proceedings of the Second Conference on Empirical Methods in Natural Language Processing,* pp. 1–10, Providence, RI, 1997.

[9] E. Charniak, "Treebank Grammars," *Proceedings of the Thirteenth National Conference on Artificial Intelligence,* pp. 1031–1036, Menlo Park, CA, 1996.

[10] M.J. Collins, "A New Statistical Parser Based on Bigram Lexical Dependencies," *Proceedings of the 34th Annual Meeting of the Associations for Computational Linguistics,* pp. 184–191, Santa Cruz, CA, 1996.

[11] C. Chelba, "A Structured Language Model," *Proceedings of ACL/EACL'97 Student Session*, pp. 498–500, Madrid, Spain, 1997.

[12] C. Chelba and F. Jelinek, "Refinement of a Structured Language Model," *Proceedings of ICAPR-98*, pp. 225–231, Plymouth, England, 1998

[13] C. Chelba and F. Jelinek, "Structured Language Modeling for Speech Recognition," *Proceedings of NLDB99*, Klagenfurt, Austria, 1999

[14] C. Chelba and F. Jelinek, "Recognition Performance of a Structured Language Model," *Proceedings of Eurospeech'99*, Vol. **4**, pp. 1567–1570, Budapest, Hungary, 1999.

[15] F. Jelinek and C. Chelba, "Putting Language into Language Modeling," *Proceedings of Eurospeech'99*, Vol. **1**, pp. KN-1–6, Budapest, Hungary, 1999.

[16] C. Chelba and P. Xu, "Richer Syntactic Dependencies for Structured Language Modeling," *Proceedings of the Automatic Speech Recognition and Understanding Workshop*, Madonna di Campiglio, Italy, 2001.

[17] P. Xu, C. Chelba, and F. Jelinek, "A Study on Richer Syntactic Dependencies for Structured Language Modeling," *Proceedings of ACL'02*, pp. 191–198, Philadelphia, 2002.

[18] D.H. Van Uystel, D. Van Compernolle, and P. Wambacq, "Naximum-Likelihood Training of the PLCG-Based Language Model," *Proceedings of the Automatic Speech Recognition and Understanding Workshop*, Madonna di Campiglio, Italy, 2001.

[19] D.H. Van Uystel, F. Van Aelten, and D. Van Compernolle, "A Structured Language Model Based on Context-Sensitive Probabilistic Left-Corner Parsing," *Proceedings of 2nd Meeting of the North American Chapter of the ACL*, pp. 223–230, Pittsburgh, 2001.

LATENT SEMANTIC LANGUAGE MODELING FOR SPEECH RECOGNITION

JEROME R. BELLEGARDA*

Abstract. Statistical language models used in large vocabulary speech recognition must properly capture the various constraints, both local and global, present in the language. While n-gram modeling readily accounts for the former, it has been more difficult to handle the latter, and in particular long-term semantic dependencies, within a suitable data-driven formalism. This paper focuses on the use of *latent semantic analysis* (LSA) for this purpose. The LSA paradigm automatically uncovers meaningful associations in the language based on word-document co-occurrences in a given corpus. The resulting semantic knowledge is encapsulated in a (continuous) vector space of comparatively low dimension, where are mapped all (discrete) words and documents considered. Comparison in this space is done through a simple similarity measure, so familiar clustering techniques can be applied. This leads to a powerful framework for both automatic semantic classification and semantic language modeling. In the latter case, the large-span nature of LSA models makes them particularly well suited to complement conventional n-grams. This synergy can be harnessed through an integrative formulation, in which latent semantic knowledge is exploited to judiciously adjust the usual n-gram probability. The paper concludes with a discussion of intrinsic trade-offs, such as the influence of training data selection on the resulting performance enhancement.

Key words. Statistical language modeling, multi-span integration, n-grams, latent semantic analysis, speech recognition.

1. Introduction. The well-known Bayesian formulation of automatic speech recognition requires a prior model of the language, as pertains to the domain of interest [34, 49]. The role of this prior is to quantify which word sequences are acceptable in a given language for a given task, and which are not: it must therefore encapsulate as much as possible of the syntactic, semantic, and pragmatic characteristics of the domain. In the past two decades, statistical n-gram modeling has steadily emerged as a practical way to do so in a wide range of applications [15]. In this approach, each word is predicted conditioned on the current context, on a left to right basis. An comprehensive overview of the subject can be found in [52], including an insightful perspective on n-grams in light of other techniques, and an excellent tutorial on related trade-offs. Prominent among the challenges faced by n-gram modeling is the inherent locality of its scope, as is evident from the limited amount of context available for predicting each word.

1.1. Scope locality. Central to this problem is the choice of n, which has implications in terms of predictive power and parameter reliability. Although larger values of n would be desirable for more predictive power, in practice, reliable estimation demands low values of n (see, for example, [38, 45, 46]). This in turn imposes an artificially local horizon to the model, impeding its ability to capture large-span relationships in the language.

*Spoken Language Group, Apple Computer Inc., Cupertino, CA 95014.

To illustrate, consider, in each of the two equivalent phrases:

(1.1) *stocks fell sharply as a result of the announcement*

(1.2) *stocks, as a result of the announcement, sharply fell*

the problem of predicting the word "*fell*" from the word "*stocks.*" In (1.1), the prediction can be done with the help of a bigram language model ($n = 2$). This is straightforward with the kind of resources currently available [50]. In (1.2), however, the value $n = 9$ would be necessary, a rather unrealistic proposition at the present time. In large part because of this inability to reliably capture large-span behavior, the performance of conventional n-gram technology has essentially reached a plateau [52].

This observation has sparked interest in a variety of research directions, mostly relying on either *information aggregation* or *span extension* [5]. Information aggregation increases the reliability of the parameter estimation by taking advantage of exemplars of other words that behave "like" this word in the particular context considered. The trade-off, typically, is higher robustness at the expense of a loss in resolution. This paper is more closely aligned with span extension, which extends and/or complements the n-gram paradigm with information extracted from large-span units (i.e., comprising a large number of words). The trade-off here is in the choice of units considered, which has a direct effect on the type of long distance dependencies modeled. These units tend to be either syntactic or semantic in nature. We now expand on these two choices.

1.2. Syntactically-driven span extension. Assuming a suitable parser is available for the domain considered, syntactic information can be used to incorporate large-span constraints into the recognition. How these constraints are incorporated varies from estimating n-gram probabilities from grammar-generated data [61] to computing a linear interpolation of the two models [36]. Most recently, syntactic information has been used specifically to determine equivalence classes on the n-gram history, resulting in so-called dependency language models [13, 48], sometimes also referred to as structured language models [14, 35, 57].

In that framework, each unit is in the form of the headword of the phrase spanned by the associated parse sub-tree. The standard n-gram language model is then modified to operate given the last $(n-1)$ *headwords* as opposed to the last $(n-1)$ *words*. Said another way, the structure of the model is no longer pre-determined: which words serve as predictors depends on the dependency graph, which is a hidden variable [52]. In the example above, the top two headwords in the dependency graph would be "*stocks*" and "*fell*" in both cases, thereby solving the problem.

The main caveat in such modeling is the reliance on the parser, and particularly the implicit assumption that the correct parse will in fact be assigned a high probability [60]. The basic framework was recently extended to operate efficiently in a left-to-right manner [14, 35], through careful optimization of both chart parsing [58] and search modules. Also noteworthy is a somewhat complementary line of research [59], which exploits the syntactic structure contained in the sentences prior to the one featuring the word being predicted.

1.3. Semantically-driven span extension. High level semantic information can also be used to incorporate large-span constraints into the recognition. Since by nature such information is diffused across the entire text being created, this requires the definition of a *document* as a semantically homogeneous set of sentences. Then each document can be characterized by drawing from a (possibly large) set of topics, usually predefined from a hand-labelled hierarchy, which covers the relevant semantic domain [33, 54, 55]. The main uncertainty in this approach is the granularity required in the topic clustering procedure [25]. To illustrate, in (1.1) and (1.2), even perfect knowledge of the general topic (most likely, "*stock market trends*") does not help much.

An alternative solution is to use long distance dependencies between word pairs which show significant correlation in the training corpus. In the above example, suppose that the training data reveals a significant correlation between "*stocks*" and "*fell*." Then the presence of "*stocks*" in the document could automatically trigger "*fell*," causing its probability estimate to change. Because this behavior would occur in both (1.1) and in (1.2), proximity being irrelevant in this kind of model, the two phrases would lead to the same result. In this approach, the pair (*stocks*, *fell*) is said to form a word trigger pair [44]. In practice, word pairs with high mutual information are searched for inside a window of fixed duration. Unfortunately, trigger pair selection is a complex issue: different pairs display markedly different behavior, which limits the potential of low frequency word triggers [51]. Still, self-triggers have been shown to be particularly powerful and robust [44], which underscores the desirability of exploiting correlations between the current word and features of the document history.

Recent work has sought to extend the word trigger concept by using a more comprehensive framework to handle the trigger pair selection [2–4, 6, 18, 28, 30]. This is based on a paradigm originally formulated in the context of information retrieval, called *latent semantic analysis* (LSA) [10, 21, 24, 26, 31, 42, 43, 56]. In this paradigm, co-occurrence analysis still takes place across the span of an entire document, but every combination of words from the vocabulary is viewed as a potential trigger combination. This leads to the systematic integration of long-term semantic dependencies into the analysis.

The concept of document assumes that the available training data is tagged at the document level, i.e., there is a way to identify article boundaries. This is the case, for example, with the ARPA North American Business (NAB) News corpus [39]. Once this is done, the LSA paradigm can be used for word and document clustering [6, 28, 30], as well as for language modeling [2, 18]. In all cases, it was found to be suitable to capture some of the global semantic constraints present in the language. In fact, hybrid n-gram+LSA language models, constructed by embedding LSA into the standard n-gram formulation, were shown to result in a substantial reduction in average word error rate [3, 4].

1.4. Organization. The focus of this paper is on semantically-driven span extension only, and more specifically on how the LSA paradigm can be exploited to improve statistical language modeling. The main objectives are: (i) to review the data-driven extraction of latent semantic information, (ii) to assess its potential use in the context of spoken language processing, (iii) to describe its integration with conventional n-gram language modeling, (iv) to examine the behavior of the resulting hybrid models in speech recognition experiments, and (v) to discuss a number of factors which influence performance.

The paper is organized as follows. In the next two sections, we give an overview of the mechanics of LSA feature extraction, as well as the salient characteristics of the resulting LSA feature space. Section 4 explores the applicability of this framework for general semantic classification. In Section 5, we shift the focus to LSA-based statistical language modeling for large vocabulary recognition. Section 6 describes the various smoothing possibilities available to make LSA-based language models more robust. In Section 7, we illustrate some of the benefits associated with hybrid n-gram+LSA modeling on a subset of the Wall Street Journal (WSJ) task. Finally, Section 8 discusses the inherent trade-offs associated with the approach, as evidenced by the influence of the data selected to train the LSA component of the model.

2. Latent semantic analysis. Let $\mathcal{V}$, $|\mathcal{V}| = M$, be some underlying vocabulary and $\mathcal{T}$ a training text corpus, comprising N articles (documents) relevant to some domain of interest (like business news, for example, in the case of the NAB corpus [39]). The LSA paradigm defines a mapping between the discrete sets $\mathcal{V}$, $\mathcal{T}$ and a continuous vector space $\mathcal{S}$, whereby each word w_i in $\mathcal{V}$ is represented by a vector $\bar{u}_i$ in $\mathcal{S}$, and each document d_j in $\mathcal{T}$ is represented by a vector $\bar{v}_j$ in $\mathcal{S}$.

2.1. Feature extraction. The starting point is the construction of a matrix (W) of co-occurrences between words and documents. In marked contrast with n-gram modeling, word order is ignored, which is of course in line with the semantic nature of the approach [43]. This makes it an instance of the so-called "bag-of-words" paradigm, which disregards collocational information in word strings: the context for each word essentially becomes the entire document in which it appears. Thus, the matrix W is accumulated from the available training data by simply keeping track of which word is found in what document.

This accumulation involves some suitable function of the word count, i.e., the number of times each word appears in each document [6]. Various implementations have been investigated by the information retrieval community (see, for example, [23]). Evidence points to the desirability of normalizing for document length and word entropy. Thus, a suitable expression for the (i, j) cell of W is:

$$w_{i,j} = (1 - \varepsilon_i) \frac{c_{i,j}}{n_j}, \tag{2.1}$$

where $c_{i,j}$ is the number of times w_i occurs in d_j, n_j is the total number of words present in d_j, and ε_i is the normalized entropy of w_i in the corpus $\mathcal{T}$. The global weighting implied by $1 - \varepsilon_i$ reflects the fact that two words appearing with the same count in d_j do not necessarily convey the same amount of information about the document; this is subordinated to the distribution of the words in the collection $\mathcal{T}$.

If we denote by $t_i = \sum_j c_{i,j}$ the total number of times w_i occurs in $\mathcal{T}$, the expression for ε_i is easily seen to be:

$$\varepsilon_i = -\frac{1}{\log N} \sum_{j=1}^{N} \frac{c_{i,j}}{t_i} \log \frac{c_{i,j}}{t_i}. \tag{2.2}$$

By definition, $0 \leq \varepsilon_i \leq 1$, with equality if and only if $c_{i,j} = t_i$ and $c_{i,j} = t_i/N$, respectively. A value of ε_i close to 1 indicates a word distributed across many documents throughout the corpus, while a value of ε_i close to 0 means that the word is present only in a few specific documents. The global weight $1 - \varepsilon_i$ is therefore a measure of the indexing power of the word w_i.

2.2. Singular value decomposition. The $(M \times N)$ word-document matrix W resulting from the above feature extraction defines two vector representations for the words and the documents. Each word w_i can be uniquely associated with a row vector of dimension N, and each document d_j can be uniquely associated with a column vector of dimension M. Unfortunately, these vector representations are unpractical for three related reasons. First, the dimensions M and N can be extremely large; second, the vectors w_i and d_j are typically very sparse; and third, the two spaces are distinct from one other.

To address these issues, one solution is to perform the (order-R) singular value decomposition (SVD) of W as [29]:

$$W \approx \hat{W} = U\, S\, V^T, \tag{2.3}$$

where U is the $(M \times R)$ left singular matrix with row vectors u_i $(1 \leq i \leq M)$, S is the $(R \times R)$ diagonal matrix of singular values $s_1 \geq s_2 \geq \cdots \geq s_R > 0$, V is the $(N \times R)$ right singular matrix with row vectors v_j $(1 \leq j \leq N)$, $R \ll \min(M, N)$ is the order of the decomposition, and ^{T}

denotes matrix transposition. As is well known, both left and right singular matrices U and V are column-orthonormal, i.e., $U^T U = V^T V = I_R$ (the identity matrix of order R). Thus, the column vectors of U and V each define an orthornormal basis for the space of dimension R spanned by the (R-dimensional) u_i's and v_j's. Furthermore, the matrix $\hat{W}$ is the best rank-R approximation to the word-document matrix W, for any unitarily invariant norm (cf., e.g., [19]). This entails, for any matrix A of rank R:

$$\min_{\{A\,:\,\mathrm{rank}(A)=R\}} \|W - A\| = \|W - \hat{W}\| = s_{R+1}\,, \tag{2.4}$$

where $\|\cdot\|$ refers to the L_2 norm, and s_{R+1} is the smallest singular value retained in the order-$(R+1)$ SVD of W. Obviously, $s_{R+1} = 0$ if R is equal to the rank of W.

Upon projecting the row vectors of W (i.e., words) onto the orthonormal basis formed by the column vectors of V, the row vector $u_i S$ characterizes the position of word w_i in the underlying R-dimensional space, for $1 \leq i \leq M$. Similarly, upon projecting the column vectors of W (i.e., documents) onto the orthonormal basis formed by the column vectors of U, the row vector $v_j S$ characterizes the position of document d_j in the same space, for $1 \leq j \leq N$. We refer to each of the M scaled vectors $\bar{u}_i = u_i S$ as a *word vector*, uniquely associated with word w_i in the vocabulary, and each of the N scaled vectors $\bar{v}_j = v_j S$ as a *document vector*, uniquely associated with document d_j in the corpus. Thus, (2.3) defines a transformation between high-dimensional discrete entities ($\mathcal{V}$ and $\mathcal{T}$) and a low-dimensional continuous vector space $\mathcal{S}$, the R-dimensional (LSA) space spanned by the u_i's and v_j's. The dimension R is bounded from above by the (unknown) rank of the matrix W, and from below by the amount of distortion tolerable in the decomposition. It is desirable to select R so that $\hat{W}$ captures the major structural associations in W, and ignores higher order effects.

2.3. Properties. By construction, the "closeness" of vectors in the LSA space $\mathcal{S}$ is determined by the overall pattern of the language used in $\mathcal{T}$, as opposed to specific constructs. Hence, two words whose representations are "close" (in some suitable metric) tend to appear in the same kind of documents, whether or not they actually occur within identical word contexts in those documents. Conversely, two documents whose representations are "close" tend to convey the same semantic meaning, whether or not they contain the same word constructs. In the same manner, from the bidiagonalization process inherent in the SVD, we can expect that the respective representations of words and documents that are semantically linked would also be "close" in the LSA space $\mathcal{S}$.

Of course, the optimality of this framework can be debated, since the L_2 norm may not be the best choice when it comes to linguistic phenomena. For example, the Kullback-Leibler divergence provides a more elegant (probabilistic) interpretation of (2.3) [31], albeit at the expense of requiring a conditional independence assumption on the words and the documents

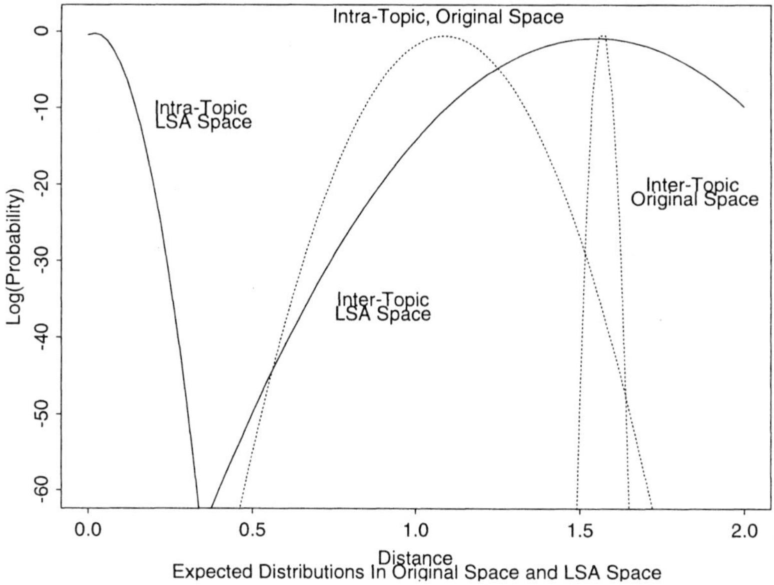

FIG. 1. *Improved Topic Separability in LSA Space (After [47]).*

[32]. This caveat notwithstanding, the correspondence between closeness in LSA space and semantic relatedness is well documented. In applications such as information retrieval, filtering, induction, and visualization, the LSA framework has repeatedly proven remarkably effective in capturing semantic information [10, 21, 24, 26, 32, 42, 43, 56].

Such behavior was recently illustrated in [47], in the context of an (artificial) information retrieval task with 20 distinct topics and a vocabulary of 2000 words. A probabilistic corpus model generated 1000 documents, each 50 to 100 words long. The probability distribution for each topic was such that 0.95 of its probability density was equally distributed among topic words, and the remaining 0.05 was equally distributed among all the 2000 words in the vocabulary. The authors of the study measured the distance[1] between all pairs of documents, both in the original space and in the LSA space obtained as above, with $R = 20$. This leads to the expected distance distributions depicted in Figure 1, where a pair of documents is considered "Intra-Topic" if the two documents were generated from the same topic and "Inter-Topic" otherwise.

It can be seen that in the LSA space the average distance between inter-topic pairs stays about the same, while the average distance between intra-topic pairs is dramatically reduced. In addition, the standard de-

[1]The relevant definition for this quantity will be discussed in detail shortly, cf. Section 3.3.

viation of the intra-topic distance distribution also becomes substantially smaller. As a result, separability between intra- and inter-topic pairs is much better in the LSA space than in the original space. Note that this holds in spite of a sharp increase in the standard deviation of the inter-topic distance distribution, which bodes well for the general applicability of the method. Analogous observations can be made regarding the distance between words and/or between words and documents.

2.4. Computational effort. Clearly, classical methods for determining the SVD of dense matrices (see, for example, [11]) are not optimal for large sparse matrices such as W. Because these methods apply orthogonal transformations (Householder or Givens) directly to the input matrix, they incur excessive fill-in and thereby require tremendous amounts of memory. In addition, they compute all the singular values of W; but here $R \ll \min(M, N)$, and therefore doing so is computationally wasteful. Instead, it is more appropriate to solve a sparse symmetric eigenvalue problem, which can then be used to indirectly compute the sparse singular value decomposition. Several suitable iterative algorithms have been proposed by Berry, based on either the subspace iteration or the Lanczos recursion method [9]. Convergence is typically achieved after 100 or so iterations.

3. LSA feature space. In the continuous vector space $\mathcal{S}$ obtained above, each word $w_i \in \mathcal{V}$ is represented by the associated word vector of dimension R, $\bar{u}_i = u_i S$, and each document $d_j \in \mathcal{T}$ is represented by the associated document vector of dimension R, $\bar{v}_j = v_j S$. This opens up the opportunity to apply familiar clustering techniques in $\mathcal{S}$, as long as a distance measure consistent with the SVD formalism is defined on the vector space. Since the matrix W embodies, by construction, all structural associations between words and documents, it follows that, for a given training corpus, $W\,W^T$ characterizes all co-occurrences between words, and $W^T\,W$ characterizes all co-occurrences between documents.

3.1. Word clustering. Expanding $W\,W^T$ using the SVD expression (2.3), we obtain (henceforth ignoring the distinction between W and $\hat{W}$):

$$W\,W^T = U\,S^2\,U^T\,. \tag{3.1}$$

Since S is diagonal, a natural metric to consider for the "closeness" between words is therefore the cosine of the angle between $u_i S$ and $u_j S$:

$$K(w_i, w_j) = \cos(\bar{u}_i, \bar{u}_j) = \frac{u_i\,S^2\,u_j^T}{\|u_i S\|\,\|u_j S\|}\,, \tag{3.2}$$

for any $1 \leq i, j \leq M$. A value of $K(w_i, w_j) = 1$ means the two words always occur in the same semantic context, while a value of $K(w_i, w_j) < 1$ means the two words are used in increasingly different semantic contexts.

Cluster 1
Andy, antique, antiques, art, artist, artist's, artists, artworks, auctioneers, Christie's, collector, drawings, gallery, Gogh, fetched, hysteria, masterpiece, museums, painter, painting, paintings, Picasso, Pollock, reproduction, Sotheby's, van, Vincent, Warhol

Cluster 2
appeal, appeals, attorney, attorney's, counts, court, court's, courts, condemned, convictions, criminal, decision, defend, defendant, dismisses, dismissed, hearing, here, indicted, indictment, indictments, judge, judicial, judiciary, jury, juries, lawsuit, leniency, overturned, plaintiffs, prosecute, prosecution, prosecutions, prosecutors, ruled, ruling, sentenced, sentencing, suing, suit, suits, witness

FIG. 2. *Word Cluster Example (After [2]).*

While (3.2) does not define a *bona fide* distance measure in the space $\mathcal{S}$, it easy leads to one. For example, over the interval $[0, \pi]$, the measure:

$$\mathcal{D}(w_i, w_j) = \cos^{-1} K(w_i, w_j)\,, \tag{3.3}$$

readily satisfies the properties of a distance on $\mathcal{S}$. At this point, it is straightforward to proceed with the clustering of the word vectors $\bar{u}_i$, using any of a variety of algorithms (see, for instance, [1]). The outcome is a set of clusters C_k, $1 \le k \le K$, which can be thought of as revealing a particular layer of semantic knowledge in the space $\mathcal{S}$.

3.2. Word cluster example. For the purpose of illustration, we recall here the result of a word clustering experiment originally reported in [2]. A corpus of $N = 21,000$ documents was randomly selected from the WSJ portion of the NAB corpus. LSA training was then performed with an underlying vocabulary of $M = 23,000$ words, and the word vectors in the resulting LSA space were clustered into 500 disjoint clusters using a combination of K-means and bottom-up clustering (cf. [4]). Two representative examples of the clusters so obtained are shown in Figure 2.

The first thing to note is that these word clusters comprise words with different part of speech, a marked difference with conventional class n-gram techniques (cf. [45]). This is a direct consequence of the semantic nature of the derivation. Second, some obvious words seem to be missing from the clusters: for example, the singular noun "*drawing*" from cluster 1 and the present tense verb "*rule*" from cluster 2. This is an instance of a phenomenon called *polysemy*: "*drawing*" and "*rule*" are more likely to appear in the training text with their alternative meanings (as in "*drawing a conclusion*" and "*breaking a rule*," respectively), thus resulting in different cluster assignments. Finally, some words seem to contribute only marginally to the clusters: for example, "*hysteria*" from cluster 1 and "*here*" from cluster 2. These are the unavoidable outliers at the periphery of the clusters.

3.3. Document clustering. Proceeding as above, the SVD expression (2.3) also yields:

$$W^T W = V S^2 V^T. \tag{3.4}$$

As a result, a natural metric to consider for the "closeness" between documents is the cosine of the angle between $v_i S$ and $v_j S$, i.e.:

$$K(d_i, d_j) = \cos(\bar{v}_i, \bar{v}_j) = \frac{v_i S^2 v_j^T}{\|v_i S\| \, \|v_j S\|}, \tag{3.5}$$

for any $1 \leq i, j \leq N$. This has the same functional form as (3.2); thus, the distance (3.3) is equally valid for both word and document clustering.[2] The resulting set of clusters D_ℓ, $1 \leq \ell \leq L$, can be viewed as revealing another layer of semantic knowledge in the space $\mathcal{S}$.

3.4. Document cluster example. An early document clustering experiment using the above measure was documented in [30]. This work was conducted on the British National Corpus (BNC), a heterogeneous corpus which contains a variety of hand-labelled topics. Using the LSA framework as above, it is possible to partition BNC into distinct clusters, and compare the sub-domains so obtained with the hand-labelled topics provided with the corpus. This comparison was conducted by evaluating two different mixture trigram language models: one built using either the LSA sub-domains, and one built using the hand-labelled topics. As the perplexities obtained were very similar [30], this validates the automatic partitioning performed using LSA.

Some evidence of this behavior is provided in Figure 3, which plots the distributions of four of the hand-labelled BNC topics against the ten document sub-domains automatically derived using LSA. While clearly not matching the hand-labeling, LSA document clustering in this example still seems reasonable. In particular, as one would expect, the distribution for the natural science topic is relatively close to the distribution for the applied science topic (cf. the two solid lines), but quite different from the two other topic distributions (in dashed lines). From that standpoint, the data-driven LSA clusters appear to adequately cover the semantic space.

[2] In fact, the measure (3.3) is precisely the one used in the study reported in Figure 1. Thus, the distances on the x-axis of Figure 1 are $\mathcal{D}(d_i, d_j)$ expressed in radians.

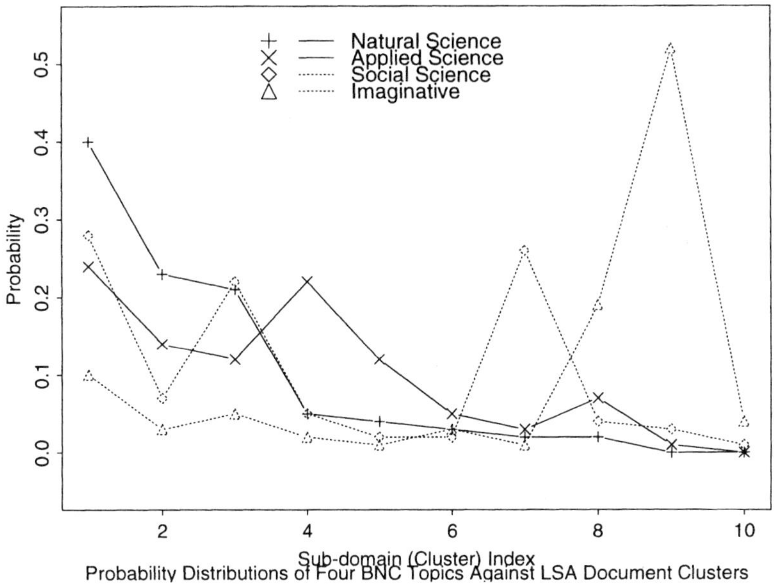

FIG. 3. *Document Cluster Example (After [30]).*

4. Semantic classification. As just seen in the previous two sections, the latent semantic framework has a number of interesting properties, including: (i) a single vector representation for both words and documents in the same continuous vector space, (ii) an underlying topological structure reflecting semantic similarity, (iii) a well-motivated, natural metric to measure the distance between words and between documents in that space, and (iv) a relatively low dimensionality which makes clustering meaningful and practical. These properties can be exploited in several areas of spoken language processing. In this section, we address the most immediate domain of application, which follows directly from the previous clustering discussion: (data-driven) semantic classification [7, 8, 12, 16, 27].

4.1. Framework extension. Semantic classification refers to the task of determining, for a given document, which one of several pre-defined topics the document is most closely aligned with. In contrast with the clustering setup discussed above, such document will not (normally) have been seen in the training corpus. Hence, we first need to extend the LSA framework accordingly. As it turns out, under relatively mild assumptions, finding a representation for a new document in the space $\mathcal{S}$ is straightforward.

Let us refer to the new document as $\tilde{d}_p$, with $p > N$, where the tilde symbol denotes the fact that the document was not part of the training data. First, we construct a feature vector containing, for each word in the underlying vocabulary, the weighted counts (2.1) with $j = p$. This feature vector $\tilde{d}_p$, a column vector of dimension M, can be thought of as an additional column of the matrix W. Thus, provided the matrices U and S do not change, the SVD expansion (2.3) implies:

$$\tilde{d}_p = U\, S\, \tilde{v}_p^T\,, \tag{4.1}$$

where the R-dimensional vector $\tilde{v}_p^T$ act as an additional column of the matrix V^T. This in turn leads to the definition:

$$\bar{\tilde{v}}_p = \tilde{v}_p\, S = \tilde{d}_p^T\, U\,. \tag{4.2}$$

The vector $\bar{\tilde{v}}_p$, indeed seen to be functionally similar to a document vector, corresponds to the representation of the new document in the space $\mathcal{S}$.

To convey the fact that it was not part of the SVD extraction, the new document $\tilde{d}_p$ is referred to as a *pseudo-document*. Recall that the (truncated) SVD provides, by definition, a parsimonious description of the linear space spanned by W. As a result, if the new document contains language patterns which are inconsistent with those extracted from W, the SVD expansion (2.3) will no longer apply. Similarly, if the addition of $\tilde{d}_p$ causes the major structural associations in W to shift in some substantial manner,[3] the parsimonious description will become inadequate. Then U and S will no longer be valid, in which case it would be necessary to recompute (2.3) to find a proper representation for $\tilde{d}_p$. If, on the other hand, the new document generally conforms to the rest of the corpus $\mathcal{T}$, then the *pseudo-document vector* $\bar{\tilde{v}}_p$ in (4.2) will be a reasonable representation for $\tilde{d}_p$.

Once the representation (4.2) is obtained, the "closeness" between the new document $\tilde{d}_p$ and any document cluster D_ℓ can then be expressed as $\mathcal{D}(\tilde{d}_p, D_\ell)$, calculated from (3.5) in the previous section.

4.2. Semantic inference. This can be readily exploited in such command-and-control tasks as desktop user interface control [7] or automated call routing [12]. Suppose that each document cluster D_ℓ can be uniquely associated with a particular action in the task. Then the centroid of each cluster can be viewed as the *semantic anchor* of this action in the LSA space. An unknown word sequence (treated as a new "document") can thus be mapped onto an action by evaluating the distance (3.3) between that "document" and each semantic anchor. We refer to this approach as *semantic inference* [7, 8]. In contrast with usual inference engines (cf.

[3]For example, suppose training was carried out for a banking application involving the word "bank" taken in a financial context. Now suppose $\tilde{d}_p$ is germane to a fishing application, where "bank" is referred to in the context of a river or a lake. Clearly, the closeness of "bank" to, e.g., "money" and "account," would be irrelevant to $\tilde{d}_p$. Conversely, adding $\tilde{d}_p$ to W would likely cause such structural associations to shift substantially, and perhaps even disappear altogether.

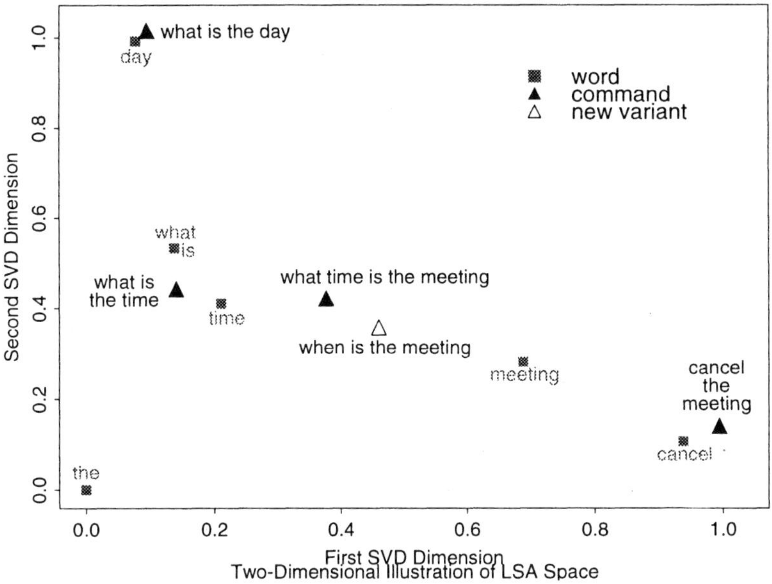

FIG. 4. *An Example of Semantic Inference for Command and Control* ($R = 2$).

[20]), semantic inference thus defined does not rely on formal behavioral principles extracted from a knowledge base. Instead, the domain knowledge is automatically encapsulated in the LSA space in a data-driven fashion.

To illustrate, consider an application with $N = 4$ actions (documents), each associated with a unique command: (i) "*what is the time*," (ii) "*what is the day*," (iii) "*what time is the meeting*," and (iv) "*cancel the meeting*." In this simple example, there are only $M = 7$ words in the vocabulary, with some interesting patterns: "*what*" and "*is*" always co-occur, "*the*" appears in all four commands, only (ii) and (iv) contain a unique word, and (i) is a proper subset of (iii). Constructing the (7×4) word-document matrix as described above, and performing the SVD, we obtain the 2-dimensional space depicted in Figure 4.

This figure shows how each word and each command is represented in the space $\mathcal{S}$. Note that the two words which each uniquely identify a command—"*day*" for (ii) and "*cancel*" for (iv)—each have a high coordinate on a different axis. Conversely, the word "*the*," which conveys no information about the identity of a command, is located at the origin. On the other hand, the semantic anchors for (ii) and (iv) fall "close" to the words which predict them best—"*day*" and "*cancel*", respectively. Similarly, the semantic anchors for (i) and (iii) fall in the vicinity of their meaningful components—"*what-is*" and "*time*" for (i) and "*time*" and "*meeting*" for (iii)—with the word "*time*," which occurs in both, indeed appearing "close" to both.

Now suppose that a user says something outside of the training setup, such as "*when is the meeting*" rather than "*what time is the meeting.*" This new word string turns out to have a representation in the space $\mathcal{S}$ indicated by the hollow triangle in Figure 4. Observe that this point is closest to the representation of command (iii). Thus, the new word string can be considered semantically most related to (iii), and the correct action can be automatically inferred. This can be thought of as a way to perform "bottom-up" natural language understanding.

By replacing the traditional rule-based mapping between utterance and action by such data-driven classification, semantic inference makes it possible to relax some of the typical command-and-control interaction constraints. For example, it obviates the need to specify rigid language constructs through a domain-specific (and thus typically hand-crafted) finite state grammar. This is turn allows the end user more flexibility in expressing the desired command/query, which tends to reduce the associated cognitive load and thereby enhance user satisfaction [12].

4.3. Caveats. Recall that LSA is an instance of the "bag-of-words" paradigm, which pays no attention to the order of words in the sentence. This is what makes it well-suited to capture semantic relationships between words. By the same token, however, it is inherently unable to capitalize on the local (syntactic, pragmatic) constraints present in the language. For tasks such as call routing, where only the broad topic of a message is to be identified, this limitation is probably inconsequential. For general command and control tasks, however, it may be more deleterious.

Imagine two commands that differ only in the presence of the word "*not*" in a crucial place. The respective vector representations could conceivably be relatively close in the LSA space, and yet have vastly different intended consequences. Worse yet, some commands may differ only through word order. Consider, for instance, the two MacOS 9 commands:

(4.3)
change popup to window
change window to popup

which are mapped onto the *exact same point* in LSA space. This makes them obviously impossible to disambiguate.

As it turns out, it is possible to handle such cases through an extension of the basic LSA framework using word agglomeration. The idea is to move from words and documents to word n-tuples and n-tuple documents, where each word n-tuple is the agglomeration of n successive words, and each (n-tuple) document is now expressed in terms of all the word n-tuples it contains. Despite the resulting increase in computational complexity, this extension is practical in the context of semantic classification because of the relatively modest dimensions involved (as compared to large vocabulary recognition). Further details would be beyond the scope of this manuscript, but the reader is referred to [8] for a complete description.

5. N-gram+LSA language modeling. Another major area of application of the LSA framework is in statistical language modeling, where it can readily serve as a paradigm for semantically-driven span extension. Because of the limitation just discussed, however, it is best applied in conjunction with the standard n-gram approach. This section describes how this can be done.

5.1. LSA component. Let w_q denote the word about to be predicted, and H_{q-1} the admissible LSA history (context) for this particular word. At best this history can only be the current document so far, i.e., up to word w_{q-1}, which we denote by $\tilde{d}_{q-1}$. Thus, in general terms, the LSA language model probability is given by:

$$\Pr(w_q|H_{q-1}, \mathcal{S}) = \Pr(w_q|\tilde{d}_{q-1}), \tag{5.1}$$

where the conditioning on $\mathcal{S}$ reflects the fact that the probability depends on the particular vector space arising from the SVD representation. In this expression, $\Pr(w_q|\tilde{d}_{q-1})$ is computed directly from the representations of w_q and $\tilde{d}_{q-1}$ in the space $\mathcal{S}$, i.e., it is inferred from the "closeness" between the associated word vector and (pseudo-)document vector in $\mathcal{S}$. We therefore have to specify both the appropriate pseudo-document representation and the relevant probability measure.

5.1.1. *Pseudo-document representation.* To come up with a pseudo-document representation, we leverage the results of Section 4.1, with some slight modifications due to the time-varying nature of the span considered. From (4.2), the context $\tilde{d}_{q-1}$ has a representation in the space $\mathcal{S}$ given by:

$$\bar{\tilde{v}}_{q-1} = \tilde{v}_{q-1}\, S = \tilde{d}_{q-1}^{T}\, U\,. \tag{5.2}$$

As mentioned before, this vector representation for $\tilde{d}_{q-1}$ is adequate under some consistency conditions on the general patterns present in the domain considered. The difference with Section 4.1 is that, as q increases, the content of the new document grows, and therefore the pseudo-document vector moves around accordingly in the LSA space. Assuming the new document is semantically homogeneous, eventually we can expect the resulting trajectory to settle down in the vicinity of the document cluster corresponding to the closest semantic content.

Of course, here it is possible to take advantage of redundancies in time. Assume, without loss of generality, that word w_i is observed at time q. Then, $\tilde{d}_{q-1}$ and $\tilde{d}_q$ differ only in one coordinate, corresponding to the index i. Assume further that the training corpus $\mathcal{T}$ is large enough, so that the normalized entropy ε_i $(1 \le i \le M)$ does not change appreciably with the addition of each pseudo-document. This makes it possible, from (2.1), to express $\tilde{d}_q$ as:

$$\tilde{d}_q = \frac{n_q - 1}{n_q}\,\tilde{d}_{q-1} + \frac{1-\varepsilon_i}{n_q}\,[0 \ldots 1 \ldots 0]^T, \tag{5.3}$$

where the "1" in the above vector appears at coordinate i. This is turn implies, from (5.2):

$$\tilde{\bar{v}}_q = \tilde{v}_q\, S = \frac{1}{n_q}\left[(n_q - 1)\,\tilde{\bar{v}}_{q-1} \;+\; (1 - \varepsilon_i)\, u_i\right]. \tag{5.4}$$

As a result, the pseudo-document vector associated with the large-span context can be efficiently updated directly in the LSA space.

5.1.2. *LSA probability.* To specify a suitable "closeness" measure, we now follow a reasoning similar to that of Section 3. Since, by construction, the matrix W embodies structural associations between words and documents, and, by definition, $W = USV^T$, a natural metric to consider for the "closeness" between word w_i and document d_j is the cosine of the angle between $u_i S^{1/2}$ and $v_j S^{1/2}$. Applying the same reasoning to pseudo-documents, we arrive at:

$$K(w_q, \tilde{d}_{q-1}) = \cos(u_q S^{1/2}, \tilde{v}_{q-1} S^{1/2}) = \frac{u_q\, S\, \tilde{v}_{q-1}^T}{\|u_q S^{1/2}\|\, \|\tilde{v}_{q-1} S^{1/2}\|}, \tag{5.5}$$

for any q indexing a word in the text data. A value of $K(w_q, \tilde{d}_{q-1}) = 1$ means that $\tilde{d}_{q-1}$ is a strong semantic predictor of w_q, while a value of $K(w_q, \tilde{d}_{q-1}) < 1$ means that the history carries increasingly less information about the current word. Interestingly, (5.5) is functionally equivalent to (3.2) and (3.5), but involves scaling[4] by $S^{1/2}$ instead of S. As before, the mapping (3.3) can be used to transform (5.5) into a real distance measure.

To enable the computation of $\Pr(w_q | \tilde{d}_{q-1})$, it remains to go from that distance measure to an actual probability measure. One solution is for the distance measure to induce a family of exponential distributions with pertinent marginality constraints. In practice, it may not be necessary to incur this degree of complexity. Considering that $\tilde{d}_{q-1}$ is only a partial document anyway, exactly what kind of distribution is induced is probably less consequential than ensuring that the pseudo-document is properly scoped (cf. Section 5.3 below). Basically, all that is needed is a "reasonable" probability distribution to act as a proxy for the true (unknown) measure.

We therefore opt to use the empirical multivariate distribution constructed by allocating the total probability mass in proportion to the distances observed during training. In essence, this reduces the complexity to a simple histogram normalization, at the expense of introducing a potential "quantization-like" error. Of course, such error can be minimized through a variety of histogram smoothing techniques. Also note that the dynamic range of the distribution typically needs to be controlled by a parameter

[4]Not surprisingly, this difference in scaling exactly mirrors the square root relationship between the singular values of W and the eigenvalues of the (square) matrices $W^T W$ and WW^T.

that is optimized empirically, e.g., by an exponent on the distance term, as discussed in [18].

Intuitively, $\Pr(w_q|\tilde{d}_{q-1})$, in marked contrast with a conventional bag-of-words (unigram) model, reflects the "relevance" of word w_q to the admissible history, as observed through $\tilde{d}_{q-1}$. As such, it will be highest for words whose meaning aligns most closely with the semantic fabric of $\tilde{d}_{q-1}$ (i.e., relevant "content" words), and lowest for words which do not convey any particular information about this fabric (e.g., "function" words like "*the*"). This behavior is exactly the opposite of that observed with the conventional n-gram formalism, which tends to assign higher probabilities to (frequent) function words than to (rarer) content words. Hence the attractive synergy potential between the two paradigms.

5.2. Integration with N-grams. Exploiting this potential requires integrating the two together. This kind of integration can occur in a number of ways, such as simple interpolation [18, 34], or within the maximum entropy framework [22, 41, 57]. Alternatively, under relatively mild assumptions, it is also possible to derive an integrated formulation directly from the expression for the overall language model probability. We start with the definition:

$$\Pr(w_q|H_{q-1}^{(n+l)}) = \Pr(w_q|H_{q-1}^{(n)}, H_{q-1}^{(l)}), \tag{5.6}$$

where H_{q-1} denotes, as before, some suitable admissible history for word w_q, and the superscripts $^{(n)}$, $^{(l)}$, and $^{(n+l)}$ refer to the n-gram component ($w_{q-1}w_{q-2}\dots w_{q-n+1}$, with $n > 1$), the LSA component ($\tilde{d}_{q-1}$), and the integration thereof, respectively.[5] This expression can be rewritten as:

$$\Pr(w_q|H_{q-1}^{(n+l)}) = \frac{\Pr(w_q, H_{q-1}^{(l)}|H_{q-1}^{(n)})}{\sum_{w_i\in\mathcal{V}} \Pr(w_i, H_{q-1}^{(l)}|H_{q-1}^{(n)})}, \tag{5.7}$$

where the summation in the denominator extends over all words in $\mathcal{V}$. Expanding and re-arranging, the numerator of (5.7) is seen to be:

$$\begin{aligned} \Pr(w_q, H_{q-1}^{(l)}|H_{q-1}^{(n)}) &= \Pr(w_q|H_{q-1}^{(n)}) \cdot \Pr(H_{q-1}^{(l)}|w_q, H_{q-1}^{(n)}) \\ &= \Pr(w_q|w_{q-1}w_{q-2}\dots w_{q-n+1}) \\ &\quad \cdot \Pr(\tilde{d}_{q-1}|w_q\, w_{q-1}w_{q-2}\dots w_{q-n+1}). \end{aligned} \tag{5.8}$$

Now we make the assumption that the probability of the document history given the current word is not affected by the immediate context preceding it. This is clearly an approximation, since $w_q w_{q-1} w_{q-2} \dots w_{q-n+1}$ may

[5]Henceforth we make the assumption that $n > 1$. When $n = 1$, the n-gram history becomes null, and the integrated history therefore degenerates to the LSA history alone, basically reducing (5.6) to (5.1).

well reveal more information about the semantic fabric of the document than w_q alone. This remark notwithstanding, for content words at least, different syntactic constructs (immediate context) can generally be used to carry the same meaning (document history). Thus the assumption seems to be reasonably well motivated for content words. How much it matters for function words is less clear [37], but we conjecture that if the document history is long enough, the semantic anchoring is sufficiently strong for the assumption to hold. As a result, the integrated probability becomes:

$$\Pr(w_q|H_{q-1}^{(n+l)}) = \frac{\Pr(w_q|w_{q-1}w_{q-2}\dots w_{q-n+1})\Pr(\tilde{d}_{q-1}|w_q)}{\sum_{w_i\in\mathcal{V}}\Pr(w_i|w_{q-1}w_{q-2}\dots w_{q-n+1})\Pr(\tilde{d}_{q-1}|w_i)}. \tag{5.9}$$

If $\Pr(\tilde{d}_{q-1}|w_q)$ is viewed as a prior probability on the current document history, then (5.9) simply translates the classical Bayesian estimation of the n-gram (local) probability using a prior distribution obtained from (global) LSA. The end result, in effect, is a modified n-gram language model incorporating large-span semantic information.

The dependence of (5.9) on the LSA probability calculated earlier can be expressed explicitly by using Bayes' rule to get $\Pr(\tilde{d}_{q-1}|w_q)$ in terms of $\Pr(w_q|\tilde{d}_{q-1})$. Since the quantity $\Pr(\tilde{d}_{q-1})$ vanishes from both numerator and denominator, we are left with:

$$\Pr(w_q|H_{q-1}^{(n+l)}) = \frac{\Pr(w_q|w_{q-1}w_{q-2}\dots w_{q-n+1})\dfrac{\Pr(w_q|\tilde{d}_{q-1})}{\Pr(w_q)}}{\sum_{w_i\in\mathcal{V}}\Pr(w_i|w_{q-1}w_{q-2}\dots w_{q-n+1})\dfrac{\Pr(w_i|\tilde{d}_{q-1})}{\Pr(w_i)}}, \tag{5.10}$$

where $\Pr(w_q)$ is simply the standard unigram probability. Note that this expression is meaningful[6] for any $n > 1$.

5.3. Context scope selection. In practice, expressions like (5.9)–(5.10) are often slightly modified so that a relative weight can be placed on each contribution (here, the n-gram and LSA probabilities). Usually, this is done via empirically determined weighting coefficients. In the present case, such weighting is motivated by the fact that in (5.9) the "prior" probability $\Pr(\tilde{d}_{q-1}|w_q)$ could change substantially as the current document unfolds. Thus, rather than using arbitrary weights, an alternative approach is to dynamically tailor the document history $\tilde{d}_{q-1}$ so that the n-gram and LSA contributions remain empirically balanced.

[6]Observe that with $n = 1$, the right hand side of (5.10) degenerates to the LSA probability alone, as expected.

This approach, referred to as context scope selection, is more closely aligned with the LSA framework, because of the underlying change in behavior between training and recognition. During training, the scope is fixed to be the current document. During recognition, however, the concept of "current document" is ill-defined, because (i) its length grows with each new word, and (ii) it is not necessarily clear at which point completion occurs. As a result, a decision has to be made regarding what to consider "current," versus what to consider part of an earlier (presumably less relevant) document.

A straightforward solution is to limit the size of the history considered, so as to avoid relying on old, possibly obsolete fragments to construct the current context. Alternatively, to avoid making a hard decision on the size of the caching window, it is possible to assume an exponential decay in the relevance of the context [3]. In this solution, exponential forgetting is used to progressively discount older utterances. Assuming $0 < \lambda \leq 1$, this approach corresponds to modifying (5.4) as follows:

$$\tilde{\bar{v}}_q = \frac{1}{n_q} \left[\lambda \, (n_q - 1) \, \tilde{\bar{v}}_{q-1} \; + \; (1 - \varepsilon_i) \, u_i \right] , \tag{5.11}$$

where the parameter λ is chosen according to the expected heterogeneity of the session.

5.4. Computational effort. From the above, the (on-line) cost incurred during recognition has three components: (i) the construction of the pseudo-document representation in $\mathcal{S}$, as generally done via (5.11); (ii) the computation of the LSA probability $\Pr(w_q | \tilde{d}_{q-1})$ in (5.1); and (iii) the integration proper, in (5.10). It can be shown (cf. [3, 4]) that the total cost of these operations, per word and pseudo-document, is $\mathcal{O}(R^2)$. This is obviously more expensive than the usual table look-up required in conventional n-gram language modeling. On the other hand, for typical values of R, the resulting overhead is, arguably, quite modest. This allows hybrid n-gram+LSA language modeling to be taken advantage of in early stages of the search [3].

6. Smoothing. Since the derivation of (5.10) does not depend on a particular form of the LSA probability, it is possible to take advantage of the additional layer(s) of knowledge uncovered earlier through word (in Section 3.1) and document (in Section 3.3) clustering. Basically, we can expect words and/or documents related to the current document to contribute with more synergy, and unrelated words and/or documents to be better discounted. Said another way, clustering provides a convenient smoothing mechanism in the LSA space [2, 3].

6.1. Word smoothing. Using the set of word clusters C_k, $1 \leq k \leq K$, produced in Section 3.1 leads to word-based smoothing. In this case, we expand (5.1) as follows:

$$\Pr(w_q|\tilde{d}_{q-1}) = \sum_{k=1}^{K} \Pr(w_q|C_k)\Pr(C_k|\tilde{d}_{q-1}), \tag{6.1}$$

which carries over to (5.10) in a straightforward manner. In (6.1), the probability $\Pr(C_k|\tilde{d}_{q-1})$ is qualitatively similar to (5.1) and can therefore be obtained with the help of (5.5), by simply replacing the representation of the word w_q by that of the centroid of word cluster C_k. In contrast, the probability $\Pr(w_q|C_k)$ depends on the "closeness" of w_q relative to this (word) centroid. To derive it, we therefore have to rely on the empirical multivariate distribution induced not by the distance obtained from (5.5), but by that obtained from the measure (3.2) mentioned in Section 3.1. Note that a distinct distribution can be inferred on each of the clusters C_k, thus allowing us to compute all quantities $\Pr(w_i|C_k)$ for $1 \le i \le M$ and $1 \le k \le K$.

The behavior of the model (6.1) depends on the number of word clusters defined in the space $\mathcal{S}$. Two special cases arise at the extremes of the cluster range. If there are as many classes as words in the vocabulary ($K = M$), then with the convention that $P(w_i|C_j) = \delta_{ij}$, (6.1) simply reduces to (5.1). No smoothing is introduced, so the predictive power of the model stays the same as before. Conversely, if all the words are in a single class ($K = 1$), the model becomes maximally smooth: the influence of specific semantic events disappears, leaving only a broad (and therefore weak) vocabulary effect to take into account. The effect on predictive power is, accordingly, limited. Between these two extremes, smoothness gradually increases, and it is reasonable to postulate that predictive power evolves in a concave fashion.

The intuition behind this conjecture is as follows. Generally speaking, as the number of word classes C_k increases, the contribution of $\Pr(w_q|C_k)$ tends to increase, because the clusters become more and more semantically meaningful. By the same token, however, the contribution of $\Pr(C_k|\tilde{d}_{q-1})$ for a given $\tilde{d}_{q-1}$ tends to decrease, because the clusters eventually become too specific and fail to reflect the overall semantic fabric of $\tilde{d}_{q-1}$. Thus, there must exist a cluster set size where the degree of smoothing (and therefore the associated predictive power) is optimal for the task considered. This has indeed been verified experimentally, cf. [2].

6.2. Document smoothing. Exploiting instead the set of document clusters D_ℓ, $1 \le \ell \le L$, produced in Section 3.3 leads to document-based smoothing. The expansion is similar:

$$\Pr(w_q|\tilde{d}_{q-1}) = \sum_{\ell=1}^{L} \Pr(w_q|D_\ell)\Pr(D_\ell|\tilde{d}_{q-1}), \tag{6.2}$$

except that the document clusters D_ℓ now replace the word clusters C_k. This time, it is the probability $\Pr(w_q|D_\ell)$ which is qualitatively similar to

(5.1), and can therefore be obtained with the help of (5.5). As for the probability $\Pr(D_\ell|\tilde{d}_{q-1})$, it depends on the "closeness" of $\tilde{d}_{q-1}$ relative to the centroid of document cluster D_ℓ. Thus, it can be obtained through the empirical multivariate distribution induced by the distance derived from (3.5) in Section 3.3.

Again, the behavior of the model (6.2) depends on the number of document clusters defined in the space $\mathcal{S}$. Compared to (6.1), however, (6.2) is more difficult to interpret at the extremes of the cluster range (i.e., $L = 1$ and $L = N$). If $L = N$, for example, (6.2) does not reduce to (5.1), because $\tilde{d}_{q-1}$ has not been seen in the training data, and therefore cannot be identified with any of the existing clusters. Similarly, the fact that all the documents are in a single cluster ($L = 1$) does not imply the degree of degenerescence observed previously, because the cluster itself is strongly indicative of the general discourse domain (which was not generally true of the "vocabulary cluster" above). Hence, depending on the size and structure of the corpus, the model may still be adequate to capture general discourse effects.

To see that, we apply $L = 1$ in (6.2), whereby the expression (5.10) becomes:

$$\Pr(w_q|H_{q-1}^{(n+l)}) = \frac{\Pr(w_q|w_{q-1}w_{q-2}\dots w_{q-n+1})\,\dfrac{\Pr(w_q|D_1)}{\Pr(w_q)}}{\displaystyle\sum_{w_i \in \mathcal{V}} \Pr(w_i|w_{q-1}w_{q-2}\dots w_{q-n+1})\,\dfrac{\Pr(w_i|D_1)}{\Pr(w_i)}}, \tag{6.3}$$

since the quantity $\Pr(D_1|\tilde{d}_{q-1})$ vanishes from both numerator and denominator. In this expression D_1 refers to the single document cluster encompassing all documents in the LSA space. In case the corpus is fairly homogeneous, D_1 will be a more reliable representation of the underlying fabric of the domain than $\tilde{d}_{q-1}$, and therefore act as a robust proxy for the context observed. Interestingly, (6.3) amounts to estimating a "correction" factor for each word, which depends only on the overall topic of the collection. This is clearly similar to what is done in the cache approach to language model adaptation (see, for example, [17, 40]), except that, in the present case, all words are treated as though they were already in the cache.

More generally, as the number of document classes D_ℓ increases, the contribution of $\Pr(w_q|D_\ell)$ tends to increase, to the extent that a more homogeneous topic boosts the effects of any related content words. On the other hand, the contribution of $\Pr(D_\ell|\tilde{d}_{q-1})$ tends to decrease, because the clusters represent more and more specific topics, which increases the chance that the pseudo-document $\tilde{d}_{q-1}$ becomes an outlier. Thus, again there exists a cluster set size where the degree of smoothing is optimal for the task considered (cf. [2]).

6.3. Joint smoothing. Finally, an expression analogous to (6.1) and (6.2) can also be derived to take advantage of both word and document clusters. This leads to a mixture probability specified by:

$$\Pr(w_q|\tilde{d}_{q-1}) = \sum_{k=1}^{K}\sum_{\ell=1}^{L} \Pr(w_q|C_k, D_\ell)\Pr(C_k, D_\ell|\tilde{d}_{q-1}), \tag{6.4}$$

which, for tractability, can be approximated as:

$$\Pr(w_q|\tilde{d}_{q-1}) = \sum_{k=1}^{K}\sum_{\ell=1}^{L} \Pr(w_q|C_k)\Pr(C_k|D_\ell)\Pr(D_\ell|\tilde{d}_{q-1}). \tag{6.5}$$

In this expression, the clusters C_k and D_ℓ are as previously, as are the quantities $\Pr(w_q|C_k)$ and $\Pr(D_\ell|\tilde{d}_{q-1})$. As for the probability $\Pr(C_k|D_\ell)$, it is qualitatively similar to (5.1), and can therefore be obtained accordingly.

To summarize, any of the expressions (5.1), (6.1), (6.2), or (6.5) can be used to compute (5.10), resulting in four families of hybrid n-gram+LSA language models. Associated with these different families are various trade-offs to become apparent below.

7. Experiments. The purpose of this section is to illustrate the behavior of hybrid n-gram+LSA modeling on a large vocabulary recognition task.[7] The general domain considered was business news, as reflected in the WSJ portion of the NAB corpus. This was convenient for comparison purposes since conventional n-gram language models are readily available, trained on exactly the same data [39].

7.1. Experimental conditions. The text corpus $\mathcal{T}$ used to train the LSA component of the model was composed of about $N = 87,000$ documents spanning the years 1987 to 1989, comprising approximately 42 million words. The vocabulary $\mathcal{V}$ was constructed by taking the 20,000 most frequent words of the NAB corpus, augmented by some words from an earlier release of the WSJ corpus, for a total of $M = 23,000$ words. The test set consisted of a 1992 test corpus of 496 sentences uttered by 12 native speakers of English. In all experiments, acoustic training was performed using 7,200 sentences of data uttered by 84 speakers (a standard corpus known as WSJ0 SI-84). On the above test data, our baseline speaker-independent, continuous speech recognition system (described in detail in [3]) produced reference error rates of 16.7% and 11.8% across the 12 speakers considered, using the standard (WSJ0) bigram and trigram language models, respectively.

We performed the singular value decomposition of the matrix of co-occurrences between words and documents using the single vector Lanczos

[7] The reader is referred to [4] for additional results in this application, and to [8] for experiments involving semantic inference.

TABLE 1
Word Error Rate (WER) Results Using Hybrid Bi-LSA and Tri-LSA Models.

Word Error Rate <WER Reduction>	Bigram $n = 2$	Trigram $n = 3$
Conventional n-Gram	16.7 %	11.8 %
Hybrid, No Smoothing	14.4 % <14 %>	10.7 % < 9 %>
Hybrid, Document Smoothing	13.4 % <20 %>	10.4 % <12 %>
Hybrid, Word Smoothing	12.9 % <23 %>	9.9 % <16 %>
Hybrid, Joint Smoothing	13.0 % <22 %>	9.9 % <16 %>

method [9]. Over the course of this decomposition, we experimented with different numbers of singular values retained, and found that $R = 125$ seemed to achieve an adequate balance between reconstruction error—minimizing s_{R+1} in (2.4)—and noise suppression—minimizing the ratio between order-R and order-$(R+1)$ traces $\sum_i s_i$. This led to a vector space $\mathcal{S}$ of dimension 125.

We then used this LSA space to construct the (unsmoothed) LSA model (5.1), following the procedure described in Section 5. We also constructed the various clustered LSA models presented in Section 6, to implement smoothing based on word clusters—word smoothing (6.1), document clusters—document smoothing (6.2), and both—joint smoothing (6.5). We experimented with different values for the number of word and/or document clusters (cf. [2]), and ended up using $K = 100$ word clusters and $L = 1$ document cluster. Finally, using (5.10), we combined each of these models with either the standard WSJ0 bigram or the standard WSJ0 trigram. The resulting hybrid n-gram+LSA language models, dubbed bi-LSA and tri-LSA models, respectively, were then used in lieu of the standard WSJ0 bigram and trigram models.

7.2. Experimental results. A summary of the results is provided in Table 1, in terms of both absolute word error rate (WER) numbers and WER reduction observed (in angle brackets). Without smoothing, the bi-LSA language model leads to a 14% WER reduction compared to the standard bigram. The corresponding tri-LSA language model leads to a somewhat smaller (just below 10%) relative improvement compared to the standard trigram. With smoothing, the improvement brought about by the LSA component is more marked: up to 23% in the smoothed bi-LSA case, and up to 16% in the smoothed tri-LSA case. Such results show that the hybrid n-gram+LSA approach is a promising avenue for incorporating large-span semantic information into n-gram modeling.

The qualitative behavior of the two n-gram+LSA language models appears to be quite similar. Quantitatively, the average reduction achieved by tri-LSA is about 30% less than that achieved by bi-LSA. This is most likely related to the greater predictive power of the trigram compared to the bigram, which makes the LSA contribution of the hybrid language model comparatively smaller. This is consistent with the fact that the latent semantic information delivered by the LSA component would (eventually) be subsumed by an n-gram with a large enough n. As it turns out, however, in both cases the average WER reduction is far from constant across individual sessions, reflecting the varying role played by global semantic constraints from one set of spoken utterances to another.

Of course, this kind of fluctuations can also be observed with the conventional n-gram models, reflecting the varying predictive power of the local context across the test set. Anecdotally, the leverage brought about by the hybrid n-LSA models appears to be greater when the fluctuations due to the respective components move in opposite directions. So, at least for $n \leq 3$, there is indeed evidence of a certain complementarity between the two paradigms.

7.3. Context scope selection. It is important to emphasize that the recognition task chosen above represents a severe test of the LSA component of the hybrid language model. By design, the test corpus is constructed with no more than 3 or 4 consecutive sentences extracted from a single article. Overall, it comprises 140 distinct document fragments, which means that each speaker speaks, on the average, about 12 different "mini-documents." As a result, the context effectively changes every 60 words or so, which makes it somewhat challenging to build a very accurate pseudo-document representation. This is a situation where it is critical for the LSA component to appropriately forget the context as it unfolds, to avoid relying on an obsolete representation. To obtain the results of Table 1, we used the exponential forgetting setup of (5.11) with a value $\lambda = 0.975$.[8]

In order to assess the influence of this selection, we also performed recognition with different values of the parameter λ ranging from $\lambda = 1$ to $\lambda = 0.95$, in decrements of 0.01. Recall from Section 5 that the value $\lambda = 1$ corresponds to an unbounded context (as would be appropriate for a very homogeneous session), while decreasing values of λ correspond to increasingly more restrictive contexts (as required for a more heterogeneous session). Said another way, the gap between λ and 1 tracks the expected heterogeneity of the current session.

Table 2 presents the corresponding recognition results, in the case of the best bi-LSA framework (i.e., with word smoothing). It can be seen that, with no forgetting, the overall performance is substantially less than the comparable one observed in Table 1 (13% compared to 23% WER reduction). This is consistent with the characteristics of the task, and underscores the role of discounting as a suitable counterbalance to frequent

[8]To fix ideas, this means that the word which occurred 60 words ago is discounted through a weight of about 0.2.

TABLE 2
Influence of Context Scope Selection on Word Error Rate.

Word Error Rate <WER Reduction>	Bi-LSA with Word Smoothing
$\lambda = 1.0$	14.5 % <13 %>
$\lambda = 0.99$	13.6 % <18 %>
$\lambda = 0.98$	13.2 % <21 %>
$\lambda = 0.975$	12.9 % <23 %>
$\lambda = 0.97$	13.0 % <22 %>
$\lambda = 0.96$	13.1 % <22 %>
$\lambda = 0.95$	13.5 % <19 %>

context changes. Performance rapidly improves as λ decreases from $\lambda = 1$ to $\lambda = 0.97$, presumably because the pseudo-document representation gets less and less contaminated with obsolete data. If forgetting becomes too aggressive, however, the performance starts degrading, as the effective context no longer has an equivalent length which is sufficient for the task at hand. Here, this happens for $\lambda < 0.97$.

8. Inherent trade-offs. In the previous section, the LSA component of the hybrid language model was trained on exactly the same data as its n-gram component. This is not a requirement, however, which raises the question of how critical the selection of the LSA training data is to the performance of the recognizer. This is particularly interesting since LSA is known to be weaker on heterogeneous corpora (see, for example, [30]).

8.1. Cross-domain training. To ascertain the matter, we went back to calculating the LSA component using the original, unsmoothed model (5.1). We kept the same underlying vocabulary $\mathcal{V}$, left the bigram component unchanged, and repeated the LSA training on non-WSJ data from the same general period. Three corpora of increasing size were considered, all corresponding to Associated Press (AP) data: (i) $\mathcal{T}_1$, composed of $N_1 = 84,000$ documents from 1989, comprising approximately 44 million words; (ii) $\mathcal{T}_2$, composed of $N_2 = 155,000$ documents from 1988 and 1989, comprising approximately 80 million words; and (iii) $\mathcal{T}_3$, composed of $N_3 = 224,000$ documents from 1988-1990, comprising approximately 117 million words. In each case we proceeded with the LSA training as described in Section 2. The results are reported in Table 3.

Two things are immediately apparent. First, the performance improvement in all cases is much smaller than previously observed (recall the corresponding reduction of 14% in Table 1). Larger training set sizes notwithstanding, on the average the hybrid model trained on AP data is about 4 times less effective than that trained on WSJ data. This suggests a relatively high sensitivity of the LSA component to the domain considered.

TABLE 3
Model Sensitivity to LSA Training Data.

Word Error Rate <WER Reduction>	Bi-LSA with No Smoothing
$\mathcal{T}_1$: $N_1 = 84,000$	16.3 % <2 %>
$\mathcal{T}_2$: $N_2 = 155,000$	16.1 % <3 %>
$\mathcal{T}_3$: $N_3 = 224,000$	16.0 % <4 %>

To put this observation into perspective, recall that: (i) by definition, content words are what characterize a domain; and (ii) LSA inherently relies on content words, since, in contrast with n-grams, it cannot take advantage of the structural aspects of the sentence. It therefore makes sense to expect a higher sensitivity for the LSA component than for the usual n-gram.

Second, the overall performance does not improve appreciably with more training data, a fact already observed in [2] using a perplexity measure. This supports the conjecture that, no matter the amount of data involved, LSA still detects a substantial mismatch between AP and WSJ data from the same general period. This in turn suggests that the LSA component is sensitive not just to the general training domain, but also to the particular style of composition, as might be reflected, for example, in the choice of content words and/or word co-occurrences. On the positive side, this bodes well for rapid adaptation to cross-domain data, provided a suitable adaptation framework can be derived.

8.2. Discussion. The fact that the hybrid n-gram+LSA approach is sensitive to composition style underscores the relatively narrow semantic specificity of the LSA paradigm. While n-grams also suffer from a possible mismatch between training and recognition, LSA leads to a potentially more severe exposure because the space $\mathcal{S}$ reflects even less of the pragmatic characteristics for the task considered. Perhaps what is required is to explicitly include an "authorship style" component into the LSA framework.[9] In any event, one has to be cognizant of this intrinsic limitation, and mitigate it through careful attention to the expected domain of use.

Perhaps more importantly, we pointed out earlier that LSA is inherently more adept at handling content words than function words. But, as is well-known, a substantial proportion of speech recognition errors come from function words, because of their tendency to be shorter, not well articulated, and acoustically confusable. In general, the LSA component will not be able to help fix these problems. This suggests that, even within a well-specified domain, syntactically-driven span extension techniques may

[9]In [47], for example, it has been suggested to define an $M \times M$ stochastic matrix (a matrix with non-negative entries and row sums equal to 1) to account for the way style modifies the frequency of words. This solution, however, makes the assumption—not always valid—that this influence is independent of the underlying subject matter.

be a necessary complement to the hybrid approach. On that subject, note from Section 5 that the integrated history (5.6) could easily be modified to reflect a headword-based n-gram as opposed to a conventional n-gram history, without invalidating the derivation of (5.10). Thus, there is no theoretical barrier to the integration of latent semantic information with structured language models such as described in [14, 35]. Similarly, there is no reason why the LSA paradigm could not be used in conjunction with the integrative approaches of the kind proposed in [53, 57], or even within the cache adaptive framework [17, 40].

9. Conclusion. Statistical n-grams are inherently limited to the capture of linguistic phenomena spanning at most n words. This paper has focused on a semantically-driven span extension approach based on the LSA paradigm, in which hidden semantic redundancies are tracked across (semantically homogeneous) documents. This approach leads to a (continuous) vector representation of each (discrete) word and document in a space of relatively modest dimension. This makes it possible to specify suitable metrics for word-document, word-word, and document-document comparisons, which in turn allows well-known clustering algorithms to be applied efficiently. The outcome is the uncovering, in a data-driven fashion, of multiple parallel layers of semantic knowledge in the space, with variable granularity.

An important property of this vector representation is that it reflects the major semantic associations in the training corpus, as determined by the overall pattern of the language, as opposed to specific word sequences or grammatical constructs. Thus, language models arising from the LSA framework are semantic in nature, and therefore well suited to complement conventional n-grams. Harnessing this synergy is a matter of deriving an integrative formulation to combine the two paradigms. By taking advantage of the various kinds of smoothing available, several families of hybrid n-gram+LSA models can be obtained. The resulting language models substantially outperform the associated standard n-grams on a subset of the NAB News corpus.

Such results notwithstanding, the LSA-based approach also face some intrinsic limitations. For example, hybrid n-gram+LSA modeling shows marked sensitivity to both the training domain and the style of composition. While cross-domain adaptation may ultimately alleviate this problem, an appropriate LSA adaptation framework will have to be derived for this purpose. More generally, such semantically-driven span extension runs the risk of lackluster improvement when it comes to function word recognition. This underscores the need for an all-encompassing strategy involving syntactically motivated approaches as well.

REFERENCES

[1] J.R. Bellegarda, *Context-Dependent Vector Clustering for Speech Recognition*, Chapter 6 in Automatic Speech and Speaker Recognition: Advanced Topics, C.-H. Lee, F.K. Soong, and K.K. Paliwal (Eds.), Kluwer Academic Publishers, NY, pp. 133–157, March 1996.

[2] J.R. Bellegarda, *A Multi-Span Language Modeling Framework for Large Vocabulary Speech Recognition*, IEEE Trans. Speech Audio Proc., Vol. **6**, No. 5, pp. 456–467, September 1998.

[3] J.R. Bellegarda, *Large Vocabulary Speech Recognition With Multi-Span Statistical Language Models*, IEEE Trans. Speech Audio Proc., Vol. **8**, No. 1, pp. 76–84, January 2000.

[4] J.R. Bellegarda, *Exploiting Latent Semantic Information in Statistical Language Modeling*, Proc. IEEE, Spec. Issue Speech Recog. Understanding, B.H. Juang and S. Furui (Eds.), Vol. **88**, No. 8, pp. 1279–1296, August 2000.

[5] J.R. Bellegarda, *Robustness in Statistical Language Modeling: Review and Perspectives*, Chapter 4 in Robustness in Language and Speech Technology, J.C. Junqua and G.J.M. van Noord (Eds.), Kluwer Academic Publishers, Dortrecht, The Netherlands, pp. 101–121, February 2001.

[6] J.R. Bellegarda, J.W. Butzberger, Y.L. Chow, N.B. Coccaro, and D. Naik, *A Novel Word Clustering Algorithm Based on Latent Semantic Analysis*, in Proc. 1996 Int. Conf. Acoust., Speech, Sig. Proc., Atlanta, GA, pp. I172–I175, May 1996.

[7] J.R. Bellegarda and K.E.A. Silverman, *Toward Unconstrained Command and Control: Data-Driven Semantic Inference*, in Proc. Int. Conf. Spoken Language Proc., Beijing, China, pp. I258–I261, October 2000.

[8] J.R. Bellegarda and K.E.A. Silverman, *Natural Language Spoken Interface Control Using Data-Driven Semantic Inference*, IEEE Trans. Speech Audio Proc., Vol. **11**, April 2003.

[9] M.W. Berry, *Large–Scale Sparse Singular Value Computations*, Int. J. Supercomp. Appl., Vol. **6**, No. 1, pp. 13–49, 1992.

[10] M.W. Berry, S.T. Dumais, and G.W. O'Brien, *Using Linear Algebra for Intelligent Information Retrieval*, SIAM Review, Vol. **37**, No. 4, pp. 573–595, 1995.

[11] M. Berry and A. Sameh, *An Overview of Parallel Algorithms for the Singular Value and Dense Symmetric Eigenvalue Problems*, J. Computational Applied Math., Vol. **27**, pp. 191–213, 1989.

[12] B. Carpenter and J. Chu–Carroll, *Natural Language Call Routing: A Robust, Self–Organized Approach*, in Proc. Int. Conf. Spoken Language Proc., Sydney, Australia, pp. 2059–2062, December 1998.

[13] C. Chelba, D. Engle, F. Jelinek, V. Jimenez, S. Khudanpur, L. Mangu, H. Printz, E.S. Ristad, R. Rosenfeld, A. Stolcke, and D. Wu, *Structure and Performance of a Dependency Language Model*, in Proc. Fifth Euro. Conf. Speech Comm. Technol., Rhodes, Greece, Vol. **5**, pp. 2775–2778, September 1997.

[14] C. Chelba and F. Jelinek, *Recognition Performance of a Structured Language Model*, in Proc. Sixth Euro. Conf. Speech Comm. Technol., Budapest, Hungary, Vol. **4**, pp. 1567–1570, September 1999.

[15] S. Chen, *Building Probabilistic Models for Natural Language*, Ph.D. Thesis, Harvard University, Cambridge, MA, 1996.

[16] J. Chu–Carroll and B. Carpenter, *Dialog Management in Vector–Based Call Routing*, in Proc. Conf. Assoc. Comput. Linguistics ACL/COLING, Montreal, Canada, pp. 256–262, 1998.

[17] P.R. Clarkson and A.J. Robinson, *Language Model Adaptation Using Mixtures and an Exponentially Decaying Cache*, in Proc. 1997 Int. Conf. Acoust., Speech, Signal Proc., Munich, Germany, Vol. **1**, pp. 799–802, May 1997.

[18] N. Coccaro and D. Jurafsky, *Towards Better Integration of Semantic Predictors in Statistical Language Modeling*, in Proc. Int. Conf. Spoken Language Proc., Sydney, Australia, pp. 2403–2406, December 1998.

[19] J.K. Cullum and R.A. Willoughby, *Lanczos Algorithms for Large Symmetric Eigenvalue Computations – Vol. 1 Theory*, Chapter 5: Real Rectangular Matrices, Brickhauser, Boston, MA, 1985.

[20] R. De Mori, *Recognizing and Using Knowledge Structures in Dialog Systems*, in Proc. Aut. Speech Recog. Understanding Workshop, Keystone, CO, pp. 297–306, December 1999.

[21] S. Deerwester, S.T. Dumais, G.W. Furnas, T.K. Landauer, and R. Harshman, *Indexing by Latent Semantic Analysis*, J. Am. Soc. Inform. Science, Vol. **41**, pp. 391–407, 1990.

[22] S. Della Pietra, V. Della Pietra, R. Mercer, and S. Roukos, *Adaptive Language Model Estimation Using Minimum Discrimination Estimation*, in Proc. 1992 Int. Conf. Acoust., Speech, Signal Processing, San Francisco, CA, Vol. **I**, pp. 633–636, April 1992.

[23] S.T. Dumais, *Improving the Retrieval of Information from External Sources*, Behavior Res. Methods, Instrum., Computers, Vol. **23**, No. 2, pp. 229–236, 1991.

[24] S.T. Dumais, *Latent Semantic Indexing (LSI) and TREC–2*, in Proc. Second Text Retrieval Conference (TREC–2), D. Harman (Ed.), NIST Pub. 500–215, pp. 105–116, 1994.

[25] M. Federico and R. De Mori, *Language Modeling*, Chapter 7 in Spoken Dialogues with Computers, R. De Mori (Ed.), Academic Press, London, UK, pp. 199–230, 1998.

[26] P.W. Foltz and S.T. Dumais, *Personalized Information Delivery: An Analysis of Information Filtering Methods*, Commun. ACM, Vol. **35**, No. 12, pp. 51–60, 1992.

[27] P.N. Garner, *On Topic Identification and Dialogue Move Recognition*, Computer Speech and Language, Vol. **11**, No. 4, pp. 275–306, 1997.

[28] D. Gildea and T. Hofmann, *Topic–Based Language Modeling Using EM*, in Proc. Sixth Euro. Conf. Speech Comm. Technol., Budapest, Hungary, Vol. **5**, pp. 2167–2170, September 1999.

[29] G. Golub and C. Van Loan, *Matrix Computations*, Johns Hopkins, Baltimore, MD, Second Ed., 1989.

[30] Y. Gotoh and S. Renals, *Document Space Models Using Latent Semantic Analysis*, in Proc. Fifth Euro. Conf. Speech Comm. Technol., Rhodes, Greece, Vol. **3**, pp. 1443–1448, September 1997.

[31] T. Hofmann, *Probabilistic Latent Semantic Analysis*, in Proc. Fifteenth Conf. Uncertainty in AI, Stockholm, Sweden, July 1999.

[32] T. Hofmann, *Probabilistic Topic Maps: Navigating Through Large Text Collections*, in Lecture Notes Comp. Science., No. 1642, pp. 161–172, Springer–Verlag, Heidelberg, Germany, July 1999.

[33] R. Iyer and M. Ostendorf, *Modeling Long Distance Dependencies in Language: Topic Mixtures Versus Dynamic Cache Models*, IEEE Trans. Speech Audio Proc., Vol. **7**, No. 1, January 1999.

[34] F. Jelinek, *Self–Organized Language Modeling for Speech Recognition*, in Readings in Speech Recognition, A. Waibel and K.F. Lee (Eds.), Morgan Kaufmann Publishers, pp. 450–506, 1990.

[35] F. Jelinek and C. Chelba, *Putting Language into Language Modeling*, in Proc. Sixth Euro. Conf. Speech Comm. Technol., Budapest, Hungary, Vol. **1**, pp. KN1–KN5, September 1999.

[36] D. Jurafsky, C. Wooters, J. Segal, A. Stolcke, E. Fosler, G. Tajchman, and N. Morgan, *Using a Stochastic Context–Free Grammar as a Language Model for Speech Recognition*, in Proc. 1995 Int. Conf. Acoust., Speech, Signal Proc., Detroit, MI, Vol. **I**, pp. 189–192, May 1995.

[37] S. Khudanpur, *Putting Language Back into Language Modeling*, presented at Workshop–2000 Spoken Lang. Reco. Understanding, Summit, NJ, February 2000.

[38] R. Kneser, *Statistical Language Modeling Using a Variable Context*, in Proc. Int. Conf. Spoken Language Proc., pp. 494–497, Philadelphia, PA, October 1996.

[39] F. Kubala, J.R. Bellegarda, J.R. Cohen, D. Pallett, D.B. Paul, M. Phillips, R. Rajasekaran, F. Richardson, M. Riley, R. Rosenfeld, R. Roth, and M. Weintraub, *The Hub and Spoke Paradigm for CSR Evaluation*, in Proc. ARPA Speech and Natural Language Workshop, Morgan Kaufmann Publishers, pp. 40–44, March 1994.

[40] R. Kuhn and R. De Mori, *A Cache-based Natural Language Method for Speech Recognition*, IEEE Trans. Pattern Anal. Mach. Intel., Vol. PAMI–**12**, No. 6, pp. 570–582, June 1990.

[41] J.D. Lafferty and B. Suhm, *Cluster Expansion and Iterative Scaling for Maximum Entropy Language Models*, in Maximum Entropy and Bayesian Methods, K. Hanson and R. Silver (Eds.), Kluwer Academic Publishers, Norwell, MA, 1995.

[42] T.K. Landauer and S.T. Dumais, *Solution to Plato's Problem: The Latent Semantic Analysis Theory of Acquisition, Induction, and Representation of Knowledge* in Psychological Review, Vol. **104**, No. 2, pp. 211–240, 1997.

[43] T.K. Landauer, D. Laham, B. Rehder, and M.E. Schreiner, *How Well Can Passage Meaning Be Derived Without Using Word Order: A Comparison of Latent Semantic Analysis and Humans*, in Proc. Conf. Cognit. Science Soc., Mahwah, NJ, pp. 412–417, 1997.

[44] R. Lau, R. Rosenfeld, and S. Roukos, *Trigger–Based Language Models: A Maximum Entropy Approach*, in Proc. 1993 Int. Conf. Acoust., Speech, Signal Proc., Minneapolis, MN, pp. II45–48, May 1993.

[45] H. Ney, U. Essen and R. Kneser, *On Structuring Probabilistic Dependences in Stochastic Language Modeling*, Computer, Speech, and Language, Vol. **8**, pp. 1–38, 1994.

[46] T. Niesler and P. Woodland, *A Variable–Length Category–Based N–Gram Language Model*, in Proc. 1996 Int. Conf. Acoust., Speech, Sig. Proc., Atlanta, GA, pp. I164–I167, May 1996.

[47] C.H. Papadimitriou, P. Raghavan, H. Tamaki, and S. Vempala, *Latent Semantic Indexing: A Probabilistic Analysis*, in Proc. 17th ACM Symp. Princip. Database Syst., Seattle, WA, 1998. Also J. Comp. Syst. Sciences, 1999.

[48] F.C. Pereira, Y. Singer, and N. Tishby, *Beyond Word n-Grams*, Computational Linguistics, Vol. **22**, June 1996.

[49] L.R. Rabiner, B.H. Juang, and C.-H. Lee, *An Overview of Automatic Speech Recognition*, Chapter 1 in Automatic Speech and Speaker Recognition: Advanced Topics, C.-H. Lee, F.K. Soong, and K.K. Paliwal (Eds.), Kluwer Academic Publishers, Boston, MA, pp. 1–30, 1996.

[50] R. Rosenfeld, *The CMU Statistical Language Modeling Toolkit and its Use in the 1994 ARPA CSR Evaluation*, in Proc. ARPA Speech and Natural Language Workshop, Morgan Kaufmann Publishers, March 1994.

[51] R. Rosenfeld, *A Maximum Entropy Approach to Adaptive Statistical Language Modeling*, Computer Speech and Language, Vol. **10**, Academic Press, London, UK, pp. 187–228, July 1996.

[52] R. Rosenfeld, *Two Decades of Statistical Language Modeling: Where Do We Go From Here*, Proc. IEEE, Spec. Issue Speech Recog. Understanding, B.H. Juang and S. Furui (Eds.), Vol. **88**, No. 8, pp. 1270–1278, August 2000.

[53] R. Rosenfeld, L. Wasserman, C. Cai, and X.J. Zhu, *Interactive Feature Induction and Logistic Regression for Whole Sentence Exponential Language Models*, in Proc. Aut. Speech Recog. Understanding Workshop, Keystone, CO, pp. 231–236, December 1999.

[54] S. ROUKOS, *Language Representation*, Chapter 6 in Survey of the State of the Art in Human Language Technology, R. Cole (Ed.), Cambridge University Press, Cambridge, MA, 1997.

[55] R. SCHWARTZ, T. IMAI, F. KUBALA, L. NGUYEN, AND J. MAKHOUL, *A Maximum Likelihood Model for Topic Classification of Broadcast News*, in Proc. Fifth Euro. Conf. Speech Comm. Technol., Rhodes, Greece, Vol. **3**, pp. 1455–1458, September 1997.

[56] R.E. STORY, *An Explanation of the Effectiveness of Latent Semantic Indexing by Means of a Bayesian Regression Model*, Inform. Processing & Management, Vol. **32**, No. 3, pp. 329–344, 1996.

[57] J. WU AND S. KHUDANPUR, *Combining Nonlocal, Syntactic and N-Gram Dependencies in Language Modeling*, in Proc. Sixth Euro. Conf. Speech Comm. Technol., Budapest, Hungary, Vol. **5**, pp. 2179–2182, September 1999.

[58] D.H. YOUNGER, *Recognition and Parsing of Context–Free Languages in Time* N^3, Inform. & Control, Vol. **10**, pp. 198–208, 1967.

[59] R. ZHANG, E. BLACK, AND A. FINCH, *Using Detailed Linguistic Structure in Language Modeling*, in Proc. Sixth Euro. Conf. Speech Comm. Technol., Budapest, Hungary, Vol. **4**, pp. 1815–1818, September 1999.

[60] X.J. ZHU, S.F. CHEN, AND R. ROSENFELD, *Linguistic Features for Whole Sentence Maximum Entropy Language Models*, in Proc. Sixth Euro. Conf. Speech Comm. Technol., Budapest, Hungary, Vol. **4**, pp. 1807–1810, September 1999.

[61] V. ZUE, J. GLASS, D. GOODINE, H. LEUNG, M. PHILLIPS, J. POLIFRONI, AND S. SENEFF, *Integration of Speech Recognition and Natural Language Processing in the MIT Voyager System*, in Proc. 1991 IEEE Int. Conf. Acoust., Speech, Signal Processing, Toronto, Canada, pp. 713–716, May 1991.

PROSODY MODELING FOR AUTOMATIC SPEECH RECOGNITION AND UNDERSTANDING*

ELIZABETH SHRIBERG† AND ANDREAS STOLCKE†

Abstract. This paper summarizes statistical modeling approaches for the use of prosody (the rhythm and melody of speech) in automatic recognition and understanding of speech. We outline effective prosodic feature extraction, model architectures, and techniques to combine prosodic with lexical (word-based) information. We then survey a number of applications of the framework, and give results for automatic sentence segmentation and disfluency detection, topic segmentation, dialog act labeling, and word recognition.

Key words. Prosody, speech recognition and understanding, hidden Markov models.

1. Introduction. Prosody has long been studied as an important knowledge source for speech understanding. In recent years there has been a large amount of computational work aimed at prosodic modeling for automatic speech recognition and understanding.[1] Whereas most current approaches to speech processing model only the words, prosody provides an additional knowledge source that is inherent in, and exclusive to, spoken language. It can therefore provide additional information that is not directly available from text alone, and also serves as a partially redundant knowledge source that may help overcome the errors resulting from faulty word recognition.

In this paper, we summarize recent work at SRI International in the area of computational prosody modeling, and results from several recognition tasks where prosodic knowledge proved to be of help. We present only a high-level perspective and summary of our research; for details the reader is referred to publications cited.

2. Modeling philosophy. Most problems for which prosody is a plausible knowledge source can be cast as statistical classification problems. By that we mean that some linguistic unit U (e.g., words or utterances) is to be classified as one of several target classes S. The role of prosody

*The research was supported by NSF Grants IRI-9314967, IRI-9618926, and IRI-9619921, by DARPA contract no. N66001-97-C-8544, and by NASA contract no. NCC 2-1256. Additional support came from the sponsors of the 1997 CLSP Workshop [7, 11] and from the DARPA Communicator project at UW and ICSI [8]. The views herein are those of the authors and should not be interpreted as representing the policies of the funding agencies.

†SRI International, 333 Ravenswood Ave., Menlo Park, CA 94025 ({ees,stolcke}@speech.sri.com). We thank our many colleagues at SRI, ICSI, University of Washington (formerly at Boston University), and the 1997 Johns Hopkins CLSP Summer Workshop, who were instrumental in much of the work reported here.

[1]Too much work in fact, to cite here without unfair omissions. We cite some specifically relevant work below; a more comprehensive list can be found in the papers cited.

is to provide us with a set of *features* F that can help predict S. In a probabilistic framework, we wish to estimate $P(S|F)$. In most such tasks it is also a good idea to use the information contained in the *word sequence* W associated with U, and we therefore generalize the modeling task to estimate $P(S|W, F)$. In fact, W and F are not restricted to pertain only to the unit in question; they may refer to the context of U as well. For example, when classifying an utterance into dialog acts, it is important to take the surrounding utterances into account.

Starting from this general framework, and given a certain labeling task, many decisions must be made to use prosodic information effectively. What is the nature of the features F to be used? How can we model the relationship between F and the target classes S? How should we model the effect of lexical information W and its interaction with prosodic properties F? In the remainder of this paper we give a general overview of approaches that have proven successful for a variety of tasks.

2.1. Direct modeling of target classes. A crucial aspect of our work, as well as that of some other researchers [6, 5] is that the dependence between prosodic features and target classes (e.g., dialog acts, phrase boundaries) is modeled directly in a statistical classifier—without the use of intermediate abstract phonological categories, such as pitch accent or boundary tone labels. This bypasses the need to hand-annotate such labels for training purposes, avoids problems of annotation reliability, and allows the model to choose the level of granularity of the representation that is best suited for the task [2].

2.2. Prosodic features. As predictors of the target classes, we extract features from a forced alignment of the transcripts (usually with phone-level alignment information), which can be based on either true words, or on (errorful) speech recognition output. Similar approaches are used by others [2]. This yields a rich inventory of "raw" features reflecting F0, pause and segment durations, and energy. From the raw features we compute a wide range of "derived" features—devised (we hope) to capture characteristics of the classes—which are normalized in various ways, conditioned on certain extraction regions, or conditioned on values of other features.

Phone-level alignments from a speech recognizer provide durations of pauses and various measures of lengthening (we have used syllable, rhyme, and vowel durations for various tasks) and speaking rate. Pitch-based features benefit greatly from a postprocessing stage that regularizes the raw F0 estimates and models octave errors [10]. As a byproduct of the postprocessing, we also obtain estimates of the speaker's F0 baseline, which we have found useful for pitch range normalizations.

Combined with F0 estimates, the recognizer output also allows computation of pitch movements and contours over the length of utterances or individual words, or over the length of windows positioned relative to a location of interest (e.g., around a word boundary). The same applies to energy-based features.

2.3. Prosodic models. Any number of statistical classifiers that can deal with a mix of categorical and real-valued features may be used to model $P(S|F,W)$. These requirements, as well as our desire to be able to inspect our models (both to understand patterns and for sanity checking), have led us to use mainly decision trees as classifiers. Decision trees have two main problems, however, which we have tried to address. First, to help overcome the problem of greediness, we wrap a feature subset selection algorithm around the standard tree growing algorithm, thereby often finding better classifiers by eliminating detrimental features up front from consideration by the tree [9]. Second, to make the trees sensitive to prosodic features in the case of highly skewed class sizes, we train on a resampled version of the target distribution in which all classes have equal prior probabilities. This approach has additional benefits. It allows prosodic classifiers to be compared (both qualitatively and quantitatively) across different corpora and tasks. In addition, classifiers based on uniform prior distributions are well suited for integration with language models, as described below.

2.4. Lexical models. Our target classes are typically cued by both lexical and prosodic information; we are therefore interested in optimal modeling and combination of both feature types. Although in principle one could add words directly as input features to a prosodic classifier, in practice this is often not feasible since it results in too large a feature space for most classifiers. Approaches for cardinality reduction (such as inferring word classes via unsupervised clustering [4]) offer promise and are an area we are interested in investigating. To date, however, we have used statistical language models (LMs) familiar from speech recognition. One or more LMs are used to effectively model the joint distribution of target classes S and words W, $P(W,S)$. With labeled training data, such models can usually be estimated in a straightforward manner. During testing on unlabeled data, we compute $P(S|W)$ to predict the possible classes and their posterior probabilities, or simply to recover the most likely target class given the words.

2.5. Model combination. The prosodic model may be combined with a language model in different ways, including

- *Posterior interpolation:* Compute $P(S|F,W)$ via the prosodic model and $P(S|W)$ via the language model and form a linear combination of the two. The weighting is optimized on held-out data. This is a weak combination approach that does not attempt to model a more fine-grained structural relationship between the knowledge sources, but it also does not make any strong assumptions about their independence.

- *Posteriors as features:* Compute $P(S|W)$ and use the LM posterior estimate as an additional feature in the prosodic classifier. This approach can capture some of the dependence between the knowledge sources. However, in practice it suffers from the fact that the LM posteriors on the training data are often strongly biased, and therefore lead the tree to over-rely on them unless extra held-out data is used for training.
- *HMM-based integration:* Compute likelihoods $P(F|S,W)$ from the prosody model and use them as observation likelihoods in a hidden Markov model (HMM) derived from the LM.[2] The HMM is constructed to encode the unobserved classes S in its state space. By associating these states with prosodic likelihoods we obtain a joint model of F, S, and W, and HMM algorithms can be used to compute the posteriors $P(S|F,W)$ that incorporate all available knowledge.
  This approach models the relationship between words and prosody at a detailed level, but it does require the assumption that prosody and words are conditionally independent given the labels S. In practice, however, this model often works very well even if the independence assumption is clearly violated.

For a detailed discussion of these approaches, and results showing their relative success under various conditions, see [12, 9, 15].

3. Applications. Having given a brief overview of the key ideas in our approach to computational prosody, we now summarize some applications of the framework.

3.1. Sentence segmentation and disfluency detection. The framework outlined was applied to the detection of sentence boundaries and disfluency interruption points in both conversational speech (Switchboard) and Broadcast News [12, 9]. The target classes S in this case were labels at each word boundary identifying the type of event: sentence boundary, various types of disfluencies (e.g., hesitations, repetitions, deletions) and fluent sentence-internal boundaries. The prosodic model was based on features extracted around each word boundary, capturing pause and phone durations, F0 properties, and ancillary features such as whether a speaker change occurred at that location.

The LM for this task was a hidden event N-gram, i.e., an N-gram LM in which the boundary events were represented by tags occurring between the word tokens. The LM was trained like a standard N-gram model from tagged training text; it thus modeled the joint probability of tags and words. In testing, we ran the LM as an HMM in which the states correspond to the unobserved (hidden) boundary events. Prosodic likelihood

[2]By equating the class distributions for classifier training, as advocated above, we obtain posterior estimates that are proportional to likelihoods, and can therefore be used directly in the HMM.

scores $P(F|S,W)$ for the boundary events were attached to these states as described above, to condition the HMM tagging output on the prosodic features F.

We tested such a model for combined sentence segmentation and disfluency detection on conversational speech, where it gave about 7% boundary classification error using correct word transcripts. The results for various knowledge sources based on true and recognized words are summarized in Table 1 (adapted from [12]). For both test conditions, the prosodic model improves the accuracy of an LM-only classifier by about 4% relative.

TABLE 1

Sentence boundary and disfluency event tagging error rates for the Switchboard corpus. The higher chance error rate for recognized words is due to incorrect word boundary hypotheses.

Model	True words	Recognized words
LM only	7.3	26.2
Prosody only	11.1	27.1
Combined	6.9	25.1
Chance	18.2	30.8

We also carried out a comparative study of sentence segmentation alone, comparing Switchboard (SWB) telephone conversations to Broadcast News (BN) speech. Results are given in Table 2 (adapted from [9]). Again the combination of word and prosodic knowledge yielded the best results, with significant improvements over either knowledge source alone.

TABLE 2

Sentence boundary tagging error rates for two different speech corpora: Switchboard (SWB) and Broadcast News (BN).

	SWB		BN	
Model	True words	Rec. words	True words	Rec. words
LM only	4.3	22.8	4.1	11.8
Prosody only	6.7	22.9	3.6	10.9
Combined	4.0	22.2	3.3	10.8
Chance	11.0	25.8	6.2	13.3

A striking result in BN segmentation was that the prosodic model alone performed better than the LM alone. This was true even when the LM was using the correct words, and even though it was trained on two orders of magnitude more data than the prosody model. Pause duration was universally the most useful feature for these tasks; in addition, SWB classifiers relied primarily on phone duration features, whereas BN classifiers made considerable use of pitch range features (mainly distance from the speaker's estimated baseline). We attribute the increased importance of pitch features in BN to the higher acoustic quality of the audio source, and the preponderance of professional speakers with a consistent speaking style.

3.2. Topic segmentation in broadcast news. A second task we looked at was locating topic changes in a broadcast news stream, following the DARPA TDT [3] framework. For this purpose we adapted a baseline topic segmenter based on an HMM of topic states, each associated with a unigram LM that models topic-specific word distributions [17]. As in the previous tagging tasks, we extracted prosodic features around each potential boundary location, and let a decision tree compute posterior probabilities of the events (in this case, topic changes). By resampling the training events to a uniform distribution, we ensured that the posteriors are proportional to event likelihoods, as required for HMM integration [9, 15].

The results on this task are summarized in Table 3. We obtained a large, 24-27% relative error reduction from combining lexical and prosodic models. Also, similar to BN sentence segmentation, the prosodic model alone outperformed the LM. The prosodic features selected for topic segmentation were similar to those for sentence segmentation, but with more pronounced tendencies. For example, at the end of topic segments, a speaker tends to pause even longer and drop the pitch even closer to the baseline than at sentence boundaries.

TABLE 3

Topic segmentation weighted error on Broadcast News data. The evaluation metric used is a weighted combination of false alarm and miss errors [3].

Model	True words	Recognized words
LM only	0.1895	0.1897
Prosody only	0.1657	0.1731
Combined	0.1377	0.1438
Chance	0.3000	0.3000

3.3. Dialog act labeling in conversational speech. The third task we looked at was dialog act (DA) labeling. In this task the goal was to classify each utterance (rather than each word boundary) into a number of types, such as statement, question, acknowledgment, and backchannel. In [7] we investigated the use of prosodic features for DA modeling, alone and in conjunction with LMs. Prosodic features describing the whole utterance were fed to a decision tree. N-gram language models specific to each DA class provided additional likelihoods. These models can be applied to DAs in isolation, or combined with a statistical dialog grammar that models the contextual effects of nearby DAs. In a 42-way classification of Switchboard utterances, the prosody component improved the overall classification accuracy of such a combined model [11]. However, we found that prosodic features were most useful in disambiguating certain DAs that are particularly confusable based on their words alone. Table 4 shows results for two such binary DA discrimination tasks: distinguishing questions from statements, and backchannels ("uh-huh", "right") from agreements ("Right!"). Again, adding prosody boosted accuracy substantially over a word-only model. The features used for these and other DA disambiguation tasks, as might be expected, depend on the DAs involved, as described in [7].

TABLE 4
Dialog act classification error on highly ambiguous DA pairs in the Switchboard corpus.

Classification task Knowledge source	True words	Rec. words
Questions vs. Statements		
LM only	14.1	24.6
Prosody only	24.0	24.0
Combined	12.4	20.2
Agreements vs. Backchannels		
LM only	19.0	21.2
Prosody only	27.1	27.1
Combined	15.3	18.3
Chance	50.0	50.0

3.4. Word recognition in conversational speech. All applications discussed so far had the goal of adding structural, semantic, or pragmatic information beyond what is contained in the raw word transcripts. Word recognition itself, however, is still far from perfect, raising the question: can prosodic cues be used to improve speech recognition accuracy? An early approach in this area was [16], using prosody to evaluate possible parses for recognized words, which in turn would be the basis for reranking word hypotheses. Recently, there have been a number of approaches that essentially condition the language model on prosodic evidence, thereby constraining recognition. The dialog act classification task mentioned above can serve this purpose, since many DA types are characterized by specific word patterns. If we can use prosodic cues to predict the DA of an utterance, we can then use a DA-specific LM to constrain recognition. This approach has yielded improved recognition in task-oriented dialogs [14], but significant improvements in large-vocabulary recognition remain elusive [11].

We have had some success using the hidden event N-gram model (previously introduced for sentence segmentation and disfluency detection) for word recognition [13]. As before, we computed prosodic likelihoods for each event type at each word boundary, and conditioned the word portion of the N-gram on those events. The result was a small, but significant 2% relative reduction in Switchboard word recognition error. This improvement was surprising given that the prosodic model had not been optimized for word recognition. We expect that more sophisticated and more tightly integrated prosodic models will ultimately make substantive contributions to word recognition accuracy.

3.5. Other corpora and tasks. We have recently started applying the framework described here to new types of data, including multiparty face-to-face meetings. We have found that speech in multiparty meetings seems to have properties more similar to Switchboard than to Broadcast News, with respect to automatic detection of target events [8]. Such data also offers an opportunity to apply prosody to tasks that have not been widely studied in a computational framework. One nice example is the modeling of turn-taking in meetings. In a first venture into this area, we have found that prosody correlates with the location and form of overlapping speech [8].

We also studied disfluency detection and sentence segmentation in the meeting domain, and obtained results that are qualitatively similar to those reported earlier on the Switchboard corpus [1]. A noteworthy result was that event detection accuracy on recognized words improved slightly when the models were trained on recognized rather than true words. This indicates that there is systematicity to recognition errors that can be partially captured in event models.

4. Conclusions. We have briefly summarized a framework for computational prosody modeling for a variety of tasks. The approach is based on modeling of directly measurable prosodic features and combination with lexical (statistical language) models. Results show that prosodic information can significantly enhance accuracy on several classification and tagging tasks, including sentence segmentation, disfluency detection, topic segmentation, dialog act tagging, and overlap modeling. Finally, results so far show that speech recognition accuracy can also benefit from prosody, by constraining word hypotheses through a combined prosody/language model.

More information about individual research projects is available at http://www.speech.sri.com/projects/hidden-events.html, http://www.speech.sri.com/projects/sieve/, and http://www.clsp.jhu.edu/ws97/discourse/.

REFERENCES

[1] D. Baron, E. Shriberg, and A. Stolcke, *Automatic punctuation and disfluency detection in multi-party meetings using prosodic and lexical cues*, in Proceedings of the International Conference on Spoken Language Processing, Denver, Sept. 2002.

[2] A. Batliner, B. Möbius, G. Möhler, A. Schweitzer, and E. Nöth, *Prosodic models, automatic speech understanding, and speech synthesis: toward the common ground*, in Proceedings of the 7th European Conference on Speech Communication and Technology, P. Dalsgaard, B. Lindberg, H. Benner, and Z. Tan, eds., Vol. **4**, Aalborg, Denmark, Sept. 2001, pp. 2285–2288.

[3] G. Doddington, *The Topic Detection and Tracking Phase 2 (TDT2) evaluation plan*, in Proceedings DARPA Broadcast News Transcription and Understanding Workshop, Lansdowne, VA, Feb. 1998, Morgan Kaufmann, pp. 223–229. Revised version available from http://www.nist.gov/speech/tests/tdt/tdt98/.

[4] P. Heeman and J. Allen, *International boundaries, speech repairs, and discourse markers: Modeling spoken dialog*, in Proceedings of the 35th Annual Meeting and 8th Conference of the European Chapter, Madrid, July 1997, Association for Computational Linguistics.

[5] J. Hirschberg and C. Nakatani, *Acoustic indicators of topic segmentation*, in Proceedings of the International Conference on Spoken Language Processing, R.H. Mannell and J. Robert-Ribes, eds., Sydney, Dec. 1998, Australian Speech Science and Technology Association, pp. 976–979.

[6] M. Mast, R. Kompe, S. Harbeck, A. Kiessling, H. Niemann, E. Nöth, E.G. Schukat-Talamazzini, and V. Warnke, *Dialog act classification with the help of prosody*, in Proceedings of the International Conference on Spoken Language Processing, H.T. Bunnell and W. Idsardi, eds., Vol. **3**, Philadelphia, Oct. 1996, pp. 1732–1735.

[7] E. Shriberg, R. Bates, A. Stolcke, P. Taylor, D. Jurafsky, K. Ries, N. Coccaro, R. Martin, M. Meteer, and C. Van Ess-Dykema, *Can prosody aid the automatic classification of dialog acts in conversational speech?*, Language and Speech, **41** (1998), pp. 439–487.

[8] E. Shriberg, A. Stolcke, and D. Baron, *Can prosody aid the automatic processing of multi-party meetings? Evidence from predicting punctuation, disfluencies, and overlapping speech*, in Proceedings ISCA Tutorial and Research Workshop on Prosody in Speech Recognition and Understanding, M. Bacchiani, J. Hirschberg, D. Litman, and M. Ostendorf, eds., Red Bank, NJ, Oct. 2001, pp. 139–146.

[9] E. Shriberg, A. Stolcke, D. Hakkani-Tür, and G. Tür, *Prosody-based automatic segmentation of speech into sentences and topics*, Speech Communication, **32** (2000), pp. 127–154. Special Issue on Accessing Information in Spoken Audio.

[10] K. Sönmez, E. Shriberg, L. Heck, and M. Weintraub, *Modeling dynamic prosodic variation for speaker verification*, in Proceedings of the International Conference on Spoken Language Processing, R.H. Mannell and J. Robert-Ribes, eds., Vol. **7**, Sydney, Dec. 1998, Australian Speech Science and Technology Association, pp. 3189–3192.

[11] A. Stolcke, K. Ries, N. Coccaro, E. Shriberg, D. Jurafsky, P. Taylor, R. Martin, C. Van Ess-Dykema, and M. Meteer, *Dialogue act modeling for automatic tagging and recognition of conversational speech*, Computational Linguistics, **26** (2000), pp. 339–373.

[12] A. Stolcke, E. Shriberg, R. Bates, M. Ostendorf, D. Hakkani, M. Plauché, G. Tür, and Y. Lu, *Automatic detection of sentence boundaries and disfluencies based on recognized words*, in Proceedings of the International Conference on Spoken Language Processing, R.H. Mannell and J. Robert-Ribes, eds., Vol. **5**, Sydney, Dec. 1998, Australian Speech Science and Technology Association, pp. 2247–2250.

[13] A. Stolcke, E. Shriberg, D. Hakkani-Tür, and G. Tür, *Modeling the prosody of hidden events for improved word recognition*, in Proceedings of the 6th European Conference on Speech Communication and Technology, Vol. **1**, Budapest, Sept. 1999, pp. 307–310.
[14] P. Taylor, S. King, S. Isard, and H. Wright, *Intonation and dialog context as constraints for speech recognition*, Language and Speech, **41** (1998), pp. 489–508.
[15] G. Tür, D. Hakkani-Tür, A. Stolcke, and E. Shriberg, *Integrating prosodic and lexical cues for automatic topic segmentation*, Computational Linguistics, **27** (2001), pp. 31–57.
[16] N.M. Veilleux and M. Ostendorf, *Prosody/parse scoring and its applications in ATIS*, in Proceedings of the ARPA Workshop on Human Language Technology, Plainsboro, NJ, Mar. 1993, pp. 335–340.
[17] J. Yamron, I. Carp, L. Gillick, S. Lowe, and P. van Mulbregt, *A hidden Markov model approach to text segmentation and event tracking*, in Proceedings of the IEEE Conference on Acoustics, Speech, and Signal Processing, Vol. **I**, Seattle, WA, May 1998, pp. 333–336.

SWITCHING DYNAMIC SYSTEM MODELS FOR SPEECH ARTICULATION AND ACOUSTICS

LI DENG*

Abstract. A statistical generative model for the speech process is described that embeds a substantially richer structure than the HMM currently in predominant use for automatic speech recognition. This switching dynamic-system model generalizes and integrates the HMM and the piece-wise stationary nonlinear dynamic system (state-space) model. Depending on the level and the nature of the switching in the model design, various key properties of the speech dynamics can be naturally represented in the model. Such properties include the temporal structure of the speech acoustics, its causal articulatory movements, and the control of such movements by the multidimensional targets correlated with the phonological (symbolic) units of speech in terms of overlapping articulatory features.

One main challenge of using this multi-level switching dynamic-system model for successful speech recognition is the computationally intractable inference (decoding with confidence measure) on the posterior probabilities of the hidden states. This leads to computationally intractable optimal parameter learning (training) also. Several versions of BayesNets have been devised with detailed dependency implementation specified to represent the switching dynamic-system model of speech. We discuss the variational technique developed for general Bayesian networks as an efficient approximate algorithm for the decoding and learning problems. Some common operations of estimating phonological states' switching times have been shared between the variational technique and the human auditory function that uses neural transient responses to detect temporal landmarks associated with phonological features. This suggests that the variation-style learning may be related to human speech perception under an encoding-decoding theory of speech communication, which highlights the critical roles of modeling articulatory dynamics for speech recognition and which forms a main motivation for the switching dynamic system model for speech articulation and acoustics described in this chapter.

Key words. State-space model, Dynamic system, Bayesian network, Probabilistic inference, Speech articulation, Speech acoustics, Auditory function, Speech recognition.

AMS(MOS) subject classifications. Primary 68T10.

1. Introduction. Speech recognition technology has made dramatic progress in recent years (cf. [30, 28]), attributed to the use of powerful statistical paradigms, availability of increasing quantities of speech data corpus, and to the development of powerful algorithms for model learning from the data. However, the methodology underlying the current technology has been founded on weak scientific principles. Not only does the current methodology require prohibitively large amounts of training data and lack robustness under mismatch conditions, its performance also falls at least one order of magnitude short of that of human speech recognition on many comparable tasks (cf. [32, 43]). For example, the best recognizers

*Microsoft Research, One Microsoft Way, Redmond, WA 98052 (deng@microsoft.com). The author wishes to thank many useful discussions with and suggestions for improving the paper presentation by David Heckerman, Mari Ostendorf, Ken Stevens, B. Frey, H. Attias, G. Ramsay, J. Ma, L. Lee, Sam Roweis, and J. Bilmes.

today still produce errors in more than one quarter of the words in natural conversational speech in spite of many hours of speech material used as training data. The current methodology has been primarily founded on the principle of statistical "ignorance" modeling. This fundamental philosophy is unlikely to bridge the performance gap between human and machine speech recognition. A potentially promising approach is to build into the statistical speech model most crucial mechanisms in human speech communication for use in machine speech recognition. Since speech recognition or perception in humans is one integrative component in the entire closed-loop speech communication chain, the mechanisms to be modeled need to be sufficiently broad — including mechanisms in both speech production and auditory perception as well as in their interactions.

Some recent work on speech recognition have been pursued along this direction [6, 18, 13, 17, 46, 47]. The approaches proposed and described in [1, 5, 49] have incorporated the mechanisms in the human auditory process in speech recognizer design. The approaches reported in [18, 21, 19, 44, 3, 54] have advocated the use of the articulatory feature-based phonological units which control human speech production and are typical of human lexical representation, breaking away from the prevailing use of the phone-sized, "beads-on-a-string" linear phonological units in the current speech recognition technology. The approaches outlined in [35, 12, 11, 14, 13] have emphasized the functional significance of the abstract, "task" dynamics in speech production and recognition. The task variables in the task dynamics are the quantities (such as vocal tract constriction locations and degrees) that are closely linked to the goal of speech production, and are nonlinearly related to the physical variables in speech production. Work reported and surveyed in [10, 15, 38, 47] have also focused on the dynamic aspects in the speech process, but the dynamic object being modeled is in the space of speech acoustics, rather than in the space of the production-affiliated variables.

Although dynamic modeling has been a central focus of much recent work in speech recognition, the dynamic object being modeled either in the space of "task" variables or of acoustic variables does not and may not be potentially able to directly take into account the many important properties in *true* articulatory dynamics. Some earlier work used [16, 22] either quantized articulatory features or articulatory data to design speech recognizers, employing highly simplistic models for the underlying articulatory dynamics. Some other earlier proposals and empirical methods exploited pseudo-articulatory dynamics or abstract hidden dynamics for the purpose of speech recognition [2, 4, 23, 45], where the dynamics of a set of pseudo-articulators is realized either by FIR filtering from sequentially placed, phoneme-specific target positions or by applying trajectory-smoothness constraints. Such approaches relied on simplistic nature in the use of the pseudo-articulators. As a result, compensatory articulation, which is a key property of human speech production and which requires modeling correlations among a set of articulators, could not be taken into account. This has drastically diminished the power of such models for potentially successful use in speech recognition.

To incorporate crucial properties in human articulatory dynamics — including compensatory articulation, target behavior, and relatively constrained dynamics (due to biomechanical properties of the articulatory organs) — in a statistical model of speech, it appears necessary to use true, multidimensional articulators, rather than the pseudo-articulators attempted in the past. Given that much of the acoustic variation observed in speech that makes speech recognition difficult can be attributed to articulatory phenomena, and given that articulation is one key component in the closed-loop human speech communication chain, it is reasonable to expect that incorporating a faithful and explicit articulatory dynamic model in the statistical structure of automatic speech recognizer will contribute to bridging the performance gap between human and machine speech recognition. Based on this motivation, a general framework for speech recognition using a statistical description of the speech articulation and acoustic processes is developed and outlined in this chapter. Central to this framework is a switching dynamic system model used to characterize the speech articulation (with its control) and the related acoustic processes, and the Bayesian network (BayesNet) representation of this model. Before presenting some details of this model, we first introduce an encoding-decoding theory of human speech perception which formalizes key roles of modeling speech articulation.

2. Roles of articulation in encoding-decoding theory of speech perception. At a global and functional level, human speech communication can be viewed as an encoding-decoding process, where the decoding process or perception is an active process consisting of auditory reception followed by phonetic/linguistic interpretation. As an encoder implemented by the speech production system, the speaker uses knowledge of meanings of words (or phrases), of grammar in a language, and of the sound representations for the intended linguistic message. Such knowledge can be made analogous to the keys used in engineering communication systems. The phonetic plan, derived from the semantic, syntactic, and phonological processes, is then executed through the motor-articulatory system to produce speech waveforms.

As a decoder which aims to accomplish speech perception, the listener uses a key, or the internal "generative" model, which is compatible with the key used by the speaker to interpret the speech signal received and transformed by the peripheral auditory system. This would enable the listener to reconstruct, via (probabilistic) analysis-by-synthesis strategies, the linguistic message intended by the speaker.[1] This encoding-decoding theory of

[1]While it is not universally accepted that listeners actually do analysis-by-synthesis in speech perception, it would be useful to use such a framework to interpret the roles

human speech communication, where the observable speech acoustics plays the role of the carrier of deep, linguistically meaningful messages, may be likened to the modulation-demodulation scheme in electronic digital communication and to the encryption-decryption scheme in secure electronic communication. Since the nature of the key used in the phonetic-linguistic information decoding or speech perception lies in the strategies used in the production or encoding process, speech production and perception are intimately linked in the closed-loop speech chain. The implication of such a link for speech recognition technology is the need to develop functional and computational models of human speech production for use as an "internal model" in the decoding process by machines. Fig. 1 is a schematic diagram showing speaker-listener interactions in human speech communication and showing the several components in the encoding-decoding theory.

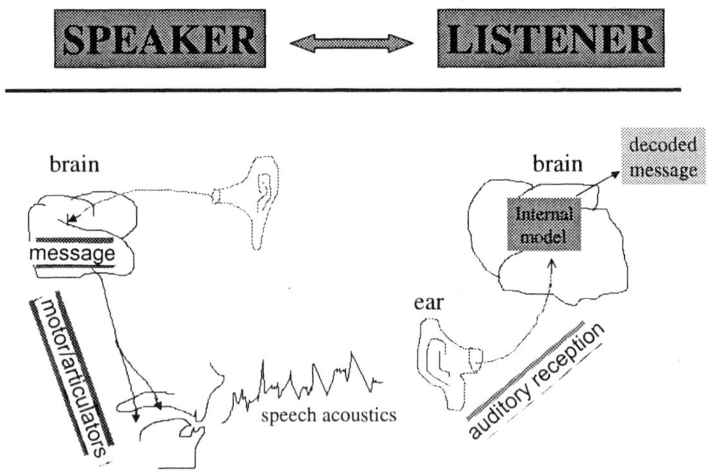

FIG. 1. *Speaker-listener interactions in the encoding-decoding theory of speech perception.*

The encoding-decoding theory of speech perception outlined above highlights crucial roles of speech articulation for speech perception. In summary, the theory consists of three basic, integrated elements: 1) approximate motor-encoding — the symbolic phonological process interfaced with dynamic phonetic process in speech production; 2) robust auditory reception — speech signal transformation prior to the cognitive process;

of articulation in speech perception.

3) cognitive decoding — optimal (by statistical criteria) matching of the auditory transformed signal with the "internal" model derived from a set of motor encoders distinct for separate speech classes. In this theory, the "internal" model in the brain of the listener is hypothesized to have been "approximately" established during the childhood speech acquisition process (or during the process of learning foreign languages in adulthood).

The speech production process as the approximate motor encoder in the above encoding-decoding theory consists of the control strategy of speech articulation, the actual realized speech articulation, and the acoustic signal as the output of the speech articulation system. On the other hand, the auditory process plays two other key roles. First, it transforms the acoustic signal of speech to make it robust against environmental variations. This provides the modified information to the decoder to make its job easier than otherwise. Second, many transient and dynamic properties in the auditory system's responses to speech help create temporal landmarks in the stream of speech to guide the decoding process [50, 53, 54]. (See more detailed discussions on the temporal landmarks in Section 4.3). As will be shown in this chapter, the optimal decoding using the switching dynamic system model as the encoder incurs exponentially growing computation. Use of the temporal landmarks generated from the auditory system's responses may successfully overcome such computational difficulties, hence providing an elegant approximate solution to the otherwise formidable computational problem in the decoding.

In addition to accounting for much of the existing human speech perception data, the computational nature of this theory, with some details described in the remaining of this chapter with special focus on statistical modeling of the dynamic speech articulation and acoustic processes, enables it to be used as the basic underpinning of computer speech recognition systems.

3. Switching state space model for multi-level speech dynamics. In this section, we outline each component of the multi-level speech dynamic model. The model serves as a computational device for the approximate encoder in the encoding-decoding theory of speech perception outlined above. We provide motivations for the construction of each model component from principles of speech science, present a mathematical description of each model component, and justify assumptions made to the mathematical description. The components in the overall model consists of a phonological model, a model for the segmental target, a model for the articulatory dynamics, and a model for the mapping from articulation to acoustics. We start with the phonological-model component.

3.1. Phonological construct. Phonology is concerned with sound patterns of speech and the nature of discrete or symbolic units that form such patterns. Traditional theories of phonology differ in the choice and interpretation of the phonological units. Early distinctive feature based theory [8] and subsequent autosegmental, feature-geometry theory [9] assumed a rather direct link between phonological features and their phonetic correlates in the articulatory or acoustic domain. Phonological rules for modifying features represented changes not only in the linguistic structure of the speech utterance, but also in the phonetic realization of this structure. This weakness has been recognized by more recent theories, e.g., articulatory phonology [7], which emphasize the importance of accounting for phonetic levels of variation as distinct from those at the phonological levels.

In the framework described here, it will be assumed that the linguistic function of phonological units is to maintain linguistic contrasts and is separate from phonetic implementation. It is further assumed that the phonological unit sequence can be described mathematically by a discrete-time, discrete-state homogeneous Markov chain. This Markov chain is characterized by its state transition matrix $\mathbf{A} = [a_{ij}]$ where $a_{ij} = P(s_k = j|s_{k-1} = i)$.

How to construct sequences of symbolic phonological units for any arbitrary speech utterance and how to built them into an appropriate Markov state (i.e., phonological state) structure will not be dealt with here. We merely mention that for effective use of the current framework in speech recognition, the symbolic units must be of multiple dimensions that overlap with each other temporally, overcoming beads-on-a-string limitations. We refer the readers to some earlier work for ways of constructing such overlapping units, either by rules or by automatic learning, which have proved effective in the HMM-like speech recognition framework [21, 19, 18, 56].

3.2. Articulatory control and targets. After a phonological model is constructed, the processes for converting abstract phonological units into their phonetic realization need to be specified. This is a central issue in speech production. It concerns the nature of invariance and variability in the processes interfacing phonology and phonetics, and specifically, whether the invariance is more naturally expressed in the articulatory or acoustic/auditory domains. Early proposals assumed a direct link between abstract phonological units and physical measurements. The "quantal theory" [53] proposed that phonological features possessed invariant acoustic correlates that could be measured directly from the speech signal. The "motor theory" [31] proposed instead that articulatory properties are associated with phonological symbols. No conclusive evidence supporting either hypothesis has been found without controversy, however.

In the current framework, a commonly held view in the phonetics literature is adopted that discrete phonological units are associated with a temporal segmental sequence of phonetic targets or goals [34, 29, 40, 41, 42]. The function of the articulatory motor control system is to achieve such targets or goals by manipulating the articulatory organs according to some control principles subject to the articulatory inertia and possibly minimal-energy constraints.

Compensatory articulation has been widely documented in the phonetics literature where trade-offs between different articulators and non-uniqueness in the articulatory-acoustic mapping allow for the possibilities that many different articulatory target configurations may be able to realize the same underlying goal, and that speakers typically choose a range of possible targets depending on external environments and their interactions with listeners [29]. In order to account for compensatory articulation, a complex phonetic control strategy need be adopted. The key modeling assumptions adopted regarding such a strategy is as follows. First, each phonological unit is associated with a number of phonetic parameters that are described by a state-dependent distribution. These measurable parameters may be acoustic, articulatory or auditory in nature, and they can be computed from some physical models for the articulatory and auditory systems. Further, the region determined by the phonetic correlates for each phonological unit can be mapped onto an articulatory parameter space. Hence the target distribution in the articulatory space can be determined simply by stating what the phonetic correlates (formants, articulatory positions, auditory responses, etc.) are for each of the phonological units (many examples are provided in [55]), and by running simulations in suitably-detailed articulatory and auditory models.

A convenient mathematical representation for the distribution of the articulatory target vector $\mathbf{t}$ is a multivariate Gaussian distribution, denoted by

$$\mathbf{t} \sim \mathcal{N}(\mathbf{t}; \mathbf{m}(s), \mathbf{\Sigma}(s)).$$

Since the target distribution is conditioned on a specific phonological unit (such as a bundle of overlapped features represented by an HMM state s) and since the target does not switch until the phonological unit changes, the statistics for the temporal sequence of the target process follows that of a segmental HMM. A most recent review of the segmental HMM can found in [26].

3.3. Articulatory dynamics. At the present state of knowledge, it is difficult to speculate how the conversion of higher-level motor control into articulator movement takes place. Ideally, modeling of articulatory dynamics and control would require detailed neuromuscular and biomechanical models of the vocal tract, as well as an explicit model of the control objectives and strategies. This is clearly too complicated to implement. A reasonable, simplifying assumption would be that the combined (non-linear) control system and articulatory mechanism behave, at a functional level, as a linear dynamic system that attempts to track the control input equivalently represented by the articulatory target in the articulatory parameter space. Articulatory dynamics can then be approximated as the response of a dynamic vocal tract model driven by a random target sequence (as a segmental HMM). (The output of the vocal tract model then produces a time-varying tract shape which modulates the acoustic properties of the speech signal as observed data.)

This simplifying assumption then reduces the generic nonlinear state equation:

$$\mathbf{z}(k+1) = \mathbf{g}_s[\mathbf{z}(k), \mathbf{t}_s, \mathbf{w}(k)]$$

into a mathematically tractable linear one:

$$\mathbf{z}(k+1) = \mathbf{\Phi}_s \mathbf{z}(k) + (\mathbf{I} - \mathbf{\Phi}_s)\mathbf{t}_s + \mathbf{w}(k), \tag{3.1}$$

where $\mathbf{z} \in \mathcal{R}^n$ is the articulatory-parameter vector, $\mathbf{I}$ is the identity matrix, $\mathbf{w}$ is the IID and Gaussian system noise ($\mathbf{w}(k) \sim \mathcal{N}[\mathbf{w}(k); \mathbf{0}, \mathbf{Q}_{s_k}]$), $\mathbf{t}_s$ is the HMM-state dependent, target vector (expressed in the articulatory domain), and $\mathbf{\Phi}_s$ is the HMM-state-dependent system matrix. The dependence of the $\mathbf{t}_s$ and $\mathbf{\Phi}_s$ parameters of the above dynamic system on the phonological state is justified by the fact that the functional behavior of an articulator depends on the particular goal it is trying to implement, and on the other articulators with which it is cooperating in order to produce compensatory articulation.

3.4. Acoustic model. While a truly consistent framework we are striving for based on explicit knowledge of speech production and perception ideally should include detailed high-order state-space models of the physical mechanisms involved, this becomes unfeasible due to excessive computational requirements. The simplifying assumption adopted is that the articulatory and acoustic state of the vocal tract can be adequately described by low-order vectors of variables representing respectively the relative positions of the major articulators, and the corresponding time-averaged spectral parameters derived from the acoustic signal (or other parameters computed from auditory models). Given further that an appropriate time scale is chosen, it will also be assumed that the relationship between articulatory and acoustic representations can be modeled by a static memoryless transformation, converting a vector of articulatory parameters into a vector of acoustic (or auditory) measurements.

This noisy static memoryless transformation can be mathematically represented by the following observation equation in the state-space model:

$$\mathbf{o}(k) = \mathbf{h}[\mathbf{z}(k)] + \mathbf{v}(k). \tag{3.2}$$

where $\mathbf{o} \in \mathcal{R}^m$ is the observation vector, $\mathbf{v}$ is the IID observation noise vector ($\mathbf{v}(k) \sim \mathcal{N}[\mathbf{v}(k); \mathbf{0}, \mathbf{R}]$) uncorrelated with the state noise $\mathbf{w}$, and $\mathbf{h}[.]$ is the static memoryless transformation from the articulatory vector to its corresponding acoustic observation vector.

There are many ways of choosing the static nonlinear function for $\mathbf{h}[\mathbf{z}]$. Let us take an example of multi-layer perceptron (MLP) with three layers (input, hidden and output). Let w_{jl} be the MLP weights from input to hidden units and W_{ij} be the MLP weights from hidden to output units, where l is the input node index, j the hidden node index and i the output node index. Then the output signal at node i can be expressed as a (nonlinear) function $\mathbf{h}(.)$ of all the input nodes (making up the input vector) according to

$$h_i(\mathbf{z}) = \sum_{j=1}^{J} W_{ij} \cdot s\left(\sum_{l=1}^{L} w_{jl} \cdot z_l \right), \quad 1 \leq i \leq I, \tag{3.3}$$

where I, J and L are the numbers of nodes at the output, hidden and input layers, respectively. $s(.)$ is the hidden unit's nonlinear activation function, taken as the standard sigmoid function of

$$s(z) = \frac{1}{1 + \exp(-z)}. \tag{3.4}$$

The derivative of this sigmoid function has the following concise form:

$$s'(z) = s(z)(1 - s(z)), \tag{3.5}$$

making it convenient for use in many computations.

Typically, the analytical forms of nonlinear functions, such as the MLP, make the associated nonlinear dynamic systems difficult to analyze and make the estimation problems difficult to solve. Approximations are frequently used to gain computational simplifications while sacrificing accuracy for approximating the nonlinear functions.

One most commonly used technique for the approximation is the truncated (vector) Taylor series expansion. If all the Taylor series terms of order two and higher are truncated, then we have the linear Taylor series approximation that is characterized by the Jacobian matrix $\mathbf{J}$ and the point of Taylor series expansion $\mathbf{z}_0$:

$$\mathbf{h}(\mathbf{z}) \approx \mathbf{h}(\mathbf{z}_0) + \mathbf{J}(\mathbf{z}_0)(\mathbf{z} - \mathbf{z}_0). \tag{3.6}$$

Each element of the Jacobian matrix $\mathbf{J}$ is partial derivative of each vector component of the nonlinear output with respect to each of the input vector components. That is,

$$\mathbf{J}(\mathbf{z}_0) = \frac{\partial \mathbf{h}}{\partial \mathbf{z}_0} = \begin{bmatrix} \frac{\partial h_1(\mathbf{z}_0)}{\partial z_1} & \frac{\partial h_1(\mathbf{z}_0)}{\partial z_2} \cdots & \frac{\partial h_1(\mathbf{z}_0)}{\partial z_n} \\ \frac{\partial h_2(\mathbf{z}_0)}{\partial z_1} & \frac{\partial h_2(\mathbf{z}_0)}{\partial z_2} \cdots & \frac{\partial h_2(\mathbf{z}_0)}{\partial z_n} \\ \vdots & \vdots & \\ \frac{\partial h_m(\mathbf{z}_0)}{\partial z_1} & \frac{\partial h_m(\mathbf{z}_0)}{\partial z_2} \cdots & \frac{\partial h_m(\mathbf{z}_0)}{\partial z_n} \end{bmatrix}. \tag{3.7}$$

As an example, for the MLP nonlinearity of Eqn. 3.3, the (i, l)-th element of the Jacobian matrix is

$$\sum_{j=1}^{J} W_{ij} \cdot s_j(y) \cdot (1 - s_j(y)) \cdot w_{jl}, \qquad 1 \leq i \leq I, \;\; 1 \leq l \leq L, \tag{3.8}$$

where $y = \sum_{l'=1}^{L} W_{jl'} z_{l'}$.

Use of the radial basis function as the nonlinearity in the general nonlinear dynamic system model, as an alternative to the MLP described above, can be found in [24].

3.5. Switching state space model. Eqns. 3.1 and 3.2 form a special version of the switching state-space model appropriate for describing multi-level speech dynamics. The top-level dynamics occurs at the discrete-state phonology, represented by the state transitions of s with a relatively long time scale. The next level is the target ($\mathbf{t}$) dynamics; it has the same time scale and provides systematic randomness at the segmental level. At the level of articulatory dynamics, the time scale is significantly shortened. This is continuous-state dynamics driven by the target process as input, which follows HMM statistics. The state equation 3.1 explicitly describes this dynamics in $\mathbf{z}$, with index of s (which takes discrete values) implicitly representing the switching process. At the lowest level of acoustic dynamics, there is no switching process. Since the observation equation 3.2 is static, this simplifying speech model assumes that acoustic dynamics results solely from articulatory dynamics.

4. BayesNet representation of the segmental switching dynamic speech model. Developed traditionally by machine-learning researchers, BayesNets have found many useful applications. A BayesNet is a graphical model that describes dependencies in the probability distributions defined over a set of variables. A most interesting class of the BayesNet, as relevant to speech modeling, is dynamic BayesNets that are specifically aimed at modeling time series statistics. For time series data such as speech vector sequences, there are causal dependencies between random variables in time. The causal dependencies give some specific, left-to-right BayesNet structures. Such specific structures either permit development of highly efficient algorithms (e.g., for the HMM) for the probabilistic inference (i.e., computation of conditional probabilities for hidden variables) and for learning (i.e., model parameter estimation), or enable the use of approximate techniques (such as variational techniques) to solve the inference and learning problems.

Both the HMM and the stationary (i.e., no switching) dynamic system model are two of the simplest examples of a dynamic BayesNet, for which the efficient algorithms developed already in statistics and in speech processing [51, 38, 20] turn out to be identical to those based on the more general principles of BayesNet theory applied to the special network structures associated with these models. However, for the more complex speech model such as the switching dynamic system model described above, no exact solutions for inference and learning are available without exponentially growing computation with the size of the data. Approximate solutions have been provided for some simple versions of the the switching dynamic system model in literatures of statistics [52], speech processing [33], and of neural computation and BayesNet [25, 39]. The BayesNet framework allows us to take a fresh view on the complex computational issues for such a model, and provides guidance and insights to the algorithm development as well as model refinement.

4.1. Basic BayesNet model. We now discuss how the particular multi-component speech model described in Section 3 can be represented and implemented by BayesNets. Fig. 2 shows one type of dependency structure (indicated by the direction of arrows) of the model, where (discrete) time index runs from left to right. The top-row random variables $s(k)$ take discrete values over the set of phonological states (overlapped feature bundles), and the remaining random variables for the targets, articulators, and acoustic vectors are continuously valued for each time index.

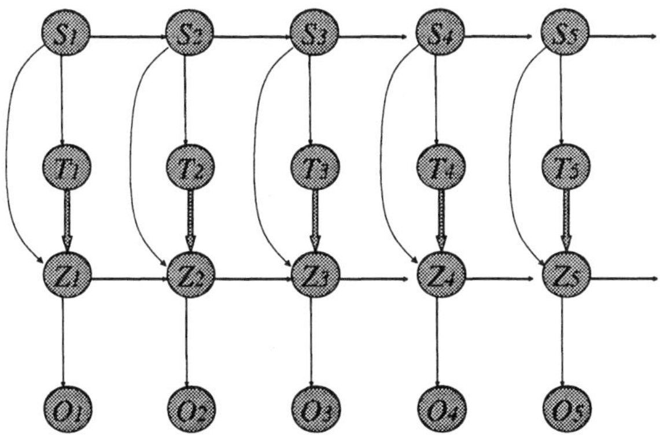

FIG. 2. *Dynamic BayesNet for a basic version of the switching dynamic system model of speech. The random variables on Row 1 are discrete, hidden linguistic states with the Markov-chain temporal structure. Those on Row 2 are continuous, hidden articulatory targets as ideal articulation. Those on Row 3 are continuous, hidden states representing physical articulation with the Markov temporal structure also. Those on Row 4 are continuous, observed acoustic/auditory vectors.*

Each dependency in the above BayesNet can be implemented by specifying the associated conditional probability. In the speech model presented in Section 3, the horizontal (temporal) dependency for the phonological (discrete) states is specified by the Markov chain transition probabilities:

$$P(s_k = j | s_{k-1} = i) = a_{ij}. \tag{4.1}$$

The vertical (level)[2] dependency for the target random variables is specified by the following conditional density function:

$$p(\mathbf{t}(k)|s_k) = \mathcal{N}(\mathbf{t}(k); \mathbf{m}(s_k), \mathbf{\Sigma}(s_k)). \tag{4.2}$$

Possible structures in the covariance matrix $\mathbf{\Sigma}(s_k)$ in the above target distribution can be explored using physical interpretations of the targets as idealized articulation. For example, the velum component is largely uncorrelated with other components; so is the glottal component. On the other hand, tongue components are correlated with each other and with the jaw component. For some linguistic units (/u/ for instance), some tongue components are correlated with the lip components. Therefore, the covariance matrix $\mathbf{\Sigma}(s_k)$ has a block diagonal structure. If we represent each component in the target vector in the BayesNet, then each target node in Fig. 2 will contain a sub-network.

The joint horizontal and vertical dependency for the articulatory (continuous) state is specified, based on state equation 3.1, by the conditional density function:

$$\begin{aligned} p_{\mathbf{z}}[\mathbf{z}(k+1)|\mathbf{z}(k), \mathbf{t}(k), s_k] &= p_{\mathbf{w}}[\mathbf{z}(k+1) - \mathbf{\Phi}_{s_k}\mathbf{z}(k) - (\mathbf{I} - \mathbf{\Phi}_{s_k})\mathbf{t}(k)] \\ &= \mathcal{N}[\mathbf{z}(k+1); \mathbf{\Phi}_{s_k}\mathbf{z}(k) + (\mathbf{I} - \mathbf{\Phi}_{s_k})\mathbf{t}(k), \mathbf{Q}_{s_k}]. \end{aligned} \tag{4.3}$$

The vertical dependency for the observation random variables is specified, based on observation equation 3.2, by the conditional density function:

$$\begin{aligned} p_{\mathbf{o}}[\mathbf{o}(k)|\mathbf{z}(k)] &= p_{\mathbf{v}}[\mathbf{o}(k) - \mathbf{h}(\mathbf{z}(k))] \\ &= \mathcal{N}[\mathbf{o}(k)); \mathbf{h}(\mathbf{z}(k)), \mathbf{R}]. \end{aligned} \tag{4.4}$$

Eqns. 4.1, 4.2, 4.4, and 4.5 then completely specify the switching dynamic model in Fig. 2 since they define all possible dependencies in its BayesNet representation. Note that while the phonological state s_k and its associated target $\mathbf{t}(k)$ in principle are at a different time scale than the phonetic variables $\mathbf{z}(k)$ and $\mathbf{o}(k)$, for simplicity purposes and as one possible implementation, Eqns. 4.1-4.5 have placed them at the same time scale.

Note also that in Eqn. 4.5 the "forward" conditional probability for the observation vector (when the corresponding articulatory vector $\mathbf{z}(k)$ is known) is Gaussian, as is the measurement noise vector's distribution. The

[2]This refers to the level of the speech production chain as the "encoder".

mean of the Gaussian is the prediction of the nonlinear function $\mathbf{h}(\mathbf{z}(k))$. However, the "inverse" or "inference" conditional probability $p[\mathbf{z}(k)|\mathbf{o}(k)]$ will not be Gaussian due to the nonlinearity of $\mathbf{h}(.)$ as well as the switching process that controls the dynamics in $\mathbf{z}(k)$. The fact that the conditional distribution for $\mathbf{z}(k)$ is not Gaussian is one major source of difficulty for the inference and learning problems associated with the nonlinear switching dynamic system model.

4.2. Extended BayesNet model. One modification and extension of the basic BayesNet model of Fig. 2 is to explicitly represent parallel streams of the overlapping phonological features and their associated articulatory dimensions. As discussed in Section 3.1, the phonological construct of the model consists of multidimensional symbols (feature bundles) overlapping in time. The BayesNet for this expanded model is shown in Fig. 3, where the individual components of the articulator vector from the parallel overlapping streams are ultimately combined to generate the acoustic vectors.

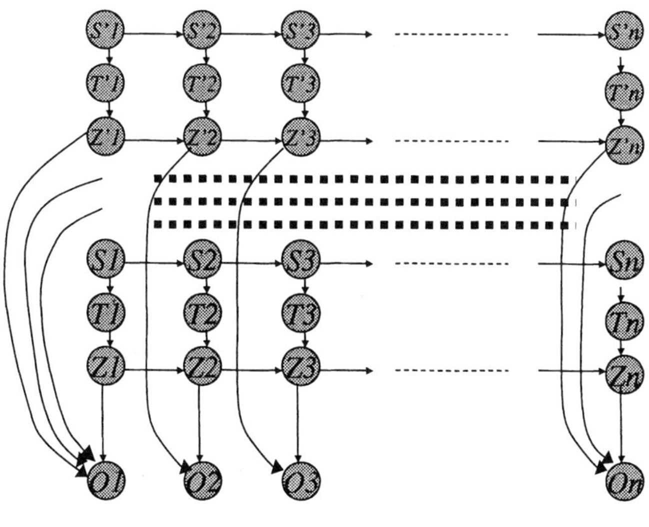

FIG. 3. *Dynamic BayesNet for an expanded version of the switching dynamic system model of speech. Parallel streams of the overlapping phonological features and their associated articulatory dimensions are explicitly represented. The articulators from the parallel streams are ultimately combined to jointly determine the acoustic vectors.*

Another modification of the basic Bayesian-net model of Fig. 2 is to incorporate the segmental constraint on the switching process for the dynamics of the target random vector $\mathbf{t}(k)$. That is, while random, $\mathbf{t}(k)$ remains fixed until the phonological state s_k switches. The switching of target $\mathbf{t}(k)$ is synchronous with that of the phonological state, and only at the time of switching, $\mathbf{t}(k)$ is allowed to take a new value according to its probability density function. This segmental constraint can be described mathematically by the following conditional probability density function:

$$p[\mathbf{t}(k)|s_k, s_{k-1}, \mathbf{t}(k-1)] = \begin{cases} \delta[\mathbf{t}(k) - \mathbf{t}(k-1)] & if \;\; s_k = s_{k-1}, \\ \mathcal{N}(\mathbf{t}(k); \mathbf{m}(s_k), \boldsymbol{\Sigma}(s_k)) & otherwise. \end{cases}$$

This adds the new dependency of random vector of $\mathbf{t}(k)$ on s_{k-1} and $\mathbf{t}(k-1)$, in addition to the existing s_k as in Fig. 2. The modified BayesNet incorporating this new dependency is shown in Fig. 4.

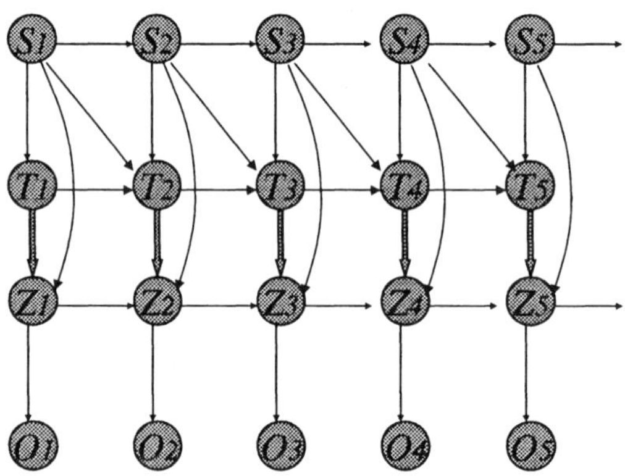

FIG. 4. *Dynamic BayesNet for the switching dynamic system model of speech incorporating the segmental constraint for the target random variables.*

4.3. Discussions. Given the BayesNet representations of switching dynamic system models for speech, rich tools for approximate inference and learning can be exploited and further developed. Since the exact inference is impossible, at least in theory, the success of applying such a model to speech recognition crucially depends on the accuracy of the approximate algorithms.

It is worth noting that while the exact optimal inference for the phonological states (the speech recognition problem) has exponential complexity in computation, once the approximate times of the switching in the phonological states become known, computational complexity can be substantially reduced. With the application of the variational technique (e.g., [27]) developed for BayesNet inference and learning to some generic, unstructured versions of the switching state-space model [39, 25]), one can separate the discrete states from the remaining portion of the network. (Recent research [37] also provides evidence that approximate methods such variational learning work well for a speech model called loosely-coupled HMM.) For the structured switching state-space model of speech dynamics as presented in this paper, this allows one to iteratively estimate the posterior distributions of the discrete phonological and continuous articulatory states. Inference on the phonological states becomes essentially a search for the state switching times with soft decisions. For example, when one uses the Gaussian mixture distribution to approximate the true posteriors in the speech model discussed so far, the E-step (needed for the recognizer's MAP decoding procedure) in the variational EM algorithm can be shown to be a solution to a set of algebraic nonlinear system of equations. Achieving efficient and accurate solutions to these closely coupled equations for the purpose of decoding the optimal phonological state sequence can be greatly facilitated when some crude estimates (e.g., within the range of several frames) of the phonological state boundaries, which we call the *landmarks*, are made available.

Interestingly, such an important role of the phonological state boundary estimates fits closely with the encoding-decoding theory of speech perception outlined in Section 2. As we discussed in Section 2, one crucial role of auditory reception for human speech perception is to provide temporal landmarks for the phonological features via the many transient neural response properties in the auditory system [50, 53, 54]. Recall that in the switching dynamic system model of speech presented in this paper, the phonological units are represented not in terms of phones that consist of a bundle of synchronously aligned features, but in terms of individual features. Therefore, the temporal landmarks associated with the individual features that may be detected by transient neural responses in the auditory system have important functional roles to play in providing the crude boundary information to facilitate the decoding of phonological states (speech perception). This common operation performed by the auditory system and by one aspect of the variational technique suggests that the variational-style decoding algorithms may be closely related to human speech perception.

5. Summary and discussions. We outlined an encoding-decoding theory of speech perception in this chapter, which highlights the importance and critical role of modeling articulatory dynamics in speech recognition. This is an integrated motor-auditory theory where the motor or production system provides the internal model for the listener's speech decoding device, while the auditory system provides sharp temporal landmarks for phonological features to constrain the decoder's search space and to minimize possible loss of decoding accuracy.

Most of current speech systems are very fragile. For further progress in the field, the author believes that it is necessary to bring in human-like intelligence of speech perception into computer systems. The switching dynamic system models discussed in this chapter offer one powerful mathematical tool for implementing the encoding-decoding mechanism of human speech communication. We have shown that the BayesNet framework allows us to take a fresh view on the complex computational issues in inference (decoding) and in learning, and to provide guidance and insights to the algorithm development.

It is hoped that the framework presented here will help integrate results from speech production and advanced machine learning within the statistical paradigms for speech recognition. An important, long-term goal will involve development of computer systems to the extent that they can be evaluated efficiently on realistic, large speech databases, collected in a variety of speaking styles (conversational styles in particular) and for a large population of speakers.

The ultimate goal of the research, whose components are described in some detail in this chapter, is to develop high-performance systems for integrated speech analysis, coding, synthesis, and recognition within a consistent statistical framework. Such a development is guided by the encoding-decoding theory of human speech communication, and is based on computational models of speech production and perception. The switching dynamic system models of speech and their BayesNet representations presented are a significant extension of the current highly simplified statistical models used in speech recognition. Further advances in this research direction will require greater integration within a statistical framework of existing research in modeling speech production, speech recognition, and advanced machine learning.

REFERENCES

[1] J. Allen. "How do humans process and recognize speech," *IEEE Trans. Speech Audio Proc.*, Vol. **2**, 1994, pp. 567–577.

[2] R. BAKIS. "Coarticulation modeling with continuous-state HMMs," *Proc. IEEE Workshop Automatic Speech Recognition*, Harriman, New York, 1991, pp. 20–21.

[3] N. BITAR AND C. ESPY-WILSON. "Speech parameterization based on phonetic features: Application to speech recognition," *Proc. Eurospeech*, Vol. **2**, 1995, pp. 1411–1414.

[4] C. BLACKBURN AND S. YOUNG. "Towards improved speech recognition using a speech production model," *Proc. Eurospeech*, Vol. **2**, 1995, pp. 1623–1626.

[5] H. BOURLARD AND S. DUPONT. "A new ASR approach based on independent processing and recombination of partial frequency bands," *Proc. ICSLP*, 1996, pp. 426–429.

[6] H. BOURLARD, H. HERMANSKY, AND N. MORGAN. "Towards increasing speech recognition error rates," *Speech Communication*, Vol. **18**, 1996, pp. 205–231.

[7] C. BROWMAN AND L. GOLDSTEIN. "Articulatory phonology: An overview," *Phonetica*, Vol. **49**, pp. 155–180, 1992.

[8] N. CHOMSKY AND M. HALLE. *The Sound Pattern of English*, New York: Harper and Row, 1968.

[9] N. CLEMENTS. "The geometry of phonological features," *Phonology Yearbook*, Vol. **2**, 1985, pp. 225–252.

[10] L. DENG. "A generalized hidden Markov model with state-conditioned trend functions of time for the speech signal," *Signal Processing*, Vol. **27**, 1992, pp. 65–78.

[11] L. DENG. "A computational model of the phonology-phonetics interface for automatic speech recognition," Summary Report, SLS-LCS, Massachusetts Institute of Technology, 1992–1993.

[12] L. DENG. "Design of a feature-based speech recognizer aiming at integration of auditory processing, signal modeling, and phonological structure of speech." *J. Acoust. Soc. Am.*, Vol. **93**, 1993, pp. 2318.

[13] L. DENG. "Computational models for speech production," in *Computational Models of Speech Pattern Processing* (NATO ASI), Springer-Verlag, 1999, pp. 67–77.

[14] L. DENG. "A dynamic, feature-based approach to the interface between phonology and phonetics for speech modeling and recognition," *Speech Communication,* Vol. **24**, No. 4, 1998, pp. 299–323.

[15] L. DENG, M. AKSMANOVIC, D. SUN, AND J. WU. "Speech recognition using hidden Markov models with polynomial regression functions as nonstationary states," *IEEE Trans. Speech Audio Proc.*, Vol. **2**, 1994, pp. 507–520.

[16] L. DENG, AND K. ERLER. "Structural design of a hidden Markov model based speech recognizer using multi-valued phonetic features: Comparison with segmental speech units," *J. Acoust. Soc. Am.*, Vol. **92**, 1992, pp. 3058–3067.

[17] L. DENG AND Z. MA. "Spontaneous speech recognition using a statistical coarticulatory model for the hidden vocal-tract-resonance dynamics," *J. Acoust. Soc. Am.*, Vol. **108**, No. 6, 2000, pp. 3036–3048.

[18] L. DENG, G. RAMSAY, AND D. SUN. "Production models as a structural basis for automatic speech recognition," *Speech Communication*, Vol. **22**, No. 2, 1997, pp. 93–111.

[19] L. DENG AND H. SAMETI. "Transitional speech units and their representation by the regressive Markov states: Applications to speech recognition," *IEEE Trans. Speech Audio Proc.*, Vol. **4**, No.4, July 1996, pp. 301–306.

[20] L. DENG AND X. SHEN. "Maximum likelihood in statistical estimation of dynamic systems: Decomposition algorithm and simulation results", *Signal Processing*, Vol. **57**, 1997, pp. 65–79.

[21] L. DENG AND D. SUN. "A statistical approach to automatic speech recognition using the atomic speech units constructed from overlapping articulatory features," *J. Acoust. Soc. Am.*, Vol. **95**, 1994, pp. 2702–2719.

[22] J. FRANKEL AND S. KING. "ASR — Articulatory speech recognition", *Proc. Eurospeech,* Vol. **1**, 2001, pp. 599–602.

[23] Y. Gao, R. Bakis, J. Huang, and B. Zhang, "Multistage coarticulation model combining articulatory, formant and cepstral features", *Proc. ICSLP*, Vol. **1**, 2000, pp. 25–28.

[24] Z. Ghahramani and S. Roweis. "Learning nonlinear dynamic systems using an EM algorithm". *Advances in Neural Information Processing Systems*, Vol. **11**, 1999, 1–7.

[25] Z. Ghahramani and G. Hinton. "Variational learning for switching state-space model". *Neural Computation*, Vol. **12**, 2000, pp. 831–864.

[26] W. Holmes. "Segmental HMMs: Modeling dynamics and underlying structure in speech," in M. Ostendorf and S. Khudanpur (eds.) *Mathematical Foundations of Speech Recognition and Processing, Volume X in IMA Volumes in Mathematics and Its Applications,* Springer-Verlag, New York, 2002.

[27] M. Jordan, Z. Ghahramani, T. Jaakkola, and L. Saul. "In introduction to variational methods for graphical models," in *Learning in Graphical Models* M. Jordon (ed.), The MIT Press, Cambridge, MA, 1999.

[28] F. Juang and S. Furui (eds.), Proc. of the IEEE (special issue), Vol. **88**, 2000.

[29] R. Kent, G. Adams, and G. Turner. "Models of speech production," in *Principles of Experimental Phonetics*, N. Lass (ed.), Mosby: London, 1995, pp. 3–45.

[30] C.-H. Lee, F. Soong, and K. Paliwal (eds.) *Automatic Speech and Speaker Recognition – Advanced Topics,* Kluwer Academic, 1996.

[31] A. Liberman and I. Mattingly. "The motor theory of speech perception revised" *Cognition*, Vol. **21**, 1985, pp. 1–36.

[32] R. Lippman. "Speech recognition by human and machines," *Speech Communication*, Vol. **22**, 1997, pp. 1–15.

[33] Z. Ma and L. Deng. "A path-stack algorithm for optimizing dynamic regimes in a statistical hidden dynamic model of speech," *Computer Speech and Language*, Vol. **14**, 2000, pp. 101–104.

[34] P. MacNeilage. "Motor control of serial ordering in speech," *Psychological Review*, Vol. **77**, 1970, pp. 182–196.

[35] R. McGowan. "Recovering articulatory movement from formant frequency trajectories using task dynamics and a genetic algorithm: Preliminary model tests," *Speech Communication*, Vol. **14**, 1994, pp. 19–48.

[36] R. McGowan and A. Faber. "Speech production parameters for automatic speech recognition," *J. Acoust. Soc. Am.*, Vol. **101**, 1997, pp. 28.

[37] H. Nock. *Techniques for Modeling Phonological Processes in Automatic Speech Recognition*, Ph.D. thesis, Cambridge University, 2001, Cambridge, U.K.

[38] M. Ostendorf, V. Digalakis, and J. Rohlicek. "From HMMs to segment models: A unified view of stochastic modeling for speech recognition" *IEEE Trans. Speech Audio Proc.*, Vol. **4**, 1996, pp. 360–378.

[39] V. Pavlovic, B. Frey, and T. Huang. "Variational learning in mixed-state dynamic graphical models," *Proc. Annual Conf. in Uncertainty in Artificial Intelligence*, 1999, UAI-99.

[40] J. Perkell, M. Matthies, M. Svirsky, and M. Jordan. "Goal-based speech motor control: a theoretical framework and some preliminary data," *J. Phonetics*, Vol. **23**, 1995, pp. 23–35.

[41] J. Perkell. "Properties of the tongue help to define vowel categories: hypotheses based on physiologically-oriented modeling," *J. Phonetics* Vol. **24**, 1996, pp. 3–22.

[42] P. Perrier, D. Ostry, and R. Laboissière. "The equilibrium point hypothesis and its application to speech motor control," *J. Speech & Hearing Research,* Vol. **39**, 1996, pp. 365–378.

[43] L. Pols. "Flexible human speech recognition," *Proceedings of the 1997 IEEE Workshop on Automatic Speech Recognition and Understanding,* Santa Barbara, 1997, pp. 273–283.

[44] M. Randolph. "Speech analysis based on articulatory behavior," *J. Acoust. Soc. Am.*, Vol. **95**, 1994, pp. 195.

[45] H. Richards, and J. Bridle. "The HDM: A segmental hidden dynamic model of coarticulation", Proc. ICASSP, Vol. **1**, 1999, pp. 357–360.

[46] R. Rose, J. Schroeter, and M. Sondhi. "The potential role of speech production models in automatic speech recognition," *J. Acoust. Soc. Am.*, Vol. **99**, 1996, pp. 1699–1709.

[47] M. Russell. "Progress towards speech models that model speech," *Proceedings of the 1997 IEEE Workshop on Automatic Speech Recognition and Understanding,* Santa Barbara, 1997, pp. 115–123.

[48] J. Schroeter and M. Sondhi. "Techniques for estimating vocal-tract shapes from the speech signal," *IEEE Trans. Speech Audio Proc.*, Vol. **2**, 1994, pp. 133–150.

[49] H. Sheikhzadeh and L. Deng. "Speech analysis and recognition using interval statistics generated from a composite auditory model," *IEEE Trans. Speech Audio Proc.*, Vol. **6**, 1998, pp. 50–54.

[50] H. Sheikhzadeh and L. Deng. "A layered neural network interfaced with a cochlear model for the study of speech encoding in the auditory system," *Computer Speech and Language*, Vol. **13**, 1999, pp. 39–64.

[51] R. Shumway and D. Stoffer. "An approach to time series smoothing and forecasting using the EM algorithm," *J. Time Series Analysis*, Vol. **3**, 1982, pp. 253–264.

[52] R. Shumway and D. Stoffer. "Dynamic linear models with switching", *J. American Statistical Association*, Vol. **86**, 1991, pp. 763–769.

[53] K. Stevens. "On the quantal nature of speech," *J. Phonetics,* Vol. **17**, 1989, pp. 3–45.

[54] K. Stevens. "From acoustic cues to segments, features and words," *Proc. ICSLP*, Vol. **1**, 2000, pp. A1–A8.

[55] K. Stevens. *Acoustic Phonetics*, The MIT Press , Cambridge, MA, 1998.

[56] J. Sun, L. Deng, and X. Jing. "Data-driven model construction for continuous speech recognition using overlapping articulatory features," *Proc. ICSLP*, Vol. **1**, 2000, pp. 437–440.

SEGMENTAL HMMS: MODELING DYNAMICS AND UNDERLYING STRUCTURE IN SPEECH

WENDY J. HOLMES*

Abstract. The motivation underlying the development of segmental hidden Markov models (SHMMs) is to overcome important speech-modeling limitations of conventional HMMs by representing sequences (or 'segments') of features and incorporating the concept of a trajectory to describe how features change over time. This paper presents an overview of investigations that have been carried out into the properties and recognition performance of various SHMMs, highlighting some of the issues that have been identified in using these models successfully. Recognition results are presented showing that the best recognition performance was obtained when combining a trajectory model with a formant representation, in comparison both with a conventional cepstrum-based HMM system and with systems that incorporated either of the developments individually.

An attractive characteristic of a formant-based trajectory model is that it applies easily to speech synthesis as well as to speech recognition, and thus can provide the basis for a 'unified' approach to both recognition and synthesis and to speech modeling in general. One practical application is in very low bit-rate speech coding, for which a formant trajectory description provides a compact means of coding an utterance. A demonstration system has been developed that typically codes speech at 600–1000 bits/s with good intelligibility, whilst preserving speaker characteristics.

Key words. Segmental HMM, dynamics, trajectory, formant, unified model, low bit-rate speech coding.

AMS(MOS) subject classifications. Primary 1234, 5678, 9101112.

1. Introduction. The acoustic-phonetic components of the most successful large-vocabulary automatic recognition (ASR) systems to date are almost exclusively based on hidden Markov models (HMMs) of some phonetically defined subword units, typically using a large inventory of context-dependent phone models that are trained on a vast quantity of speech data. The models themselves tend to be gender-dependent but are otherwise 'speaker-independent', although there is often on-line adaptation to any particular speaker. Using this type of approach, impressive performance has been achieved on recognition of read speech. For example, in the 1998 US Defense Advanced Research Projects Agency (DARPA) evaluations using broadcast news material, for the 'planned' portion of the test set (read speech in quiet conditions) a word error rate of 7.8% was obtained [21]. However, the general level of performance drops if the recording conditions and speaking style are less controlled, with the lowest error rate on the spontaneous portion of the 1998 test set being 14.4% [21]. Conversational speech is particularly challenging, especially when the conversations are between individuals who know each other very well, when the percentage of recognition errors may be several times that for read speech, even if

*20/20 Speech Limited, Malvern Hills Science Park, Geraldine Road, Malvern, Worcs., WR14 3SZ, UK.

the vocabulary size is much smaller. For example, on the CallHome corpus of telephone conversations between family members, typical word error rates exceed 30% [7]. Although there are many aspects to the problems associated with this type of recognition task, the extent of the drop in performance when moving beyond constrained domains suggests that there may be inherent deficiencies in the acoustic modeling paradigm.

HMMs provide a framework which is broadly appropriate for modeling speech patterns, accommodating both variability in timescale and short-term spectral variability. However these models are simply general statistical pattern matchers, and do not take advantage of the constraints inherent in the speech production process and make certain assumptions that conflict with what is known about the nature of speech production and its relationship with acoustic realization. In particular, the following three assumptions which are made by the HMM formalism are clearly inappropriate for modeling speech patterns:

- *Piece-wise stationarity.* It is assumed that a speech pattern is produced by a piece-wise stationary process, with instantaneous transitions between stationary states.
- *The independence assumption.* The probability of a given acoustic vector corresponding to a given state depends only on the vector and the state, and is otherwise independent of the sequence of acoustic vectors preceding and following the current vector and state. The model therefore takes no account of the dynamic constraints of the physical system which has generated a particular sequence of acoustic data, except inasmuch as these can be incorporated in the feature vector associated with a state. In a typical speaker-independent HMM recognizer where each modeling unit is represented by a multi-modal Gaussian distribution to include all speakers, the model in effect treats each frame of data as if it may have been spoken by a different speaker.
- *State duration distribution.* A consequence of the independence assumption is that the probability of a model staying in the same state for several frames is determined only by the 'self-loop' transition probability. Thus the state duration in an HMM conforms to a geometric probability distribution which assigns maximum probability to a duration of one frame and exponentially decreasing probabilities to longer durations.

The impact of these inappropriate assumptions can be reduced by, for example, using a generous allocation of states which allows a sequence of piece-wise stationary segments to better approximate the dynamics and also makes a duration of one frame per state more appropriate. A popular way of mitigating the effects of the independence assumption is to use an acoustic feature vector which includes information over a time span of several frames. This is most usually achieved by including the first and sometimes also the second time-derivative of the original static features.

However, although such techniques have been shown to be of practical benefit in improving recognition performance, the independence assumption actually becomes even less valid because the observed data for any one frame are used to contribute to a time span of several feature vectors.

Rather than trying to modify the data to fit the model, it should be better to adopt a model that is more appropriate for the characteristics of speech signals. The inappropriate assumptions of HMMs are linked with their frame-synchronous characteristic, whereby states are associated with single acoustic feature vectors. In order to improve the underlying model, it is necessary somehow to incorporate the concept of modeling frame sequences rather than individual frames, with the aim of providing an accurate model of speech dynamics across an utterance in a way which takes into account predictable factors such as speaker continuity. At the same time, it is important to retain the many desirable attributes of the HMM framework, such as rigorous data-driven training, optimal decoding and delayed decision making. The desire to overcome the above limitations of HMMs by modeling frame sequences has motivated the development of a variety of extensions, modifications and alternatives to HMMs (see [20] for a review). Models of this type are usually referred to as "segment models" [20], or as "segmental HMMs" [16].

2. Segment models. A "segmental HMM" (SHMM) can be defined in general terms as a Markov model where segments, rather than frames, are the homogeneous units which are treated as probabilistic functions of the model states. The relationship between successive acoustic feature vectors representing a sub-phonemic speech segment can be approximated by some form of trajectory through the feature space.

Several workers have suggested segment models that incorporate the notion of a trajectory (e.g. [3, 11, 16]). The model described in [16] is a particular form of segmental HMM that was investigated by the author, working with Prof. M.J. Russell. The main characteristic that distinguishes the model is the representation of feature variability, which includes the concept of a 'probabilistic' trajectory.

A probabilistic-trajectory SHMM (PTSHMM) for a speech sound provides a representation of the range of possible underlying trajectories for that sound, where the trajectories are of variable duration and each duration has a state-dependent probability. To accommodate the fact that an observed sequence of feature vectors will in general not follow any underlying trajectory exactly, a trajectory is modeled as 'noisy'. A segment is thus described by a stochastic process whose mean changes as a function of time according to the parameters of the trajectory. The model therefore makes a distinction between two types of variability: the first is *extra-segmental* variation in the underlying trajectory, and the second is *intra-segmental* variation of the observations around any one trajectory. Intuitively, extra-segmental variations represent general factors, such as differences between speakers or chosen pronunciation for a speech sound, which would lead to different trajectories for the same sub-phonemic unit. Intra-segmental variations can be regarded as representing the much smaller frame-to-frame variation that exists in the realization of a particular pronunciation in a given context by any one speaker. For reasons of mathematical tractability and of trainability, all variability associated with PTSHMMs is modeled with Gaussian distributions assuming diagonal covariance matrices, and only parametric trajectory models are considered.

The theory and implementation of PTSHMMs have been presented and discussed in detail in [14] and [16]. The first part of the current paper will concentrate on giving an overview of *linear* trajectory models, including the linear PTSHMM and other segmental HMMs that can be viewed as simplified forms of it. The main aim of this overview will be to show some of the ways in which the conventional HMM can be extended, and also some of the practical issues involved in achieving success with these extended models. The second part of the paper will concentrate on the choice of features on which the trajectory model is based, and on some of the potential advantages that can be gained from choosing features that have a closer relationship with the underlying speech production system than the currently more popular features such as mel-frequency cepstrum coefficients (MFCCs).

3. Linear-trajectory segmental HMMs. A simple model of dynamics is one in which it is assumed that the underlying trajectory vector changes linearly over time. Previous studies of trajectory representations of MFCCs [14] have suggested that a linear trajectory model is sufficient to capture typical time-evolving characteristics (at least when using three segments per phone). The adequacy of a linear model is also supported by Gish and Ng [11], who found that a linear trajectory was sufficient for most sounds, even when only using one segment per phone.

3.1. Theory. A linear trajectory $f_{(m,c)}$ is defined by its slope m and the segment mid-point value c, such that $f_{(m,c)}(t) = c + m(t - \frac{1}{2}T)$ if the trajectory has a duration of $T + 1$ frames. In a linear PTSHMM, extra-segmental variability takes the form of variation in the slope and mid-point parameters. The distributions of the two trajectory parameters for a given state can be defined by Gaussian distributions $N_{\mu,\gamma}$ and $N_{\nu,\eta}$ for the slope and mid-point respectively. Intra-segment variation can be represented by a Gaussian distribution with fixed variance τ. All distributions are assumed to have diagonal covariance matrices, and for notational simplicity all observation sequences are therefore assumed to be one-dimensional. The joint probability of a sequence of observations $y = y_0, ..., y_T$ and any particular values of the slope m and mid-point c is thus defined as follows:

$$P(y, m, c) = N_{\mu,\gamma}(m) N_{\nu,\eta}(c) \prod_{t=0}^{T} N_{f_{m,c}(t),\tau}(y_t).$$

The PTSHMM emission probability $P(y)$ can be computed by integrating $P(y, m, c)$ over the unknown trajectory parameters m and c:

$$P(y) = \int_{-\infty}^{\infty} \int_{-\infty}^{\infty} N_{\mu,\gamma}(m) N_{\nu,\eta}(c) \prod_{t=0}^{T} N_{f_{m,c}(t),\tau}(y_t) dc dm. \tag{3.1}$$

In [14] it was shown that the integral in the above expression can be evaluated to give a tractable expression that does not depend on the values of m or c, thus:

$$\begin{aligned} P(y) = & \sqrt{\frac{\tau}{(T+1)\eta + \tau}} \sqrt{\frac{\tau}{q\gamma + \tau}} \left(\frac{1}{\sqrt{2\pi\tau}} \right)^{(T+1)} \\ & \times \exp\left(-\frac{1}{2} \left(\frac{(T+1)(\nu - c'(y))^2}{(T+1)\eta + \tau} + \frac{q(\mu - m'(y))^2}{q\gamma + \tau} \right.\right. \\ & \left.\left. + \frac{1}{\tau} \left(\sum_{t=0}^{T} y^2 - (T+1) c'(y)^2 - qm'(y)^2 \right) \right) \right), \end{aligned} \tag{3.2}$$

where

$$q = \left(\sum_{t=0}^{T} \left(t - \frac{T}{2} \right)^2 \right).$$

In Equation 3.2 above,

$$m'(y) = \frac{\sum_{t=0}^{T} (t - \frac{T}{2}) y_t}{\sum_{t=0}^{T} (t - \frac{T}{2})^2}, \qquad c'(y) = \frac{\sum_{t=0}^{T} y_t}{T+1}$$

define the slope and mid-point value respectively of the linear trajectory which provides the least-squared error best fit to the sequence of observations y. HMMs are usually trained by a parameter re-estimation procedure called the Baum–Welch algorithm [1], which is an example of the more general method known as the expectation-maximization (EM) algorithm[2]. Extended Baum–Welch parameter re-estimation formulae were derived for linear-trajectory PTSHMMs [14] [16], following the approach of Liporace [19], whereby an auxiliary function was introduced and new values were calculated for the model parameters in order to maximize that auxiliary function.

Returning to the linear-PTSHMM expression for $P(y)$ given above, it is interesting to consider the characteristics of the models that arise if certain parameters are fixed to have values of zero. If the slope mean μ and the slope variance γ are both fixed at zero, the trajectory is defined by just the mid-point distribution and is thus constant over time. The expression for $P(y)$ becomes:

$$P(y) = \int_{-\infty}^{\infty} N_{\nu,\eta}(c) \prod_{t=0}^{T} N_{c,\tau}(y_t) dc. \tag{3.3}$$

This model is a static PTSHMM, equivalent to the model suggested in [8]. Static PTSHMMs distinguish between extra-segmental and intra-segmental variability, and therefore impose some degree of continuity constraint between successive observations within a segment. This type of model cannot, however, capture local frame-to-frame dynamic characteristics.

If the mid-point variance η is also fixed at zero, the expression for $P(y)$ is further simplified as follows:

$$P(y) = \prod_{t=0}^{T} N_{\nu,\tau}(y_t), \tag{3.4}$$

which is equivalent to the standard-HMM probability calculation (although here a maximum segment duration will be applied, and there is also the possibility of easily including a realistic duration model, as in the approach of [26] for example).

A standard HMM can be regarded as a static PTSHMM with zero extra-segment variance. Similarly, by taking a linear PTSHMM and setting the extra-segment variances η and γ of the mid-point and slope both to zero, the mid-point and slope values will always be equal to the model means and therefore define a single linear trajectory, thus:

$$P(y) = \prod_{t=0}^{T} N_{f_{\mu,\nu}(t),\tau}(y_t). \tag{3.5}$$

This model represents a linear "fixed-trajectory" segmental HMM (FT-SHMM), equivalent to linear trajectory-based segment models such as those suggested by [11] and by [3].

If only the slope variance γ is set to zero, the probability expression becomes:

$$P(y) = \int_{-\infty}^{\infty} N_{\nu,\eta}(c) \prod_{t=0}^{T} N_{f_{\mu,c}(t),\tau}(y_t)dc. \tag{3.6}$$

This model is a constrained form of linear PTSHMM, whereby the mid-point of the trajectory can vary across examples of a model unit, but the slope is always fixed at the model mean value. In future discussion it will be convenient to refer to this type of linear PTSHMM as having "constrained slope", while the linear PTSHMM with variability in all its parameters and represented in Equation (3.1) has "flexible slope".

The following sections summarize the findings of experimental investigations into linear PTSHMMs with full flexibility in all their parameters and also into the range of simplified cases described above.

3.2. Digit recognition experiments. The recognition task was selected with the aim of providing the basis for meaningful comparisons to be made between different segmental HMMs and conventional HMMs, while also enabling the properties of the segmental model to be investigated. A speaker-independent connected-digit recognition task using phone-based models was chosen, because it requires a connected-word recognition algorithm such that segmentation occurs simultaneously with recognition but is a simple task with a small vocabulary so that analysis of recognition errors is relatively straightforward. In addition, the small vocabulary offers a faster experiment turn-around time than is possible with larger vocabularies. The increase in computational load associated with segment models is such that it was considered important to begin with a small task when investigating the properties of PTSHMMs.

3.2.1. Speech data. The test data were four lists of 50 digit triples spoken by each of 10 male speakers. The training data were from 225 different male speakers, each reading 19 four-digit strings taken from a vocabulary of 10 strings. The available speech data had been sampled at 19.98 kHz and analyzed using a 27-channel critical-band filterbank spanning the range 0–10 kHz, producing output channel amplitudes quantized in units of 0.5 dB at a rate of 100 frames/s. A cosine transform was applied, and the first eight cosine coefficients together with an average amplitude parameter were used as the feature set.

3.2.2. Experimental framework. The experiments described here used continuous-density Gaussian HMMs with a single-Gaussian output distribution per state and diagonal covariance matrices. The emphasis has been on *relative* performance of different models rather than on achieving the best possible absolute performance. Therefore, although the systems are intended to provide a sufficiently good level of baseline performance to make comparisons meaningful, no attempt was made to optimize details of the feature analysis, model inventory and so on.

Three-state context-independent monophone models and four single-state non-speech models were used. The simple left–right model structure that is typically used in most HMM systems was adopted. Self-loop transitions were allowed in the standard HMMs. For the segmental HMMs representing speech sounds, only transitions from a state on to the immediately following state were allowed and each state was assigned a maximum segment duration of 10 frames. This model structure thus imposes a maximum phone duration of 300 ms, which was considered adequate for most speech sounds in connected speech. The self-loops were retained for the non-speech models, to provide a simple way of accommodating any long periods of silence. Because these experiments were intended to evaluate different models of acoustics, rather than duration-modeling differences, all transitions from a state were assigned the same transition probability, and for the segmental HMMs all the segment durations were assigned equal probability divided between the allowed duration range. None of the transition probabilities or duration probabilities were re-estimated.

The model means for the standard HMMs were initialized based on a small quantity of hand-annotated training data, and all model variances were initialized to the same arbitrary value. The model means and variances were then trained with five iterations of Baum–Welch re-estimation. After five training iterations, these standard HMMs gave a word error rate of 8.6% on the test data. These models provided the starting point for training different sets of segmental HMMs, all of which were trained with five iterations of the appropriate Baum–Welch type re-estimation procedure. For performance comparisons, the standard HMMs were also subjected to a further five iterations of re-estimation.

3.2.3. Segmental HMM training. The simplest segmental models were those represented in Equation (3.4), which used the standard-HMM probability calculation but incorporated the constraint on maximum duration. These simple static segmental HMMs and the linear FTSHMMs (shown in Equation (3.5)) were both initialized directly from the trained standard-HMM parameters, with the intra-segment variances being copied from the standard-HMM variances and the extra-segment variances being set to zero. The slope means of the FTSHMMs were initialized to zero, so that the starting point was the same as for the simplest segmental HMMs, with the difference being that when training the FTSHMMs the slope mean was allowed to deviate away from its initial value.

Initial estimates for the different sets of PTSHMMs were computed by first using the trained standard HMMs to segment the complete set of training data, and using the statistics of these data to estimate the various parameters of each set of PTSHMMs according to the relevant modeling assumptions. In all cases the means and variances of the midpoints were initialized from the distributions of the sample means for the individual segments. The initialization strategies that were used for the other parameters were different for the different types of PTSHMM and are summarized in Table 1. Each set of segmental models was trained with five iterations of the appropriate Baum–Welch-type re-estimation procedure [14] [16].

3.2.4. Results. The connected-digit recognition results are shown in Table 2 for the different sets of segmental models, compared with the baseline HMMs after both five and 10 training iterations. The main findings are summarized in the following paragraphs.

The simple segmental HMMs with a maximum segment duration of 10 frames gave an error rate of 6.6%, which is lower than that of the conventional HMMs even when further training had been applied (8.4%). Thus, in these experiments there were considerable advantages in constraining the maximum segment duration, which acts to prevent unrealistically long occupancies for the speech model states.

The lowest word error rate achieved with the static PTSHMMs was 7.5%, which is not quite as good as the 6.6% obtained with the simplest

TABLE 1
Initialization strategy for various model parameters of different sets of PTSHMMs.

Model set	Slope parameters	Intra-segment variance
static PTSHMM (Eq. (3.3))	mean and variance both fixed at 0	variance of observations about segment mean
linear PTSHMM (flex. slope) (Eq. (3.1))	mean and variance of slopes of best-fit linear trajectories	variance of observations about best-fit linear trajectories
linear PTSHMM (constr. slope) (Eq. (3.6))	mean and variance both set to 0, but mean can vary in training	variance of observations about segment mean (*i.e.* line with zero slope)

segmental HMMs. Both sets of models appeared to be adequately trained after five iterations of re-estimation, as performing a further five iterations did not reduce the word error rate. It therefore appears that, for a static trajectory assumption, there is no benefit from adopting the PTSHMM approach of separating out intra- from extra-segmental variability.

The linear FTSHMMs gave a word error rate of 4.9%, which is an improvement over the 6.6% error rate achieved with the simplest segmental HMMs. This result demonstrates the benefits of incorporating a linear trajectory representation to describe how features change over time.

The best performance achieved with the linear PTSHMMs was an error rate of 2.9%, which represents a reduction in error rate of 40% over the result with the linear FTSHMMs. Considerable further advantage was thus gained by separating out extra- from intra-segmental variability, in addition to the benefits of the linear trajectory description.

The linear PTSHMMs with constrained slope gave the best recognition performance, whereas the linear PTSHMMs with flexible slope performed worse than the baseline standard HMMs. This finding suggests that linear PTSHMMs provide better discrimination when they represent extra-segmental variability in the mid-point but not in the slope parameters.

3.2.5. Comparisons with HMMs using time-derivative features. The experiments described so far have demonstrated recognition performance improvements by incorporating a linear model of temporal dynamics within a segment-based framework. However, successful conventional HMM-based recognizers almost always include some representation of dynamic characteristics within the acoustic feature vectors themselves. Comparisons were therefore made with models using conventional HMM probability calculations with time-derivative features, computed for each frame using the typical approach of applying linear regression over a window of five frames centred on the current frame. Both HMMs and then simple segmental HMMs were trained using an acoustic feature set which

TABLE 2
Connected-digit recognition results for different sets of segmental HMMs compared with conventional HMMs.

Model type	% Sub.	% Del.	% Ins.	% Err.
HMM (five training iterations)	6.2	1.5	0.9	8.6
HMM (10 training iterations)	6.0	1.6	0.8	8.4
Simple segmental HMM	5.2	0.7	0.7	6.6
Static PTSHMM	5.2	2.2	0.1	7.5
Linear FTSHMM	3.8	0.5	0.6	4.9
Linear PTSHMM (flex. slope)	4.9	4.0	0.0	9.0
Linear PTSHMM (constr. slope)	2.0	0.8	0.1	2.9

included time-derivative features of the original nine instantaneous features, to give a total of eighteen features. The performance of these models was significantly improved over that which had been obtained using only instantaneous features, to give error rates of 1.5% and 1.6% for the HMMs and simple segmental HMMs respectively. Thus, when derivative features were included, the maximum-duration constraints provided by the simple segmental HMM did not give any advantage over the standard HMM. The conventional HMMs with time-derivative features have given an error rate of only 1.5%, whereas the best error rate achieved with the linear PTSHMMs (using only instantaneous features) was 2.9%. The result of this comparison is disappointing, but can be explained by differences in the extent to which the two models are able to represent dynamics. Although the use of derivative features only provides implicit modeling of dynamics, some representation of change is provided for every frame. However, the segmental models studied here have been limited to representing dynamics within any one segment, so further performance advantages may be obtained by using derivative features with the segmental models, as has been found by other researchers, for example [5]. Given that the error rate was already very low with conventional HMMs when including time-derivative features, it was not considered worthwhile trying this approach for the digit recognition task. However, the next section describes some experiments on a much more challenging task, which have been carried out both with and without time-derivative features.

3.3. Phone classification experiments. Phone classification involves determining the identity of speech segments with specified phonetic boundaries, so providing a means to investigate and compare phonetic-modeling capabilities for different speech sounds. Studying classification rather than recognition has computational advantages, but also allows for the investigation of description and discrimination abilities separately from segmentation properties. A useful set of data for evaluating phonetic classification performance is the DARPA TIMIT acoustic-phonetic continuous-speech database of American English [10], for which all the utterances have been phonetically transcribed, segmented and labeled. TIMIT was designed to provide broad phonetic coverage, and is therefore particularly appropriate for comparing approaches to acoustic-phonetic modeling.

When classifying the data segments, all phones were treated as equally likely (no language model was used to constrain the allowed phone sequences). This approach was considered appropriate for investigating improved acoustic-phonetic modeling, but it does make the task very difficult. As with the digit experiments, the emphasis here is on the *relative* performance of the different segmental HMMs.

3.3.1. Speech data and model sets. The experiments reported here used the TIMIT designated training set and the core test set, using data only from the male speakers. The available data had been analyzed by applying a 20 ms Hamming window to the 16 kHz-sampled speech at a rate of 100 frames/s and computing a fast Fourier transform. The output had been converted to a mel scale with 20 channels, and a cosine transform had been applied. The first 12 cosine coefficients together with an average amplitude feature formed the basic feature set, but some experiments also included time-derivative features computed for each frame by applying linear regression over a five-frame window centred on the current frame.

The inventory of model units was defined as the set of 61 symbols which are used in the time-aligned phonetic transcriptions provided with TIMIT. However, the two different silence symbols used in these transcriptions were represented by a single silence model, to give 60 model units. In common with most other work using TIMIT, when scoring recognition output the 60-symbol set was reduced to the 39-category scoring set given in [18]. Experiments were carried out with context-independent (monophone) models, and also with right-context-dependent biphones, which depend on only the immediately-following phoneme context.

The basic model structure for both the conventional and the segmental HMMs was the same as the one used for the digit experiments, with three states per speech model and single-state non-speech models. However, this structure imposes a minimum duration of three frames for every speech unit, whereas some of the labeled phone segments are shorter than three frames. In order to accommodate these very short utterances, the structure of all the speech models was extended to allow transitions from the initial state to all emitting states with a low probability.

3.3.2. Training procedure. First, a set of standard-HMM monophone models was initialized and trained with five iterations of Baum–Welch re-estimation. These models were then used to initialize different sets of monophone segmental HMMs, using the same approach as the one adopted in the digit-recognition experiments. The discussion here will focus on comparing the performance of linear FTSHMMs and constrained-slope linear PTSHMMs with that of simple segmental HMMs (models using conventional HMM probabilities but with a maximum-duration constraint). For all sets of segment models, five iterations of the appropriate Baum–Welch-type re-estimation were applied, the trained monophone models were used to initialize biphone models and three further iterations of re-estimation were then carried out.

3.3.3. Classification results. Table 3 shows the classification error for different segmental HMMs, for both monophone and biphone models, with and without time-derivative features. From these results it can be seen that, under all the different experimental conditions, the linear FTSHMMs gave some performance improvement over the simple segmental HMMs, and there were further performance benefits from linear PTSHMMs that incorporated a model of variability in the mid-point parameter.

TABLE 3

Classification results for the male portion of the TIMIT core test set using biphone models. The percentage of phone errors is shown for different types of linear-trajectory segmental HMMs compared with the baseline provided by simple segmental HMMs.

Model type	Cepstrum features only	Include delta features
Simple SHMM	43.0	29.4
Linear FTSHMM	39.0	27.4
Linear PTSHMM	38.2	26.8

The relative performance of the linear PTSHMMs and the simple segmental HMMs was analyzed as a function of phone class, in order to determine whether the segmental model of dynamics was more beneficial for some types of sound than others. The results of this analysis are shown in Table 4 for a selection of sound classes. The linear PTSHMMs improved performance for all the phone classes, but were most beneficial for the diphthongs and the semivowels and glides. These sounds are characterized by highly dynamic continuous changes, for which the trajectory model should be particularly advantageous. The performance gain was smallest for the stops, which have rather different characteristics involving abrupt changes between relatively steady-state regions. Some model of dynamics across segment boundaries may be the best way to represent these changes.

3.4. Discussion. The experiments described above have shown some improvements in recognition performance through modeling trajectories of mel-cepstrum features, although for best performance it was necessary to also include time-derivatives in the feature set. These experiments all used a fixed model structure with three segments per phone. There are various aspects of the acoustic modeling which could be developed to take greater account of the characteristics of speech patterns. For example, rather than using three segments to model every phone, it would seem more appropriate to represent each phone with the minimum number of segments to describe typical trajectories in order to maximize the benefit from the constraints

TABLE 4
Classification performance of linear PTSHMMs relative to that of simple segmental HMMs, shown for different phone classes (using biphone models with mel-cepstrum features).

Phone class	No. examples	SHMM (% err.)	PTSHMM (% err.)	% PTSHMM improvement
Stops (p, t, dx, k, b, d, g)	566	56.7	54.8	3.4
Fricatives (f, v, th, dh, s, z, sh, hh)	710	41.7	38.9	6.8
Semivowels and glides (l, r, y, w)	497	39.2	33.2	15.4
Vowels (iy, ih, eh, ae, ah, uw, uh, er)	1178	53.8	48.9	9.1
Diphthongs (ey, ay, oy, aw, ow)	376	48.9	41.2	15.8

provided by the segment model. With a linear trajectory model, three segments are probably necessary for certain sounds (such as diphthongs), but for many other sounds (e.g. nasals, voiceless fricatives and short vowels) one segment should be sufficient. It may also be beneficial to employ phone-dependent constraints on the range of allowed segment durations.

Another aspect of the modeling concerns the choice of acoustic features. While it has been shown that recognition performance can be improved by representing trajectories of mel-cepstrum features, the motivation for modeling dynamics in speech comes from the continuous dynamic nature of speech production. It may therefore be better to apply the trajectory model to features that are more directly related to the mechanisms of speech production. A useful functional representation of speech production is provided by the vocal tract resonances, or formants. Experiments using formant-trajectory models together with a phone-dependent model structure are described in the next section.

4. Recognition using formant trajectories. Although it is well known that the frequencies of the formants are extremely important for determining the phonetic content of speech sounds, formant frequencies are not normally used as features for ASR due to a number of practical difficulties that tend to arise when attempting to extract and use formant information in recognition. For example, the formants are often not clearly apparent as distinct peaks in the short-term spectrum. In the extreme, formants do not provide the required information for making certain distinctions, such as identifying silence. Furthermore, formant-labeling difficulties can arise even when the spectrum shape shows a clear simple resonance structure because, for example, two formants that are close together may merge to form a single spectral peak. A consequence of all these factors is that it is difficult to identify formants independently from the recognition process that decides on the phone identities.

A method of formant analysis has been developed [13] that includes techniques to largely overcome the difficulties normally associated with extracting and using formant information. In addition, improvements in recognition performance were demonstrated by incorporating formant information in a standard-HMM recognition system [13]. Work on applying the formant-based system to linear-trajectory segmental-HMM recognition is described below.

4.1. Formant analysis. The formant analyzer is described in some detail in [13], but a brief overview is given here. The system is based on a codebook of spectral cross-sections and associated formant labelings that has been previously set up by a human expert. Given an input speech signal, short-term spectral cross-sections are matched against this codebook to find the entries that give the best spectral match and hence derive possible formant labelings for each frame of speech. Continuity constraints and other constraints are employed to eliminate many possibilities but, in those cases where there is still uncertainty about how the formants should be allocated to spectral peaks, alternative sets of formant frequencies are offered to the recognition process. To indicate cases where the formants are not well defined by the spectrum shape, an empirical degree-of-confidence measure is provided for each estimated formant frequency. The confidence measure is calculated using information about formant amplitude and spectrum curvature. When the amplitude is low or the formant structure is not well defined (as in many voiceless fricatives), the confidence will be much lower than when there is a peaky spectrum shape (typical of vowels).

Figure 1 shows a spectrogram with superimposed formant tracks indicating the output of the formant analyzer. It can be seen that the analyzer has tracked the formants quite accurately with smooth trajectories that capture dynamics such as the movement of F2 in the /aI/ diphthong of the word "nine". In the word "two", there are two alternative formant trajectory choices that have been suggested and the recognizer would be required to select between these two choices.

4.2. Using the formant analyzer output in recognition. In addition to the formant frequencies, features giving some measure of spectrum level and spectrum shape are needed to perform recognition. Low-order cepstrum features are a convenient way of providing this information. To make use of the formant alternatives and confidence measures that are produced by the analyzer, some modifications to the recognition calculations are required. When the analyzer offered alternative formant allocations, for each model state the set was chosen that gave the highest probability of generating the observations. The choice between alternatives was made on

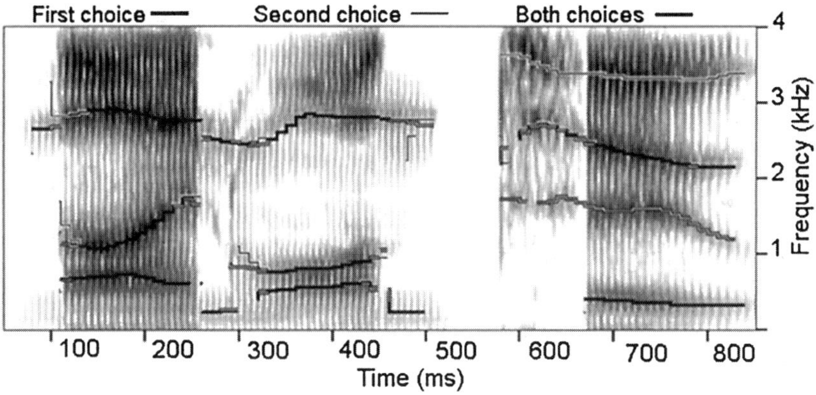

FIG. 1. *Spectrogram of an utterance of the words "nine one two", with superimposed formant tracks showing alternative formant allocations offered by the analyzer for F1, F2 and F3. Tracks are not plotted when there is no confidence in their accuracy.*

a frame-by-frame basis when performing standard-HMM recognition, and on a segment-by-segment basis when using segmental HMMs.

To use the confidence measure in recognition, it was represented as the variance of a notional Gaussian distribution of the true formant frequency about the estimated value. By representing the formant confidence estimates as variances, it is straightforward to incorporate them in the HMM probability calculations in a way that can be justified theoretically (see [9] for details). In this interpretation, the formant analyzer emits the parameters of a normal distribution representing its belief about the position of each formant. When the confidence is high, the variance is low to represent strong belief in the estimate. Conversely, when the confidence is low, the variance will be large, so representing almost equal belief in all possible frequencies. With this Bayesian interpretation, the confidences are incorporated simply into the recognition calculations by adding the appropriate confidence variance to that of the model state output distribution. The same approach can be applied to extend the probability calculations for FTSHMMs. The situation is more complicated for PTSHMMs due to the probabilistic nature of the trajectory model, and so far all segmental HMM experiments with formants have used the fixed trajectory model.

4.3. Digit recognition experiments.

4.3.1. Method. Using the same connected-digit recognition task that was used in the experiments described in Section 3.2, and with both standard and segmental HMMs, recognition performance when using formant features to describe fine spectral detail was compared with that obtained when using more typical mel-cepstrum features. In order to assess the usefulness of the formants directly, the same total number of features was used for both feature sets and exactly the same low-order cepstrum features were used for describing general spectrum shape. The output of an excitation-synchronous FFT was therefore used both to estimate formant fequencies with associated confidence measures and to compute a mel-cepstrum. One feature set for the experiments comprised the first eight cepstrum coefficients and an overall energy feature, while for the other feature set cepstrum coefficients 6, 7 and 8 were replaced by three formant features.

4.3.2. Model sets. For these segmental HMM experiments, each phone was modeled by an appropriate number of segments in order to describe its spectral characteristics using linear trajectories, with the number of segments assigned based on phonetic knowledge. Three segments were used to model voiceless stops, affricates and some diphthongs, with two segments being used to represent voiced stops, most diphthongs and a few long monophthongs. When using linear trajectories, one segment was considered sufficient for nasals, fricatives, semivowels and most monophthongal vowels. For each segment, a minimum and maximum segment duration was set to allow a plausible range of durations for each phone. Linear-trajectory segmental HMMs were compared with the simplest type of 'segmental' models, which used the same state allocation and maximum-duration constraints, but with standard-HMM emission probability calculations. Further comparisons were carried out with standard HMMs using the phone-dependent state allocation, and with standard HMMs using the more conventional allocation of three states per phone. All experimental conditions used single-state non-speech models.

4.3.3. Results. The results are summarized in Table 5. The baseline is provided by the standard HMMs using mel-cepstrum features and three states per phone. These models gave an error rate of 3.5%.[1] Incorporating the variable-state allocation designed for the linear-trajectory model made the performance of the standard HMMs worse, which can be explained by the lower number of states that were then available to capture the acoustic characteristics of many of the phones. This disadvantage of the smaller number of states was overcome by introducing a segmental structure with segment-duration constraints. A considerable further advantage was gained by incorporating the linear trajectory model. For all the model sets, there was a small but consistent advantage gained from including formants in the feature vector, for the same total number of features. The best overall performance is provided by the formant-based linear-trajectory segmental HMMs. The total number of free parameters in this model set is in fact fewer than the number in the standard HMMs with three states per phone. Although the linear trajectory requires additional parameters, this addition is more than compensated for by the reduction in the total number of states.

[1]These results are not directly comparable with the earlier results that were shown in Table 2 for a number of reasons, including the use of a different set of features and also some other minor differences in the experimental framework.

TABLE 5

Connected-digit recognition results for different sets of standard and segmental HMMs. Word error rates for a feature set comprising the first eight MFCCs (and an overall energy feature) are compared with those for a feature set in which the three highest MFCCs are replaced by three formant features.

Model type	8 MFCCs	5 MFCCs + 3 formants
Standard HMM with three states per phone	3.5	2.5
Standard HMM with variable state allocation	6.4	5.9
Simple segmental HMMs	3.2	2.9
Linear FTSHMMs	2.6	2.3

4.3.4. Discussion. The results of the experiments described above have indicated that using segmental HMMs to model formant trajectories seems to be a promising approach to ASR. Another attractive aspect of modeling formant trajectories is that such a model naturally lends itself to speech synthesis as well as to recognition. There is thus the possibility of using the same model for both recognition and synthesis [25], which in turn leads to a compact model for low bit-rate speech coding [15]. The application of the formant-based linear trajectory model to speech coding is discussed in the next section.

5. Recognition-based speech coding. Successful speech coding at low data rates of a few hundred bits/s requires a compact, low-dimensional representation of the speech signal, which is generally applied to variable-length 'segments' of speech. Automatic speech recognition is potentially a powerful way of identifying useful segments for coding. If the segments are meaningful in phonetic terms, knowledge of segment identity can be used to guide the coding. In the extreme, very low data rates can be achieved by transmitting only phoneme identity information. A number of recognition-based coders have been suggested that use HMMs (e.g. [22, 17, 23, 27]). In all of these systems, an utterance is coded in terms of phone-based recognition units and relevant duration information. The main differences are in the schemes that are used to reconstruct the utterance at the receiver. One possibility is to use the HMMs themselves, but simple use of the HMM state means will tend to lead to inappropriate discontinuities in the synthesized speech. [27] used a more elaborate scheme which produced smoother sequences by also using information from time-derivative features. However, the underlying assumptions of piecewise-stationarity and of independence are such that HMMs are inherently limited as speech production models. Another limitation when using HMMs for synthesis is that typical feature sets such as LPC coefficients [22] or mel-frequency cepstral coefficients [27] impose limits on the coded speech quality. Other systems have regenerated the utterances using completely separate systems, such as time-normalized versions of complete segments [23] or a synthesis-by-rule system [17].

A formant-trajectory segment model of the type described in the previous section can naturally lend itself to speech synthesis as well as to speech recognition. This type of model can therefore provide the basis for a 'unified' approach to speech coding in which the same (appropriate) model of speech recognition is used as the basis for both the recognition step and the synthesis step. In this way it is possible to address the issues associated with achieving successful recognition-synthesis coding at low bit-rates [15].

5.1. A general framework for a 'unified' speech coding model. A good model for recognition-based coding needs to provide a compact representation of speech, while offering both accurate recognition performance and high quality synthesis. Such a model could be used for coding at a range of data rates, by trading bits against retention of speaker characteristics. With no limitations on vocabulary size, at the lowest bit rates speech could be generated from a phoneme sequence. At higher data rates, the coding could be applied directly to speech production parameters. The principles of operation for such a 'unified' model for speech recognition and synthesis, and its application to speech coding, are illustrated in Figure 2. The approach requires an accurate model for speech dynamics that preserves the distinctions between different speech sounds and also a suitable representation of speech production mechanisms.

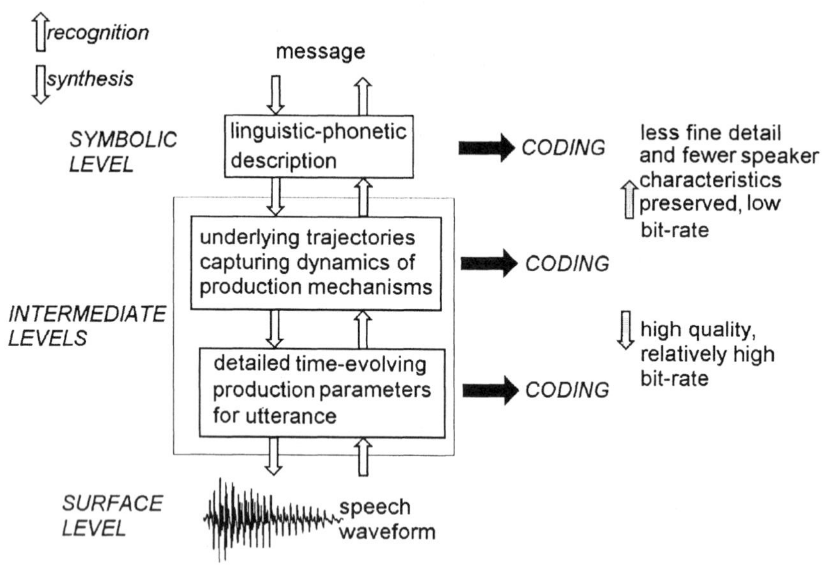

FIG. 2. *Schematic representation of a unified model for speech recognition and synthesis, showing its application to speech coding across a range of data rates.*

5.2. A simple coding scheme to illustrate the approach. Work so far has concentrated on demonstrating the principle of recognition-synthesis coding using the same linear formant-trajectory model for both recognition and synthesis. The coding is applied to analyzed formant trajectories, and so is at the high bit-rate end of the range of coding schemes discussed above. The main stages in this simple coding scheme are illustrated in Figure 3. The recognition uses the system described in the previous section, with formant-based linear-trajectory segmental HMMs, and the synthesis uses the JSRU parallel-formant synthesizer [12]. The speech is first analyzed in terms of formant trajectories and other information that is needed for synthesis (formant amplitudes, fundamental frequency and degree of voicing). The recognizer is used to determine the segment identities and the locations of the boundaries of the segments. All the synthesizer control parameters are then coded as linear trajectories for the segments that have been identified in recognition. The method of coding and the bit allocation for the different parameters was chosen to be reasonably economical, but was not optimized for maximum efficiency.

A particular aim of this work has been to use information from the recognition process to assist in the formant coding. For example, when a sequence involving a vowel followed by another vowel or a sonorant was recognized, at the coding stage a continuity constraint was imposed on the formant trajectories across the segment boundary (in future there is the possibility of also incorporating continuity constraints within the recognition process itself). When the recognizer selected between two alternative formant trajectories offered by the initial formant analysis, the trajectory chosen in recognition was the one that was used in the coding. More detail about the coding method can be found in [15].

Although formants have been used as the basis for other speech coding schemes (e.g. [6, 28]), the use of trajectory-based recognition is an important distinguishing feature of the approach being pursued here.

5.3. Coding experiments. The coding method has been tested on the speaker-independent connected-digit recognition task that was used in the speech recognition experiments described above, and also on a speaker-dependent task of recognizing spoken airborne reconnaissance mission (ARM) reports using a 500-word vocabulary. For each utterance coded, the bit rate was calculated (see [15] for details) and the quality of the coding was evaluated by informal listening tests. The segment-coded utterances were generally perceived as somewhat stylized in comparison with the original utterances. However, speaker characteristics were retained for all the variety of speakers tested, and the speech was generally highly intelligible.

For the digit data, typical coding rates were 600–800 bits/s. For the ARM task, the rates tended to be higher at about 800–1000 bits/s. These rates reflect the nature of the speech material: with this coding scheme, the bit rate does not simply depend on the vocabulary size, but it does

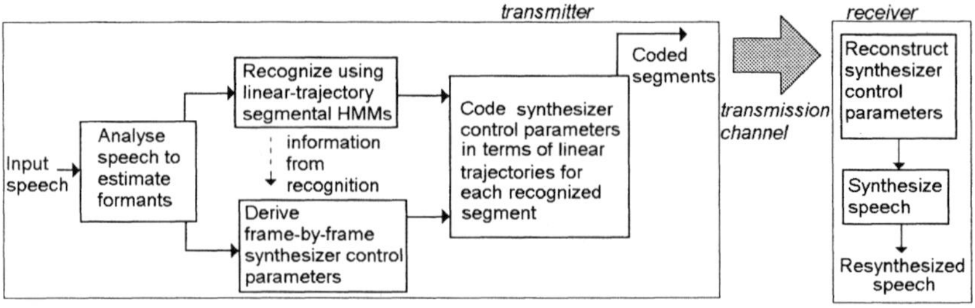

FIG. 3. *Block diagram showing the main stages in a simple recognition-synthesis coding scheme using linear formant trajectories in both recognition and synthesis.*

depend on the number of segments identified per second of speech. The higher bit rates for the ARM task arose because this data set included more acoustically complex words and the reports were spoken rather more quickly than the digit strings that were used in the other task.

6. Summary and discussion. This paper has given an overview of a range of studies that have been carried out investigating the application of trajectory-based segmental HMMs to ASR and speech coding, and more generally to modeling the characteristics of speech signals. Improvements in recognition performance have been demonstrated by modeling linear trajectories of both mel-cepstrum and formant features, with the overall best performance being obtained by adopting the trajectory model and including formants in the feature set. Although the formant analyzer used in these experiments includes a number of special characteristics to reduce problems due to formant analysis errors, there were still some errors from the formant analyzer that were propagated to the recognizer. Any system that performs formant analysis as a first stage prior to recognition will tend to suffer from these problems, which are likely to become worse under more difficult environmental conditions. The difficulties involved in extracting and using formant (or articulatory) information have led a number of workers (e.g. [25, 24, 4]) to suggest that this information is best incorporated in a multiple-level framework, incorporating some production-related model of dynamics as an intermediate level between the abstract phonological units and the observed acoustic features. The aim is for the trajectory model to enforce production-related constraints on possible acoustic realizations without requiring explicit extraction of articulatory or formant information. This type of extension can be applied to the linear-trajectory segmental HMM described here (see [25]), which should enable the model to enforce more powerful constraints while not suffering from analysis errors. Such a model should be especially useful in difficult environmental conditions and whenever the acoustic signal is degraded.

REFERENCES

[1] L. BAUM, *An inequality and associated maximization technique in statistical estimation for probabilistic functions of Markov processes*, Inequalities, III (1972), pp. 1–8.

[2] A. DEMPSTER, N. LAIRD, AND D. RUBIN, *Maximum likelihood from incomplete data via the em algorithm*, Journal of the Royal Statistical Society, Series B, **39** (1977), pp. 1–38.

[3] L. DENG, M. AKSMANOVIC, D. SUN, AND J. WU, *Speech recognition using hidden Markov models with mixtures of trend functions*, IEEE Trans Speech and Audio Processing, **2** (1994), pp. 507–520.

[4] L. DENG AND J. MA, *Spontaneous speech recognition using a statistical coarticulatory model for the vocal-tract-resonance dynamics*, Journal of the Acoustical Society of America, **108** (2000), pp. 1–13.

[5] V. DIGALAKIS, *Segment-based stochastic models of spectral dynamics for continuous speech recognition*, PhD thesis, Boston University, 1992.

[6] B.C. DUPREE, *Formant coding of speech using dynamic programming*, Electronics Letters, **20** (1980), pp. 279–280.

[7] J. FISCUS, W. FISHER, A. MARTIN, M. PRZYBOCKI, AND D. PALLETT, *2000 nist evaluation of conversational speech recognition over the telephone: English and mandarin performance results*, in Proceedings of the 2000 Speech Transcription Workshop, University of Maryland, 2000.

[8] M.J. GALES AND S.J. YOUNG, *Segmental hidden Markov models*, in EUROSPEECH, Berlin, 1993, pp. 1611–1614.

[9] P. GARNER AND W. HOLMES, *On the robust incorporation of formant features into hidden Markov models for automatic speech recognition*, in ICASSP, Seattle, 1998, pp. 1–4.

[10] J. GAROFOLO, L.F. LAMEL, W. FISHER, J. FISCUS, D. PALLETT, AND N. DAHLGREN, *The darpa timit acoustic-phonetic continuous speech corpus cdrom*, ntis order number pb01-100354, available from ldc, 1993.

[11] H. GISH AND K. NG, *A segmental speech model with applications to word spotting*, in ICASSP, Minneapolis, 1993, pp. 447–450.

[12] J.N. HOLMES, *A parallel-formant synthesizer for machine voice output*, in Computer Speech Processing, 1985.

[13] J.N. HOLMES, W.J. HOLMES, AND P.N. GARNER, *Using formant frequencies in speech recognition*, in EUROSPEECH, Rhodes, 1997.

[14] W.J. HOLMES, *Modelling segmental variability for automatic speech recognition*, PhD thesis, University of London, 1997.

[15] ———, *Towards a unified model for low bit-rate speech coding using a recognition-synthesis approach*, in ICSLP, Sydney, 1998.

[16] W.J. HOLMES AND M.J. RUSSELL, *Probabilistic-trajectory segmental HMMs*, Computer Speech and Language, **13** (1999), pp. 3–37.

[17] M. ISMAIL AND K.M. PONTING, *Between recognition and synthesis – 300 bits/second speech coding*, in EUROSPEECH, Rhodes, 1997, pp. 441–444.

[18] K.F. LEE AND H.W. HON, *Speaker-independent phone recognition using hidden Markov models*, ASSP, **37** (1989), pp. 1641–1648.

[19] L.A. LIPORACE, *Maximum likelihood estimation for multivariate observations of Markov sources*, IT, **28** (1982), pp. 729–734.

[20] M. OSTENDORF, V.V. DIGALAKIS, AND O.A. KIMBALL, *From HMM's to segment models: A unified view of stochastic modeling for speech recognition*, IEEE Trans Speech and Audio Processing, **4** (1996), pp. 360–378.

[21] D. PALLETT, J. FISCUS, J. GAROFOLO, A. MARTIN, AND M. PRZYBOCKI, *1998 broadcast news benchmark test results: English and non-english word error rate performance measures*, in Proceedings of the DARPA Broadcast News Workshop, Virginia, 1999.

[22] J. Picone and G. Doddington, *A phonetic vocoder*, in ICASSP, Glasgow, 1989, pp. 580–583.

[23] C. Ribeiro and I. Trancoso, *Improving speaker recognisability in phonetic vocoders*, in ICSLP, Sydney, 1998, pp. 2611–2614.

[24] H.B. Richards and J.S. Bridle, *The hdm: a segmental hidden dynamic model of coarticulation*, in ICASSP, Phoenix, 1999.

[25] M.J. Russell and W.J. Holmes, *Progress towards a unified model for speech pattern processing*, Proc. IOA, **20** (1998), pp. 21–28.

[26] M.J. Russell and R.K. Moore, *Explicit modelling of state occupancy in hidden Markov models for automatic speech recognition*, in ICASSP, 1985, pp. 5–8.

[27] K. Tokuda, T. Masuko, J. Hiroi, T. Kobayahi, and T. Kitamura, *A very low bit rate speech coder using HMM-based speech recognition/synthesis techniques*, in ICASSP, Seattle, 1998, pp. 609–612.

[28] P. Zolfaghari and T. Robinson, *A segmental formant vocoder based on linearly varying mixture of Gaussians*, in EUROSPEECH, Rhodes, 1997, pp. 425–428.

MODELLING GRAPH-BASED OBSERVATION SPACES FOR SEGMENT-BASED SPEECH RECOGNITION

JAMES R. GLASS*

Abstract. Most speech recognizers use an observation space which is based on a temporal sequence of spectral "frames." There is another class of recognizer which further processes these frames to produce a segment-based network, and represents each segment by a fixed-dimensional "feature." In such feature-based recognizers the observation space takes the form of a temporal graph of feature vectors, so that any single segmentation of an utterance will use a subset of all possible feature vectors. In this work we describe a maximum *a posteriori* decoding strategy for feature-based recognizers and derive two normalization critera useful for a segment-based Viterbi or A^* search. We show how a segment-based recognizer is able to obtain good results on the tasks of phonetic and word recognition.

Key words. Segment-based speech recognition, phonetic recognition.

1. Introduction. The fundamental goal of automatic speech recognizers is to identify the spoken words in a speech waveform. This process generally begins with a signal processing stage which converts the recorded waveform to some form of acoustic representation. Subsequently, one or more search stages, which incorporate linguistic constraints such as acoustic and language models, attempt to identify the most likely word hypotheses. For many years the prototypical acoustic representation has consisted of a time-frequency representation, which is computed at regular intervals over the speech signal (e.g., every 10 ms). This sequence of spectral observations, or frames, is intended to capture the salient time-frequency dynamics of the underlying phonological units present in the speech waveform. Most acoustic modelling techniques use the spectral frame as the input to pattern classifiers which attempt to determine the probability that a frame was produced by a particular linguistic unit.

Over the past two decades, first-order hidden Markov models (HMMs) have emerged as the dominant stochastic model for speech recognition [25]. HMMs use classifiers, such as Gaussian mixtures or artificial neural networks, to emit a state-dependent likelihood or posterior probability on a frame-by-frame basis. In contrast to HMMs and other frame-based processing techniques, the SUMMIT speech recognizer developed by our group uses a segment-based framework for its acoustic-phonetic representation of the speech signal [7]. In this framework, acoustic feature vectors are extracted both over hypothesized segments, and at their boundaries, for phonetic analysis. The resulting graph-based observation space differs considerably from the more conventional frame-based sequential observations.

*MIT Laboratory for Computer Science, 200 Technology Square, Cambridge, MA 02139, USA (glass@mit.edu).

Our interest in exploring segment-based approaches is based on our experience with phonetic classification. One of our findings has been that a homogeneous acoustic representation can compromise performance. For example, the classification performance of stop consonants, /ptkbdg/, and nasal consonants, /mnŋ/, have opposite dependencies on the duration of the spectral analysis window [11]. While the ability to discriminate between the stops improves with shorter analysis windows, the nasal consonant performance improves as the analysis window is lengthened. Another motivation for heterogeneous measurements is the observation that different temporal basis vectors can significantly change phonetic classification performance. For example, Halberstadt found that smoothly varying cosine-basis functions are better suited for discriminating between vowels and nasal consonants, while piecewise-constant basis vectors (e.g., averages) do better for discriminating fricatives and stop consonants [10]. These results support the notion that combining information sources could reduce overall error.

By combining different sources of information and combining classifiers we have been able to achieve state-of-the-art phonetic classification performance. On the TIMIT acoustic phonetic corpus [5], using standard training and test sets, we have been able to achieve an 18.3% context-independent error rate on a 39 phone classification problem [12]. These results are the best that we have seen on this task. More generally, we have found that a segment-based approach, whereby acoustic modelling is performed over an entire segment, has provided us with a powerful framework for exploring a variety of ways in which to incorporate acoustic-phonetic information into the speech recognizer.

In this paper we describe the probabilistic formulation we currently use for our segment-based recognizer. This recognizer differs considerably from most frame-based decoding techniques, as well as from many other segment-based approaches. In the following section we provide a brief survey of other segment-based approaches which have been explored in the speech recognition literature. This is followed by a derivation of the MAP decoding techniques we have developed for our recognizer. Finally, we report some current results on phonetic and word recognition experiments.

2. Background. There have been many segment-based approaches which have been explored in the speech community [22]. In HMM-based approaches, which retain many similarities to HMM-based techniques, researchers have explored variable frame-rate analysis [24], segment-based HMMs [19], and the segmental HMM [15, 30]. In trajectory-based modelling approaches, there have been stochastic segment models [23], parametric trajectory models [6], as well as stochastic and statistical trajectory models [9]. These methods typically try to model dynamic behavior over the course of a phonetic segment, although the modelling is ultimately performed at the individual frame level. Finally, there are a class of approaches

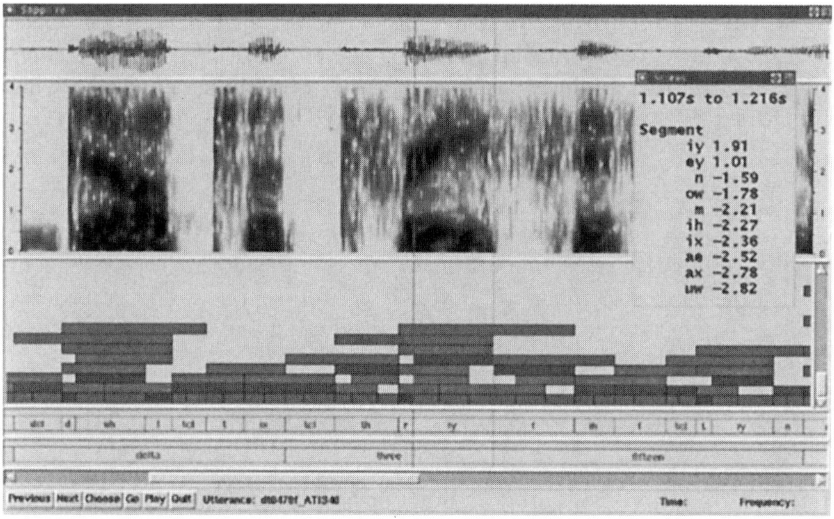

FIG. 1. *This figure contains several displays from the* SUMMIT *segment-based speech recognizer. The top two displays contain the speech waveform and associated spectrogram, respectively. Below them, a segment-network display shows hypothesized segments; each segment spans a time range. The darker colored segments show the segmentation which achieved the highest score during search. The two transcriptions below the segment-network contain the best-scoring phonetic, and word sequences, respectively. The pop-up menu shows the ranked log-likelihood ratio scores for the [*i*] *in the word "three".*

which could be called feature-based. Examples in this area include the FEATURE system [4], the SUMMIT system described in this paper, as well the LAFF system [31].

The SUMMIT segment-based system developed in our group uses a segment- and landmark-based framework to represent the speech signal. Acoustic or probabilistic landmarks form the basis for a phonetic network, or graph, as shown in Figure 1. Acoustic features are extracted over hypothesized phonetic segments and at important acoustic landmarks. Gaussian mixture classifiers are then used to provide phonetic-level probabilities for both segmental and landmark features. Words are represented as pronunciation graphs whereby phonemic baseforms are expanded by a set of phonological rules [14]. Phone probabilities are determined from training data. A probabilistic MAP decoding framework is used. A modified Viterbi beam search finds the best path through both acoustic-phonetic and pronunciation graphs while incorporating language constraints. All graph representations are based on weighted finite-state transducers [8, 21]. One of the properties of the segment-based framework is that, as will be described later, the models must incorporate both positive and negative examples of lexical units. Finally, a secondary A^* search can provide word-graph or N-best sentence outputs for further processing.

3. MAP decoding. In most probabilistic formulations of speech recognition the goal is to find the sequence of words $W^* = w_1, \ldots, w_N$, which has the maximum *a posteriori* (MAP) probability $P(W|A)$, where A is the set of acoustic observations associated with the speech waveform:

$$W^* = \arg\max_W P(W|A). \tag{3.1}$$

In most speech recognizers, decoding is accomplished by hypothesizing (usually implicitly) a segmentation, S, of the waveform into a sequence of sub-word states or linguistic units, U. A full search would consider all possible segmentations for each hypothesized word and sub-word sequence, so that Equation 3.1 can be rewritten as:

$$W^* = \arg\max_{S,U,W} \sum_{\forall S,U} P(S,U,W|A). \tag{3.2}$$

If some form of dynamic programming or graph search, such as Viterbi or A^*, is used to find the best path, this expression can be simplified to

$$S^*, U^*, W^* = \arg\max_{S,U,W} P(S,U,W|A). \tag{3.3}$$

In subsequent equations, S^* and U^* will be dropped for notational simplicity. Using Bayes rule, the term $P(S,U,W|A)$ can be decomposed:

$$P(S,U,W|A) = \frac{P(A|S,U,W)P(S|U,W)P(U|W)P(W)}{P(A)}. \tag{3.4}$$

Since $P(A)$ is independent of S, U, and W, it will not affect the outcome of the search, and is usually ignored unless it is being used as a normalization mechanism. The term $P(W)$ is estimated by a language model which predicts the *a priori* probability of a particular sequence of words being spoken. The term $P(U|W)$ can be considered to be a phonological model which predicts the probability of a sequence of sub-word units being generated by the given word sequence W. In an HMM this term might correspond to the state sequence corresponding to a word sequence, and could be completely deterministic. Some speech recognizers incorporate a stochastic component at this level to attempt to model phonological variations in the way a word can be realized in fluent speech (e.g., "did you" being realized as the phoneme sequence /dɪǰu/) [14, 26]. The term $P(S|U,W)$ models the probability of the segmentation itself, and typically depends only on U. In HMMs for example, this term corresponds to the likelihood of a particular state sequence, and is generated by the state transition probabilities. More generally, this term can be considered as a duration model which predicts the probability of individual segment durations. Many researchers have explored the use of more explicit duration models for recognition, especially in the context of segment-based recognition [18, 23]. The remaining term in Equation 3.4, $P(A|S,U,W)$, relates to

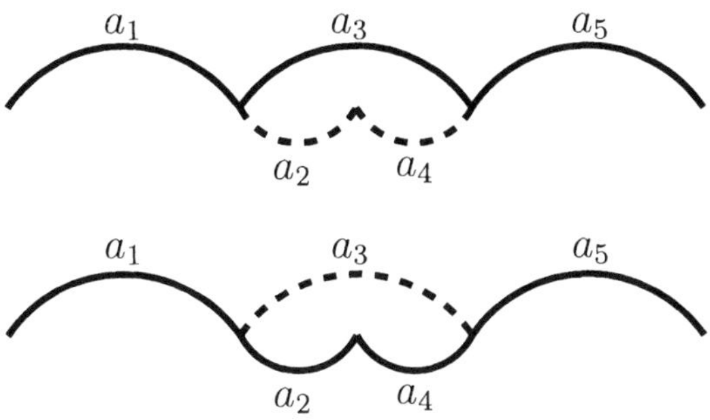

FIG. 2. *Two segmentations (in solid lines) through a simple five segment graph with acoustic observations* $\{a_1, \ldots, a_5\}$. *The top segmentation is associated with observations* $\{a_1, a_3, a_5\}$, *while the bottom segmentation is associated with* $\{a_1, a_2, a_4, a_5\}$.

the ASR acoustic modelling component, and is the subject of this paper. For simplicity we will assume that acoustic likelihoods are conditionally independent of W so that $P(A|S, U, W) = P(A|S, U)$. This assumption is also standard in conventional HMM-based systems.

In conventional speech recognizers, the acoustic observation space, A, corresponds to a temporal sequence of acoustic frames (e.g., spectral slices). Each hypothesized segment, s_i, is represented by the series of frames computed between segment start and end times. Thus, the acoustic likelihood $P(A|S, U)$, is derived from the same observation space for all word hypotheses. In feature-based recognition, as illustrated in Figure 2, each segment, s_i, is represented by a single feature vector, a_i. Given a particular segmentation, S, (a contiguous sequence of segments spanning the entire utterance), A consists of X, the feature vectors associated with the segments in S, as well as Y, the feature vectors associated with segments not in S, such that $A = X \cup Y$ and $X \cap Y = \emptyset$. In order to compare different segmentations it is necessary to predict the likelihood of both X and Y, since $P(A|S, U) = P(X, Y|S, U)$. Thus, in addition to the observations X associated with the segments in S, we must consider all other possible observations in the space Y, corresponding to the set of all other possible segments, R. In the top path in Figure 2, $X = \{a_1, a_3, a_5\}$, and $Y = \{a_2, a_4\}$. In the bottom path, $X = \{a_1, a_2, a_4, a_5\}$, and $Y = \{a_3\}$. The total observation space A, contains *both* X and Y, so for MAP decoding it is necessary to estimate $P(X, Y|S, U)$. Note that since X implies S we can say $P(X, Y|S, U) = P(X, Y|U)$. The following sections describe two methods we have developed to account for the entire observation space in our segment-based recognizer.

3.1. Modelling non-lexical units. One possible method for modelling Y is to assign them to a non-lexical unit, or *anti-phone*, $\bar{\alpha}$ (e.g., too big or too small). Given a segmentation, S, assign feature vectors in X to valid linguistic units, and all others to Y to the non-unit, $\bar{\alpha}$. Since $P(X, Y|\bar{\alpha})$ is a constant for any given segment graph, we can write $P(X, Y|U)$ assuming independence between X and Y:

$$P(X,Y|U) = P(X|U)P(Y|\bar{\alpha})\frac{P(X|\bar{\alpha})}{P(X|\bar{\alpha})} \propto \frac{P(X|U)}{P(X|\bar{\alpha})}. \tag{3.5}$$

Thus, we need only consider segments in S during search

$$W^* = \arg\max_{S,U,W} \prod_{i=1}^{n} \frac{P(x_i|u_i)}{P(x_i|\bar{\alpha})} P(s_i|u_i)P(U|W)P(W). \tag{3.6}$$

Where u_i is the linguistic unit assigned to segment s_i, and its associated feature vector observation x_i.

The advantage of the anti-phone normalization framework for decoding is that it models the entire observation space, using both positive and negative examples. Log likelihood scores are normalized by the anti-phone so that good scores are all positive, while bad scores will be negative. In Figure 1 for example, only the top two log likelihood ratio's for the /i/ segment were positive. Invalid segments, which do not correspond to a valid phone, are likely to have negative scores for all phonetic hypotheses. Note that the anti-phone is not used for lexical access, but purely for normalization purposes. In general, it is useful for pattern matching problems with graph-based observation spaces.

3.2. Near-miss modelling. Anti-phone modelling divides the observation space into essentially two parts; subsets of segments which are either on or off a hypothesized segmentation S. The anti-phone model is quite general since it must model all examples of observations which are not phones. A larger inventory of anti-phones might better model segments which are near misses of particular phones. One such method, called *near-miss* modelling, was developed by Chang [1, 2]. Near-miss modelling partitions the observation space into a set of near-miss subsets, where there is one near-miss subset, A_i, associated with each segment, s_i, in the segment network (although A_i could be empty). During recognition, observations in a near-miss subset are mapped to the near-miss model of the hypothesized phone. The net result can be represented as:

$$W^* = \arg\max_{S,U,W} \prod_{i=1}^{n} P(x_i|u_i)P(A_i|u_i)P(s_i|u_i)P(U|W)P(W) \tag{3.7}$$

where $P(A_i|u_i)$ is computed as the product of observations in A_i being generated by the near-miss model associated with u_i (i.e., $\bar{u}_i$).

Near-miss models can be an anti-phone, but can potentially be more sophisticated. The challenge is to partition the observation space during the search so that all observations in A are included. Chang developed an effective criterion for creating near-miss subsets. By definition, the near-miss subsets associated with any segmentation S must be mutually exclusive, and collectively exhaustive so that $A = \bigcup(x_i \cup A_i) \quad \forall s_i \in S$ and $A_i \cap A_j = \emptyset \quad \forall s_i, s_j \in S, i \neq j$. Note that for any given segmentation, S, $X = \bigcup x_i$, and Y, $Y = \bigcup A_i$. Chang recognized that a temporal criterion could be used to guarantee proper near-miss subset creation. This is because any particular segmentation through a segment network accounts for all times exactly once. Thus, segments which all span the same time naturally form a near-miss set. Using a common reference point, such as the segment mid-point, appropriate near-miss subsets can be defined which satisfy the necessary near-miss conditions [1].

3.3. Modelling landmarks. In addition to modelling segments, it is also possible to model phonetic transitions at hypothesized landmarks or phonetic boundaries. If we represent landmark-based observations by Z, it is necessary to determine $P(X, Y, Z|S, U)$. Assuming conditional independence between the segmental and landmark-based observations, we can say $P(X, Y, Z|S, U) = P(X, Y|S, U)P(Z|S, U)$. Further, if we assume conditional independence between landmark-based observations, we can represent $P(Z|S, U)$ by

$$P(Z|S, U) = \prod_{i=1}^{m} P(z_i|SU) \tag{3.8}$$

where z_i is the feature vector extracted at all m hypothesized landmarks in the speech waveform. Since every segmentation accounts for every landmark there is no need for the normalization procedures discussed for segmental observations. Note that depending on the hypothesized segmentation, some landmarks will be considered to be transitions between lexical units, whereas others will be considered to be internal to a unit.

Landmark-based models are able to capture the relative dynamics which occur between phones. These models have been used to effectively generate an initial segmentation graph for our segment-based recognizer by using them in a forward Viterbi pass followed by a backwards A^* search [2]. Block processing enables the segment network to be computed with pipelined computation [17].

4. Experiments. We have evaluated our segmental framework in both phonetic and word recognition experiments. The experiments described in the next two sections have been presented in more detail elsewhere [12, 32]. All experiments have made use of the SUMMIT segment-based speech recognition system [7]. This recognizer combines segment- and landmark-based classifiers. Feature extraction is typically based on averages and derivatives of Mel-frequency cepstral coefficients (MFCCs), plus additional information such as energy. Principal component analysis is used to normalize and whiten the feature space. Acoustic models are based on mixtures of diagonal Gaussians.

4.1. Phonetic recognition. Phonetic recognition experiments were carried out the the TIMIT acoustic-phonetic corpus [5]. This corpus has been used by many researchers to report phonetic classification and recognition experiments. In our experiments, we used the standard 462 speaker training set, and 24 speaker core-test set. By convention, phonetic recognition errors were reduced to the common 39 classes. Segmental and landmark representations were based on five variations of averages and derivatives of 12 Mel-Frequency and PLP cepstral coefficients, plus energy and duration [12]. 4-fold aggregation was used to improve the robustness of the Gaussian mixtures [13]. The language model used in all experiments was a phone bigram based on the training data.

The phonetic error rates obtained on the core test set varied from 30.1% to 24.4% depending on the sophistication of the acoustic measurements [12]. The best results were obtained using an anti-phone model and a committee-based classifier to combine the outputs of five different segmental, and landmark models. Table 1 compares this result with the best results reported in the literature.

TABLE 1
Reported phonetic recognition error rates on the TIMIT *core test set.*

Method	% Error
Triphone CDHMM [16]	27.1
Recurrent Neural Network [27]	26.1
Bayesian Triphone HMM [20]	25.6
Anti-Phone, Heterogeneous classifiers [12]	24.4

4.2. Word recognition. Word recognition experiments have been performed on a spontaneous-speech, telephone-based, conversational interface task in the weather domain [34]. For these experiments, a 50,000 utterance training set was used, and an 1806 utterance test set was used containing in-domain queries. The recognizer for this task used a vocabulary of 1957 words, as well as a class bigram and trigram language model [8, 32]. As shown in Table 2, the word error rate (WER) obtained by the system using only context-dependent segmental models was 9.6%. When landmark models were used, the WER decreased to 7.6%. The overall performance decreased to 6.1% when both models were combined.

TABLE 2
Word recognition error rates for a weather domain task.

Method	% Error
Segment models	9.6
Landmark models	7.6
Combined	6.1

5. Discussion. The results shown in Table 1 compare a number of published results on phonetic recognition which have been based on the TIMIT core test set. There are still differences regarding the complexity of the acoustic and language models, thus making a direct comparison somewhat difficult. Nevertheless, we believe our results are competitive with those obtained by others, and that our performance will improve when we increase the complexity of our models.

The framework we have outlined in this paper provides flexibility to explore the relative advantages of segment versus landmark representations. As we have shown, it is possible to use only segment-based feature vectors, or landmark-based feature vectors (which could reduce to frame-based processing), or a combination of both.

The anti-phone normalization criterion can be interpreted as a likelihood ratio. In this way it has similarities with techniques being used in word-spotting, which compare acoustic likelihoods with those of "filler" models [28, 29, 33]. The likelihood or odds ratio was also used by Cohen to use HMMs for segmenting speech [3].

One of the most questionable assumptions made in this work was the independence assumption made between the observations in X, and those in Y. Segments that temporally overlap with each other are clearly related to each other to some degree. In the future, it would be worthwhile examining alternative methods for modelling the joint X, Y space.

6. Summary. In this paper we have described a probabilistic decoding framework for decoding a graph-based observation space. This method is particularly appropriate for segment-based speech recognizers which transform the observation space from a sequence of frames, to a graph of features. Graph-based observation spaces allow for a wide-variety of alternative modelling methods to be explored, than could be achieved with frame-based approaches. We have developed two techniques for decoding graph-based observation spaces, based on anti-phone or near-miss modelling, and have achieved good results on phonetic recognition. We have also observed improved performance when combining segmental models into a landmark-based word recognizer.

Acknowledgements. There are a number of colleagues, past and present, who have contributed to this work including Jane Chang, Andrew Halberstadt, T.J. Hazen, Lee Hetherington, Michael McCandless, Nikko Ström, and Victor Zue. The author would also like to thank Mari Ostendorf for providing many useful comments that helped improve the paper. This research was supported by DARPA under contract N66001-99-C-1-8904 monitored through the Naval Command, Control and Ocean Surveillance Center.

REFERENCES

[1] J. Chang. Near-miss modeling: A segment-based approach to speech recognition. Ph.D. thesis, EECS, MIT, June 1998.

[2] J. Chang and J. Glass. Segmentation and modeling in segment-based recognition. In *Proc. Eurospeech*, pages 1199–1202, Rhodes, Greece, October 1997.

[3] J. Cohen. Segmenting speech using dynamic programming. *Journal of the Acoustic Society of America*, **69**(5):1430–1438, May 1981.

[4] R. Cole, R. Stern, M. Phillips, S. Brill, A. Pilant, and P. Specker. Feature-based speaker-independent recognition of isolated letters. In *Proc. ICASSP*, pages 731–733, Boston, MA, April 1983.

[5] J. Garofolo, L. Lamel, W. Fisher, J. Fiscus, D. Pallet, and N. Dahlgren. The DARPA TIMIT acoustic-phonetic continuous speech corpus CDROM. NTIS order number PB91-505065, October 1990.

[6] H. Gish and K. Ng. A segmental speech model with applications to word spotting. In *Proc. ICASSP*, pages 447–450, Minneapolis, MN, April 1993.

[7] J. Glass, J. Chang, and M. McCandless. A probabilistic framework for feature-based speech recognition. In *Proc. ICSLP*, pages 2277–2280, Philadelphia, PA, October 1996.

[8] J. Glass, T. Hazen, and L. Hetherington. Real-time telephone-based speech recognition in the Jupiter domain. In *Proc. ICASSP*, pages 61–64, Phoenix, AZ, March 1999.

[9] W. Goldenthal. Statistical trajectory models for phonetic recognition. Technical report MIT/LCS/TR-642, MIT Lab. for Computer Science, August 1994.

[10] A. Halberstadt. Heterogeneous acoustic measurements and multiple classifiers for speech recognition. Ph.D. thesis, MIT Dept. EECS, November 1998.

[11] A. Halberstadt and J. Glass. Heterogeneous measurements for phonetic classification. In *Proc. Eurospeech*, pages 401–404, Rhodes, Greece, September 1997.

[12] A. Halberstadt and J. Glass. Heterogeneous measurements and multiple classifiers for speech recognition. In *Proc. ICSLP*, pages 995–998, Sydney, Australia, December 1998.

[13] T. Hazen and A. Halberstadt. Using aggregation to improve the performance of mixture Gaussian acoustic models. In *Proc. ICASSP*, pages 653–656, Seattle, WA, May 1998.

[14] L. Hetherington. An efficient implementation of phonological rules using finite-state transducers. In *Proc. Eurospeech*, pages 1599–1602, Aalborg, Denmark, September 2001.

[15] W. Holmes and M. Russell. Modeling speech variability with segmental HMMs. In *Proc. ICASSP*, pages 447–450, Atlanta, GA, May 1996.

[16] L. Lamel and J.L. Gauvain. High performance speaker-independent phone recognition using CDHMM. In *Proc. Eurospeech*, pages 121–124, Berlin, Germany, September 1993.

[17] S. Lee and J. Glass. Real-time probabilistic segmentation for segment-based speech recognition. In *Proc. ICSLP*, pages 1803–1806, Sydney, Australia, December 1998.

[18] K. Livescu and J. Glass. Segment-based recognition on the PhoneBook task: Initial results and observations on duration modeling. In *Proc. Eurospeech*, pages 1437–1440, Aalborg, Denmark, September 2001.

[19] J. Marcus. Phonetic recognition in a segment-based HMM. In *Proc. ICASSP*, pages 479–482, Minneapolis, MN, April 1993.

[20] J. Ming and F. Smith. Improved phone recognition using bayesian triphone models. In *Proc. ICASSP*, pages 409–412, Seattle, WA, May 1998.

[21] M. Mohri. Finite-state transducers in language and speech processing. *Computational Linguistics*, **23**(2):269–311, June 1997.

[22] M. Ostendorf, V. Digilakis, and O. Kimball. From HMM's to segment models: a unified view of stochastic modelling for speech recognition. *IEEE Trans. SAP*, **4**(5):360–378, September 1996.

[23] M. Ostendorf and S. Roucos. A stochastic segment model for phoneme-based continuous speech recognition. *IEEE Trans. ASSP*, 37(12):1857–1869, December 1989.

[24] K. Ponting and S. Peeling. The use of variable frame rate analysis in speech recognition. *Computer Speech and Language*, **5**:169–179, 1991.

[25] L. Rabiner. A tutorial on hidden Markov models and selected applications in speech recognition. *Proceedings of the IEEE*, 1989.

[26] M. Riley and A. Ljolje. Lexical access with a statistically-derived phonetic network. In *Proc. Eurospeech*, pages 585–588, Genoa, Italy, September 1991.

[27] A. Robinson. An application of recurrent nets to phone probability estimation. *IEEE Trans. Neural Networks*, **5**(2):298–305, March 1994.

[28] J. Rohlicek, W. Russell, S. Roucos, and H. Gish. Continuous hidden Markov modelling for speaker-independent word spotting. In *Proc. ICASSP*, pages 627–630, Glasgow, Scotland, May 1989.

[29] R. Rose and D. Paul. A hidden Markov model based keyword recognition system. In *Proc. ICASSP*, pages 129–132, Albuquerque, NM, April 1990.

[30] M. Russell. A segmental HMM for speech pattern modelling. In *Proc. ICASSP*, pages 499–502, Minneapolis, MN, 1993.

[31] K. Stevens. Lexical access from features. In *Workshop on speech technology for man-machine interaction*, Bombay, India, 1990.

[32] N. Ström, L. Hetherington, T. Hazen, E. Sandness, and J. Glass. Acoustic modelling improvements in a segment-based speech recognizer. In *Proc. IEEE Automatic Speech Recognition and Understanding Workshop*, pages 139–142, Keystone, CO, 1999.

[33] J. Wilpon, L. Rabiner, C.H. Lee, and E. Goldman. Automatic recognition of keywords in unconstrained speech using hidden Markov models. *IEEE Trans. ASSP*, **38**(11):1870–1878, November 1990.

[34] V. Zue, S. Seneff, J. Glass, J. Polifroni, C. Pao, T. Hazen, and L. Hetherington. Jupiter: A telephone-based conversational interface for weather information. *IEEE Trans. Speech and Audio Proc.*, **8**(1):85–96, January 2000.

TOWARDS ROBUST AND ADAPTIVE SPEECH RECOGNITION MODELS

HERVE BOURLARD*, SAMY BENGIO†, AND KATRIN WEBER*‡

Abstract. In this paper, we discuss a family of new Automatic Speech Recognition (ASR) approaches, which somewhat deviate from the usual ASR approaches but which have recently been shown to be more robust to nonstationary noise, without requiring specific adaptation or "multi-style" training. More specifically, we will motivate and briefly describe new approaches based on multi-stream and subband ASR. These approaches extend the standard hidden Markov model (HMM) based approach by assuming that the different (frequency) streams representing the speech signal are processed by different (independent) "experts", each expert focusing on a different characteristic of the signal, and that the different stream likelihoods (or posteriors) are combined at some (temporal) stage to yield a global recognition output. As a further extension to multi-stream ASR, we will finally introduce a new approach, referred to as HMM2, where the HMM emission probabilities are estimated via state specific feature based HMMs responsible for merging the stream information and modeling their possible correlation.

Key words. Robust speech recognition, hidden Markov models, subband processing, multi-stream processing.

1. Introduction. Current automatic speech recognition systems are based on (context-dependent or context-independent) phone models described in terms of a sequence of hidden Markov model (HMM) states, where each HMM state is assumed to be characterized by a stationary probability density function. Furthermore, time correlation, and consequently the dynamics of the signal, inside each HMM state are also usually disregarded (although the use of delta and delta-delta features may capture some of this correlation). Consequently, apart from the dependencies captured via the topology of the HMM, most temporal dependencies are usually very poorly modeled.[1] Ideally, we want to design a particular HMM that is able to accommodate multiple time-scale characteristics so that we can capture phonetic properties, as well as syllable structures, which seem to have many attractive properties [9], including invariants that are more

*Dalle Molle Institute for Perceptual Artificial Intelligence (IDIAP), and Swiss Federal Institute of Technology at Lausanne (EPFL), 4, Rue du Simplon, CH-1920 Martigny, Switzerland (bourlard@idiap.ch).

†Dalle Molle Institute for Perceptual Artificial Intelligence (IDIAP), 4, Rue du Simplon, CH-1920 Martigny, Switzerland (bengio@idiap.ch).

‡weber@idiap.ch.

[1]This problem is not specific to the fact that phone models are generally used. Whole word models, or syllable models, built up as sequences of HMM states will suffer from exactly the same drawbacks, the only potential advantage of moving towards "larger" units being that one can then have more word (or syllable) specific distributions (usually resulting in more parameters and an increased risk of undersampled training data). Consequently, building an ASR system simply based on syllabic HMMs will not alleviate the limitations of the current recognizers since those models will still be based on the short-term piecewise stationary assumptions mentioned above.

robust to noise. For example, acoustic features such as the modulation spectrogram[2] exhibit some correlation with syllabic features and can be used to improve state-of-the-art ASR systems [33]. It is, however, clear that those different time-scale features will also exhibit different levels of stationarity and will require different HMM topologies to capture their dynamics.

There are many potential advantages to such a multi-stream approach, including:

1. The definition of a principled way to merge different temporal knowledge sources such as acoustic and visual inputs, even if the temporal sequences are not synchronous and do not have the same data rate – see [28] and [29] for further discussion about this.
2. The possibility to incorporate multiple time resolutions (as part of a structure with multiple unit lengths, such as phone and syllable).
3. Multiband-based ASR [6, 14] involving the independent processing and combination of partial frequency bands is a very particular case of multi-stream recognition. Although this will not be explicitly discussed here, there are many potential advantages to this multi-band approach, including (i) better robustness to speech impaired by narrowband noise, and (ii) the possibility of applying different time/frequency tradeoffs and different recognition strategies in the subbands.

In the following, we will not discuss the underlying algorithms ("complex" variants of Viterbi decoding, if one wants to take the possible asynchrony into account), nor detailed experimental results (see [11] for recent results). Instead, we will mainly discuss different combination strategies pointing towards the same formalism.

2. Psycho-acoustic evidence.

It seems to me that what can happen in the future is... that experiments get harder and harder to make, more and more expensive... and scientific discovery gets slower and slower. (Richard Feynman, 1918–1988, *The Character of Physical Law*, Cambridge, MA, p. 172.)

2.1. Product of errors rule and its interpretation.

The work of Fletcher and his colleagues (see the insightful review of his work in [1]) suggests that human decoding of the linguistic message is based on decisions within narrow frequency subbands that are processed quite independently of each other. Empirical evidence suggests that the combination of decisions from these subbands is done at some intermediate level and in such a way that the global error rate is equal to the product of error rates in the subbands. In other words, if we have two frequency bands (streams) c_1 and c_2, and each of them is respectively yielding a probability of error (error rate) $e(q_j|x^1)$ and $e(q_j|x^2)$ for a particular class q_j and an input

[2] Initially proposed as a way to assess room acoustics [16].

pattern $x = \{x^1, x^2\}$, where x^1 and x^2 represent the output features of the two frequency streams[3], the total error rate $e(q_j|x^1, x^2)$ resulting from the simultaneous use of the two streams is given by:

$$e(q_j|x^1, x^2) = e(q_j|x^1)\, e(q_j|x^2) \,. \tag{2.1}$$

Although this conclusion is often questioned by the scientific community[4], it is probably not worth arguing too long about it since it is pretty clear that (2.1) is anyway the optimal rule to obtain the best performance out of a possibly noisy multi-stream system (but it requires *perfect* knowledge about which stream, if any, is noisy). Moreover, a similar rule can usually explain some of the empirical observations in audio-visual processing (see, e.g., [24] and [20]).

Although pretty simple, rule (2.1) is not always easy to interpret (and even less for engineers!). So let us have a closer look at it. Since the probability of being correct whenever we assign a particular observation x to a class q is equal to the *a posteriori probability* $P(q|x)$ (i.e., the probability of error is equal to $1 - P(q|x)$, see [8], page 12)[5], rule (2.1) can also be written as:

$$\begin{aligned} e(q_j|x^1, x^2) &= (1 - P(q_j|x^1))(1 - P(q_j|x^2)) \\ &= 1 - \sum_{k=1}^{2} P(q_j|x^k) + \prod_{k=1}^{2} P(q_j|x^k) \,, \end{aligned} \tag{2.2}$$

where $P(q_j|x^k)$ denotes the class posterior probabilities obtained for the k-th input stream. Rewriting (2.2) in terms of (total) correct recognition probability ($P(q_j|x^1, x^2) = 1 - e(q_j|x^1, x^2)$), we have:

$$P(q_j|x^1, x^2) = \sum_{k=1}^{2} P(q_j|x^k) - \prod_{k=1}^{2} P(q_j|x^k) \,. \tag{2.3}$$

In the case of K streams, the above expression will have $2^K - 1$ terms, containing all possible stream combinations.

These expressions are quite reasonable since they also reflect a standard property of probabilities of joint events.[6] Actually, this *product of*

[3]Since we decided not to deal with the temporal constraints, this notation is oversimplified. In the case of temporal sequences, x^1 and x^2 will be sequences (possibly of different lengths and different rates) of features, and q_j will be a sequence of HMM states.

[4]Since the relevant Fletcher experiments were done (i) with nonsense syllables only, and (ii) using high-pass or low-pass filters (i.e., two streams) only, it is not clear whether or not this is an accurate statement for disparate bands in continuous speech.

[5]See Section 3 in this article for further evidence.

[6]The probability of union of two events $P(A \text{ or } B) = P(A) + P(B) - P(A, B)$, which is also equal to $P(A) + P(B) - P(A)P(B)$ if events A and B are independent. Indeed, in estimating the proportions of a sequence of trials in which A and B occur, respectively, one counts twice those trials in which both occur.

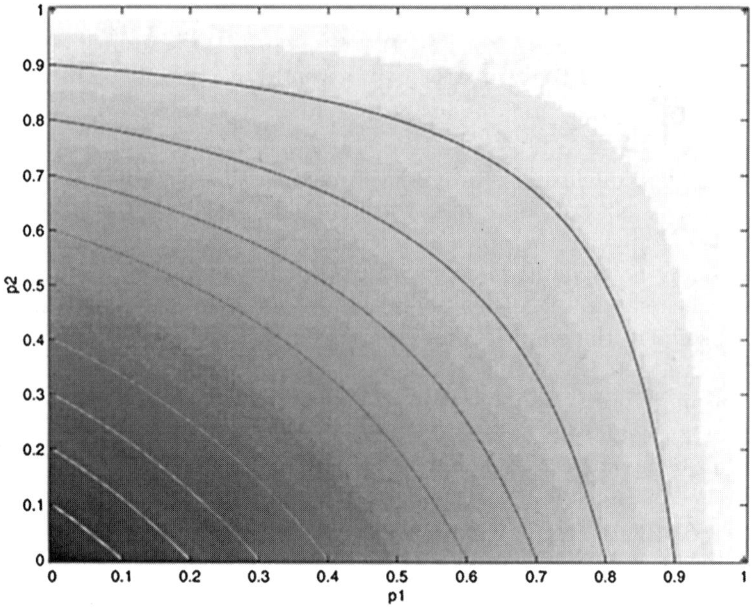

FIG. 1. *"Optimal" classification strategy based on two (independent) observation streams yielding posterior probabilities $P(q_j|x^1)$ and $P(q_j|x^2)$. The grey level represents the "total" probability of correct recognition (with white corresponding to the maximum probability), and the different curves represent the equal recognition probability curves (as a function of $P(q|x^1)$ and $P(q|x^2)$) above which the probability of correct recognition will be higher than a prescribed value.*

errors rule tells us that the probability of correct classification on human full-band hearing is equal to the probability that there is correct (human) classification in *any* subband. Consequently, this also means that human hearing seems capable of processing numerous bands and selecting the one that gives correct recognition.

The resulting (very simple but nonlinear) product of errors function is illustrated in Figure 1 for all possible values of $P(q_j|x^1)$ (horizontal axis) and $P(q_j|x^2)$ (vertical axis). From this figure, it is interesting to note how much flexibility an "optimal" multi-stream system potentially has in keeping the (total) probability of correct recognition above a certain threshold, even if one of the streams is extremely noisy (and yields high error rates). This can indeed be measured by the area above a given equal recognition rate curve. For example, for $P(q_j|x^1, x^2) = 0.9$, nearly one third of the space is available! It is clear that this property cannot be achieved by using the usual product of likelihoods, where if one of the likelihoods is poorly estimated, the whole product deteriorates.[7]

[7]In addition to the fact that it is usually difficult to compare/combine likelihoods

This conclusion remains valid for more than two streams. Actually, it can even be shown that the area above a given (multi-dimensional) equal error rate surface grows exponentially with the number of streams. To make the link easier with what follows in the remainder of this article, observe that, in the case of three input streams, (2.3) becomes:

$$P(q_j|x^1,x^2,x^3) = \sum_{k=1}^{3} P(q_j|x^k) + \prod_{k=1}^{3} P(q_j|x^k) - \sum_{\ell>k=1}^{3} P(q_j|x^k)P(q_j|x^\ell) \, . \tag{2.4}$$

Obviously, this reflects a "perfect" world. In actual engineering systems though the posterior probabilities $P(q_j|x^k)$ will have to be estimated on the basis of a set of parameters Θ, and, in the case of two streams, (2.3) should be written:

$$P(q_j|x^1,x^2,\Theta) = \sum_{k=1}^{2} P(q_j|x^k,\Theta) - \prod_{k=1}^{2} P(q_j|x^k,\Theta) \, . \tag{2.5}$$

Figure 1 does not change, but the position in the space depends on Θ, as well as on the different stream features. Ideally, robust training and adaptation should be performed in the Θ space to guarantee that $P(q_j|x^1,x^2,\Theta)$ is always above a certain threshold, or to directly maximize (2.5). In the following, we discuss approaches going in that direction.

2.2. Discussion. The above analysis allows us to draw a few conclusions and to design the features of an "optimal" ASR system:

1. Human hearing performs combination of frequency streams according to the product of errors rule discussed above. In this case (and assuming that the subbands are independent, which is false), correct classification of any subband is empirically equivalent to correct full-band classification. *In subband-based ASR systems, this means that we should design the system and the training criterion to maximize the classification performance on subbands, while also making sure that the subband errors are independent.*
2. As a direct consequence of the above, it is also obvious that the more subbands we use, the higher the full-band correct classification rate will be. As done in human hearing, *ASR systems should thus use a large number of subbands* to have a better chance to increase recognition rates. It is interesting to note here that this trend has recently been followed in [15].

---

computed from features in different spaces, possibly of different dimensions (since likelihoods, as usually computed (assuming Gaussian densities with diagonal covariance matrices), are "dimensional", i.e., dependent on the dimension of the feature space).

3. In order to estimate the reliability of each stream, *ASR systems should be able to estimate subband posteriors as accurately as possible.* We will show in the next section that this is not impossible.
4. If ASR systems can reliably estimate local posteriors, we can implement the product of errors rule, which should guarantee the minimum of errors (if the above conditions are satisfied). Furthermore, each time we improve the classification rate in *any* subband, the recognition rate should improve.

3. Estimating posteriors. *The purpose of models is not to fit the data but to sharpen the questions.* (Samuel Karlin, 1923-, 11th R.A. Fisher Memorial Lecture, Royal Society, 20 April 1983.)

From the discussion above, it seems clear that we should work on the basis of a posteriori probabilities[8]. Given that we often work in the framework of hybrid HMM/ANN systems [5] (using artificial neural networks (ANN) for estimating local posterior probabilities which are transformed into scaled likelihoods used as HMM emission probabilities), and although some of the arguments below will also be valid for likelihood-based systems, we will focus our discussion on posteriors.

As initially reported in [5], Figure 2 illustrates the fact that an ANN can reliably estimate local posterior probabilities $P(q_j|x)$. Indeed, recalling the properties of posterior probabilities discussed in the previous section, good estimates of posterior probabilities should also be a measure of the fraction of correct classification. Consequently, when representing the correct classification rate as a function of the posterior probabilities as estimated at the output of a neural network, the ideal Bayes (posterior-based) classifier would yield a diagonal, which is quite the case for both the training data and the cross-validation data (not used for training, but for which correct classification was known).

Dividing these local posterior probabilities by the prior probabilities $P(q_j)$ as estimated on the training set, yields scaled local likelihoods that can be used to compute [12]

$$\frac{P(M|X)}{P(M)} = \frac{P(X|M)}{P(X)}, \tag{3.1}$$

where M represents a complete HMM (modeling a specific subword unit, a word, or a sentence) composed of several units computing $\frac{p(q_j|x)}{p(q_j)}$, and X an observation sequence associated with M. This can then be simply multiplied (as in usual HMMs) by $P(M)$ to include external knowledge sources (such as a language model).

4. Multi-stream and mixture of experts. From an engineering perspective, one way to introduce the multi-stream formalism in a pattern

[8]Which are known, anyway, to yield the minimum error rate solution.

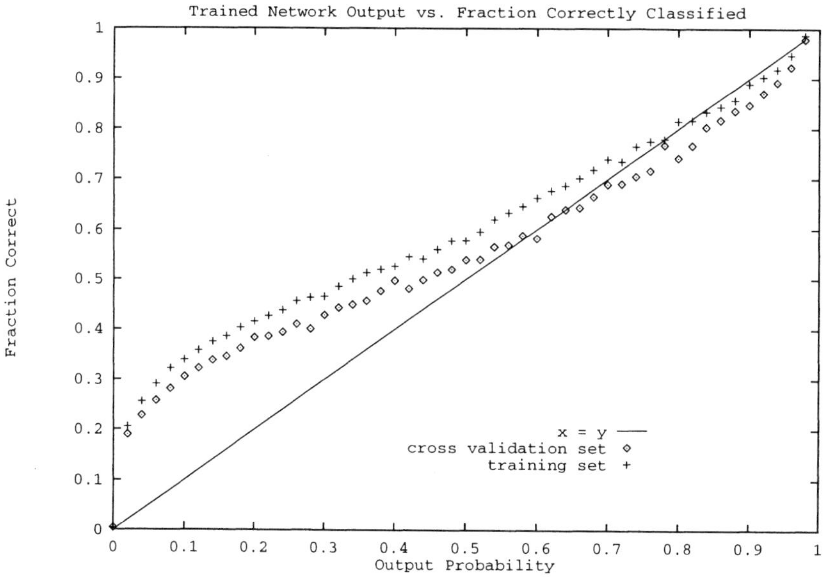

FIG. 2. *It is possible to generate "good" posterior probabilities out of a neural network, and these are indeed good measures of the probability of being correct. This plot was generated on real speech data by collecting statistics over the acoustic parameters from 1750 Resource Management speaker-independent training sentences and 500 cross-validation sentences (not used for training, but for which correct classification was known).*

classification task (such as ASR) is to use the approach of mixture of experts, as proposed in the framework of neural networks [4]. The general idea of a mixture of experts is to process the (same) input x according to different linear or nonlinear (neural network) functions ("experts"), and to combine the outputs of each expert according to a weighted sum, and where the weights also result from a (linear or nonlinear) function of the input pattern x.

Typically, this approach (as for HMMs) can be formulated in terms of latent variables, where the missing variable is the set of experts who are reliable at each step. As illustrated in Figure 3, let M represent the hypothesized model (HMM) associated with an input sequence X. If $\mathcal{E} = \{E_1, \ldots, E_k, \ldots, E_K\}$ represents a set of *mutually exclusive and exhaustive* experts[9] (and where $P(E_k)$ is defined as the probability that E_k is the most reliable expert), then $P(M|X)$ can be estimated as:

[9]As discussed later, the initial multi-stream approach (Section 5.1) was not using strictly exhaustive experts since they did not cover all possible stream combinations. The full combination approach, as discussed in Section 5.3, will actually use all possible combinations.

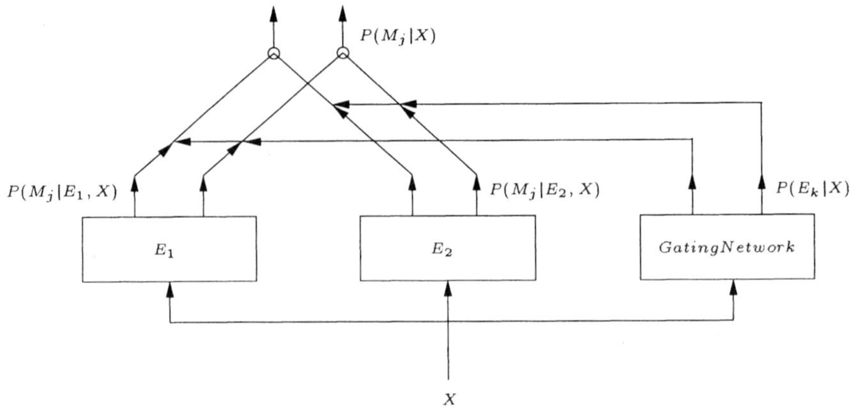

FIG. 3. *Posterior-based mixture of experts. Experts (e.g., neural networks) are extracting their own posterior estimates, which are then combined through weights also estimated (by the "gating network") from the data. These weights could also be adapted online.*

$$
\begin{aligned}
P(M|X) &= \sum_{k=1}^{K} P(M, E_k|X) \\
&= \sum_{k=1}^{K} P(M|E_k, X) P(E_k|X) \\
&\simeq \sum_{k=1}^{K} P(M^k|X^k) P(E_k|X)
\end{aligned}
\tag{4.1}
$$

where X^k represents the respective inputs of expert/function E_k [10], M^k the model for the speech unit M used to process X^k, and $P(E_k|X)$ the (relative) reliability of expert E_k given the whole input.[11] The approximations in (4.1) result from the assumptions that (i) the probability of a model M given a particular expert E_k is only estimated from the submodel M^k associated with the expert, and (ii) that expert-specific model is only looking at its specific input features. As briefly recalled in Section 3, segment-based posteriors in (4.1) can easily be estimated when using artificial neural networks.

Ideally, as discussed in [6, 14, 29] and illustrated in Figure 4, the expert combination presented above should take place at the level of M, i.e., at the level of the particular (non-emitting) states denoted "$\otimes$". However, this is not trivial and will often require a significant adaptation of the recognizer.

[10] In the case of multi-stream inputs, X^k will typically be a subset of X (containing the features relative to E_k).

[11] Since, as illustrated in Figure 4, each sequence X^k will be processed with a different/specific HMM.

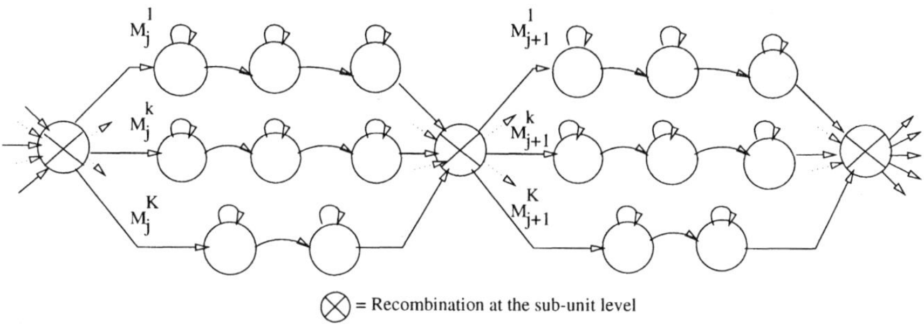

FIG. 4. *General form of a K-streams recognizer with anchor points between speech units (to force synchrony between the different streams). Note that the model topology is not necessarily the same for the different sub-systems.*

It is only in the case of segment likelihoods combination (by products) that one can develop a tractable solution to this optimization problem. In this case, it can be shown that the product of segment-based, expert-specific, likelihoods can be expressed as local likelihood products of an equivalent 1st order HMM. In this case, some adaptation of the transition probabilities [32] may be required. This algorithm, referred to as "HMM combination", is then implemented as a straightforward adaptation of the HMM decomposition algorithm presented in [30].

In the case of more complex combination criteria, as in the case of a mixture of experts or the approach discussed is Section 5 (related to the mixture of experts model and the psycho-acoustic evidence discussed in Section 2), HMM combination/decomposition is no longer a tractable solution. In this case, other approaches have to be used, e.g., using a two level dynamic programming algorithm or using (4.1) to rescore an N-best list of hypotheses (providing us with a set of possible segmentation/anchor points).

Although it is clear that:

1. The empirical results discussed in Section 2 were obtained on the basis of segments (non-sense syllables),
2. only the segment level combination can allow for asynchrony between the streams[12],

we will mainly focus, in the remainder of this paper, on the combination at the state level.

5. Multiband-based ASR with latent variables.

5.1. General formalism. As a particular case of multi-stream processing, we have been investigating an ASR approach based on independent

[12]Although not using the nonlinear (optimal?) combination functions discussed in this paper, preliminary results presented in [6, 14] suggested that asynchrony was not a major factor — see, though, [21] and [29] for further discussion about this.

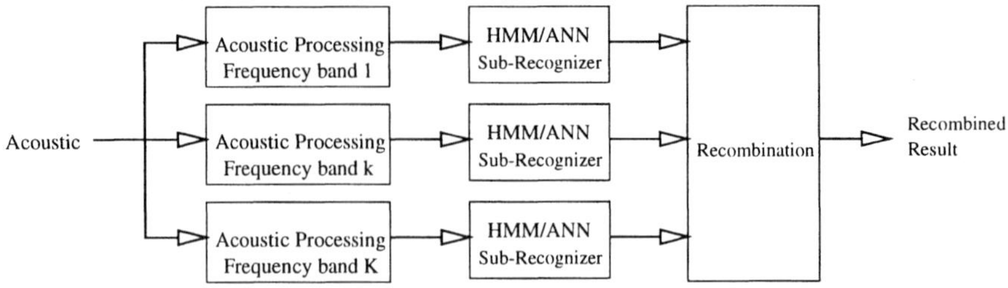

FIG. 5. *Typical multiband-based ASR architecture. In multiband speech recognition, the frequency range is split into several bands, and information in the bands is used for phonetic probability estimation by independent modules. These probabilities are then combined for recognition later in the process at some segmental level.*

processing and combination of frequency subbands. The general idea, as illustrated in Fig. 5, is to split the whole frequency band (represented in terms of critical bands) into a few subbands on which different recognizers are independently applied. The resulting probabilities are then combined for recognition later in the process at some segmental level (here we consider the state level). Starting from critical bands, acoustic processing is now performed independently for each frequency band, yielding K input streams, each being associated with a particular frequency band.

In this case, each of the K sub-recognizers (streams) uses the information contained in a specific frequency band $X^k = \{x_1^k, x_2^k, \ldots, x_n^k, \ldots, x_N^k\}$, where each x_n^k represents the acoustic (spectral) vector at time n in the k-th stream. In (4.1), $P(M^k|X^k)$ represents the a posteriori probability of a sub-model M^k (k-th frequency band model for M) and can be estimated from local posteriors $P(q_j^k|x_n^k)$ (e.g., estimated at the output of an ANN), where q_j^k denotes a state j of model M^k). $P(E_k|X)$ represents the "reliability" of expert E_k, working on the k-th frequency band, and can be estimated in different ways (e.g., based on SNR).

As discussed in the previous section, combination at the segment level according to the criteria discussed here is not easy. However, combination at the HMM-state level, by combining local posteriors $P(q_j^k|x_n^k)$, can be done in many ways [6], including untrained or trained linear functions (e.g., as a function of automatically estimated local SNR), as well as trained nonlinear functions (e.g., by using neural networks). This is quite simple to implement and amounts to performing a standard Viterbi decoding in which local (log) probabilities are obtained from a linear or nonlinear combination of the local subband probabilities. For example, in our initial subband-based ASR, local posteriors $P(q_j|x_n)$ (or scaled likelihoods) were estimated according to:

$$P(q_j|x_n) = \sum_{k=1}^{K} w_k P(q_j|x_n^k, \Theta_k) \tag{5.1}$$

where, in our case, each $P(q_j|x_n^k, \Theta_k)$ is computed with a band-specific ANN of parameters Θ_k and with x_n^k (possibly with temporal context) at its input. The weighting factors can be assigned a uniform distribution (already performing very well [6]) or be proportional to the estimated SNR. Over the last few years, several results were reported showing that such a simple approach was usually quite robust to band limited noise.

In Section 5.3 below, we discuss a new approach that was recently developed at IDIAP, and presented in [3, 11, 23], and show (i) how it significantly enhances the baseline multiband approach, and (ii) how it relates to the above discussions (and psycho-acoustic evidence).

5.2. Motivations and drawbacks. The multiband approach has several potential advantages, which are briefly discussed here.

Better robustness to band-limited noise — The speech signal may be impaired (e.g., by noise, stream characteristics, reverberation,...) only in some specific frequency bands. When recognition is based on several independent decisions from different frequency subbands, the decoding of a linguistic message need not be severely impaired, as long as the remaining clean subbands supply sufficiently reliable information. This was confirmed by several experiments (see, e.g., [6]). Surprisingly, even when the combination is simply performed at the HMM state level, it is observed that the multiband approach yields better performance and noise robustness than a regular full-band system.[13]

Similar conclusions also hold in the framework of the missing feature theory [19, 22]. In this case, it was shown that, *if one knows the position of the noisy features*, significantly better classification performance could be achieved by disregarding the noisy data (using marginal distributions) or by integrating over all possible values of the missing data conditionally on the clean features — See Section 5.3 for further discussion about this. In the multiband approach, we do not try to explicitly identify the noisy band (and to disregard it). Instead, we process all the subbands independently (to avoid "spreading" the noise across all components of the feature vector or in the local probability estimate) and recombine them according to a particular weighting scheme that should de-emphasize (or cancel out) the noisy bands.

Better modeling — As for a regular full-band system, it was shown in [6] that all-pole modeling significantly improved the performance of multiband

[13]It could however be argued that, in this case, the multiband approach boils down to a regular full-band recognizer in which several likelihoods of (assumed) independent features are estimated and multiplied together to yield local likelihoods (since, in likelihood based systems, expected values for the full-band is the same than the concatenated expected values of subbands). This is however not true when using posterior based systems (such as hybrid HMM/ANN systems) where the subbands are presented to different nets that are independently trained in a discriminant way on each individual subband. Finally, as discussed in this paper, we also believe that the combination criterion should be different than a simple product of (scaled) likelihoods or posteriors.

systems. However, as an additional advantage of the subband approach, it can be shown or argued that:

1. This all-pole modeling may be more robust if performed on several subbands (low dimensional spaces) than on the full-band signal [27].
2. Since the dimension of each (subband) feature space is smaller, it is easier to estimate reliable statistics (resulting in a more robust parameterization).

Stream asynchrony — Transitions between more or less stationary segments of speech (corresponding to an HMM state) do not necessarily occur at the same time for different frequency bands [21], which makes the piecewise stationary assumption more fragile for HMMs. The subband approach may have the potential of relaxing the synchrony constraint inherent in current HMM systems.

Stream specific processing and modeling — Different recognition strategies might ultimately be applied in different subbands. For example, different time/frequency resolution tradeoffs could be chosen (e.g., time resolution and width of analysis window depending on the frequency subband). Finally, some subbands may be inherently better for certain classes of speech sounds than others.

Major objections and drawbacks — There are a few, related, drawbacks to this multiband approach [21]:

1. One of the common objections to this separate modeling of each frequency band has been that important information in the form of correlation between bands may be lost. Although this may be true, several studies [21], as well as the good recognition rates achieved on small frequency bands [10, 15], tend to show that most of the phonetic information is contained in each frequency band (possibly provided that we have enough temporal information)[14].
2. To define and independently process frequency bands, it is obviously necessary to start from spectral coefficients (critical bands), which, however, are not orthogonal and do not permit competitive performance for clean speech. In standard ASR systems, these coefficient are typically orthogonalized using a DCT (cepstral) transformation. Even in the case of ANN probability estimation (where ANN is supposed to extract and model the correlation across coefficients), it has been observed that orthogonalization of the features still helped a bit. However, in the case of narrowband additive noise, we obviously want to subtract as much as possible of the noise before the DCT transform to avoid spreading the noise across

[14]And, indeed, the discussion in Section 2, as well as many other psycho-acoustic experiments, seem to suggest that human hearing can actually extract a lot of phonetic/syllabic information from band limited signals.

all the feature components. For subband ASR systems, a partial but effective solution to this problem consists in performing an independent DCT in each subband [6, 26].
Alternative solutions to this problem have recently been proposed in which an attempt is made to decorrelate as much as possible the (temporally) successive output energies of each individual filter-bank — see, e.g., [18, 7, 25]. This is usually obtained by performing some kind of temporal filtering (and, consequently, spreading the possible noise over time instead of over frequency) or frequency filtering (and consequently spreading the possible noise over a limited frequency range only).

3. As opposed to the empirical evidence discussed in Section 2, the initial subband-based ASR system presented in [6] did not make use of all possible subband combinations. This will be fixed by the method presented next.

5.3. Full combination subband ASR. Following the developments and discussions above, it seems reasonable to assume that a subband ASR system should simultaneously deal with all the $L = 2^K$ possible subband combinations S^ℓ (with $\ell = 1, \ldots, L$, including the empty set[15]) resulting from an initial set of K frequency (critical) bands x^k. However, while it is pretty easy to quickly estimate any subband likelihood or marginal distribution when working with Gaussian or multi-Gaussian densities [19], this is harder when using ANN to estimate posterior probabilities. In this latter case, indeed, it would be necessary to train (and run, during recognition) 2^K neural networks, which would become very quickly intractable.

In the following, we briefly present the solution recently proposed in [11] and [23], and discuss its relationships with the themes developed in the current paper.

Ideally, we would thus like to compute the posterior probabilities for each of the $L = 2^K$ possible combinations S_n^ℓ (including all possible single bands, pairs of bands, triples, etc) of the K subbands x_n^k. Indeed, since we do not know a priori where the noise is located, we should integrate over all possible positions[16]. Using the formalism of mixture of experts, we can thus write:

$$P(q_j|x_n, \Theta) = \sum_{\ell=1}^{L} P(q_j, E_\ell|x_n, \Theta)$$

[15]Which would correspond to the case where all the bands are unreliable. In this case, the best posterior estimate is the prior probability $P(q_j)$, and one of the L terms in the following equations will contain only this prior information.

[16]This amounts to assuming that the position of the noise or, in other words, the position of the reliable frequency bands, is a hidden (latent) variable on which we will integrate to maximize the posterior probabilities (in the spirit of the EM algorithm).

$$
\begin{aligned}
&= \sum_{\ell=1}^{L} P(q_j|E_\ell, x_n, \Theta)P(E_\ell|x_n) \\
&= \sum_{\ell=1}^{L} P(q_j|S_n^\ell, \Theta_\ell)P(E_\ell|x_n)
\end{aligned} \tag{5.2}
$$

where Θ represents the whole parameter space, while Θ_ℓ denotes the set of (ANN) parameters used to compute the subband posteriors. Of course, implementation of (5.2) requires the training of L neural networks to estimate all the posteriors $P(q_j|S_n^\ell, \Theta_\ell)$ that have to be combined according to a weighted sum, with each weight representing the relative reliability of a specific set of subbands. In the case of stationary interference, this reliability could be estimated on the basis of the average (local) SNR in the considered set. Alternatively, it could also be estimated as the probability that the local SNR is above a certain threshold, and where the threshold has been estimated to guarantee a prescribed recognition rate (e.g., lying above a certain equal recognition rate curve in Figure 1) [3].

Typically, training of the L neural nets would be done once and for all on clean data, and the recognizer would then be *adapted* online simply by adjusting the weights $P(E_\ell|x_n)$ (still representing a limited set of L weights) to increase the global posteriors. This adaptation could be performed by online estimation of the SNR or by an online version of the EM (deleted-interpolation) algorithm. Although this approach is not really tractable, it has the advantage of avoiding the independence assumption between the subbands of a same set, as well as allowing any DCT transformation of the combination before further processing. Consequently, this combination, referred to as *Full Combination*, was actually implemented [10] for the case of four frequency subbands (each containing several critical bands), thus requiring the training of 16 neural nets, and used as an "optimal" reference point.

An interesting approximation to this "optimal" solution though consists in simply train one neural network per subband for a total of K models, and to approximate all the other subband combination probabilities directly from these. In other words, re-introducing the independence assumption[17] between subbands, subband combination posteriors would be estimated as [10, 11]:

$$
P(q_j|S_n^\ell, \Theta_\ell) = P(q_j) \prod_{k \in S^\ell} \frac{P(q_j|x_n^k, \Theta_k)}{P(q_j)} \tag{5.3}
$$

Experimental results obtained from this approximated Full Combination approach in different noisy conditions are reported in [10, 11], where

[17] Actually, it is shown in [10, 11] that we only need to introduce a weak (conditional) independence assumption.

the performance of this approximation above was also compared to the "optimal" estimators (5.2). Interestingly, it was shown that this independence assumption did not hurt us much and that the resulting recognition performance[18] was similar to the performance obtained by training and recombining all possible L nets (and significantly better than the original single subband approach). In both cases, the recognition rate and the robustness to noise were greatly improved compared to the initial subband approach (5.1). This further confirms that we do not seem to lose "critically" important information when neglecting the correlation between bands.

Finally, it is particularly interesting to note here that using (5.3) in (5.2) yields something very similar to the "optimal" product of errors rule (2.4) observed empirically:

$$P(q_j|x_n,\Theta) = \sum_{\ell=1}^{L} \frac{P(E_\ell|x_n)}{C_\ell} \prod_{k\in S^\ell} P(q_j|x_n^k,\Theta_k) \tag{5.4}$$

with $C_\ell = P(q_j)^{(n_\ell - 1)}$, and n_ℓ being the number of subbands in S^ℓ. In [10], it is shown that this normalization factor is important to achieve good performance. This Full Combination rule thus takes exactly the same form as the product of errors rule [such as (2.4) or (2.5)], apart from the fact that the weighting factors are different. In (5.4), the weighting factors can be interpreted as (scaled) probabilities estimating the relative reliability of each combination, while in the product of errors rule these are simply equal to $+1$ or -1. Another difference is that the product of errors rule involves $2^K - 1$ terms while the Full Combination rule involves 2^K terms, one of them representing the contribution of the prior probability.

In the next section, we discuss a further extension of this approach where the segmentation into subbands is no longer done explicitly, but is achieved dynamically over time, and where the integration over all possible frequency segmentations is part of the same formalism.

6. HMM2: Mixture of HMMs. All HMM emission probabilities discussed in the previous models are typically modeled through Gaussian mixtures or artificial neural networks. Also, in the multiband based recognizers discussed above, we have to decide *a priori* the number and position of the subbands being considered. As also briefly discussed above, it is not always clear what the "optimal" recombination criterion should be. In the following, we introduce a new approach, referred to as HMM2, where the emission probabilities of the HMM (now referred to as "temporal HMMs") are estimated through a secondary, state-dependent, HMM (referred to as "feature HMMs") specifically working along the feature vector. As briefly discussed below (see references such as [2] and [31] for further detail), this

[18]Obtained on the Numbers'95 database, containing telephone-based speaker independent free format numbers, on which NOISEX noise was added.

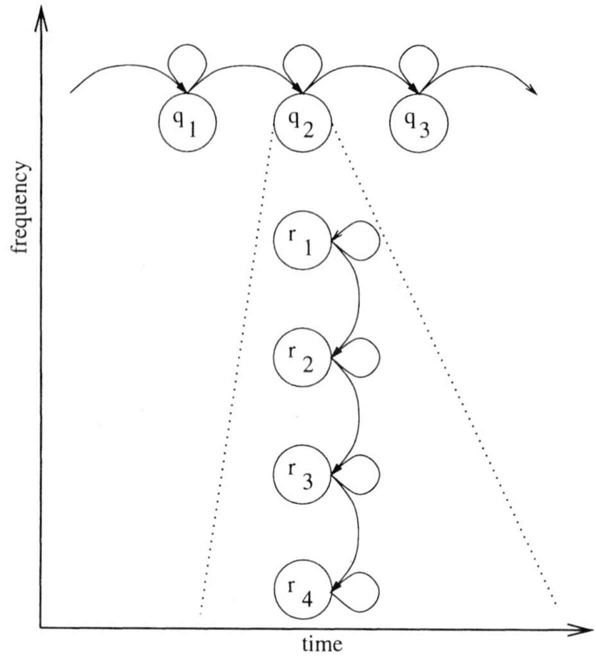

FIG. 6. *HMM2: the emission distributions of the temporal HMM are estimated by secondary, state-specific, feature HMMs.*

model will then allow for dynamic (time and state dependent) subband (frequency) segmentation as well as "optimal" recombination according to a standard maximum likelihood criterion (although other criteria used in standard HMMs could also be used).

In HMM2, as illustrated in Figure 6, each temporal feature vector x_n is considered as a fixed length sequence of S components $x_n = (x_n^1, .., x_n^S)$, which is supposed to have been generated at time n by a specific feature HMM associated with a specific state q_j of the temporal HMM. Each feature HMM state r_l is thus emitting individual feature components x_n^s, whose distributions are modeled by, e.g., one dimensional Gaussian mixtures. The feature HMM thus looks at all possible subband segmentations and automatically performs the combination of the likelihoods to yield a single emission probability. The resulting emission probability can then be used as emission probability of the temporal HMM. As an alternative, we can also use the resulting feature segmentation in multiband systems or as additional acoustic features features in a standard ASR system. Indeed, if HMM2 is applied to the spectral domain, it is expected that the feature HMM will "segment" the feature vector into piecewise stationary spectral regions, which could thus follow spectral peaks (formants) and/or spectral valley regions.

In the example illustrated in Figure 6, the HMM2 is composed of a temporal HMM that handles sequences of features through time, and feature HMMs assigned to the different temporal HMM states. The temporal HMM is composed of 3 left-to-right connected states (q_1, q_2 and q_3), while the state-specific feature HMM is composed of 4 ("top-down") states (r_1, r_2 r_3 and r_4). Although not reflected in Figure 6, each feature HMM $\{r_1, r_2, r_3, r_4\}$ is specific to a temporal HMM state (emission probability distribution), with different parameters, and possibly different HMM topologies. More formally, as done in [2] and [17], the feature state should have been denoted r_j^k, with k representing the associated temporal state index and j the feature state index.

Of course, the topology of the feature HMM, extracting the correlation information within feature vectors, could take many forms, including ergodic HMMs and/or topologies with a number of states larger the number of feature components, in which case "high-order" correlation information could be modeled. In the following though, we constrained the feature HMM to a strictly "top-down" topology. Moreover, since we were interested in extracting information in the spectral domain and in possible relationships with multiband ASR systems, we considered features in the spectral domain. Each of the feature HMM states is then supposed to model one of the K frequency bands, where the positions and bandwidths of these bands are determined dynamically.

In [2], we introduced an EM algorithm to jointly train all the parameters of such an HMM2 in order to maximize the data likelihood. This derivation is based on the fact that an HMM is a special kind of mixture of distributions, and therefore HMM2, as a mixture of HMMs, can be considered as a more general kind of mixture distribution. During decoding, the Viterbi algorithm is used to find the path through the HMM2 which best explains the input data. Local state likelihoods of the temporal HMM can however be estimated using either Viterbi or the complete likelihood calculation, summing over all possible paths through the feature HMM:

$$p(x_n|q_j) = \sum_R P(r_0|q_j) \prod_{s=1}^{S} p(x_n^s|r_l, q_j) P(r_l|r_{l-1}, q_j) \tag{6.1}$$

where q_j is the temporal HMM state at time n, r_l the feature HMM state at feature s, R the set of all possible paths through the feature HMM, $P(r_0|q_t)$ the initial state probability of the feature HMM, $p(x_n^s|r_l, q_j)$ the probability of emitting feature component x_n^s while in feature HMM state r_l of temporal state q_j, and $P(r_l|r_{l-1}, q_j)$ the transition probabilities of the feature HMM in temporal state q_j.

We believe that HMM2 (which includes the classical mixture of Gaussian HMMs as a particular case) has several potential advantages, including:

1. Better feature correlation modeling through the feature HMM topology (e.g., working in the frequency domain). Also, the com-

plexity of this topology and the probability density function associated with each state easily control the number of parameters.

2. Automatic non-linear spectral warping. In the same way the conventional HMM does time warping and time integration, the feature-based HMM performs frequency warping and frequency integration.
3. Dynamic formant trajectory modeling. As further discussed below, the HMM2 structure has the potential to extract some relevant formant structure information, which is often considered as important to robust speech recognition.

To illustrate these advantages and the relationship of HMM2 with dynamic multi-band ASR, we trained all parameters of an HMM2, using frequency filtered filterbank features [25]. We employed the HMM2 topology as shown in Figure 6. Training was done with the EM algorithm, and decoding was performed using the Viterbi algorithm for both the temporal and the frequency HMM. Figure 7 illustrates (on unseen test data) the temporal and frequency segmentation obtained as a by-product from Viterbi, plotted onto a spectrogram of our features. At each time step, we kept the 3 positions where the feature HMM changed its state during decoding (for instance, at the first time frame, the feature HMM goes from state r_1 to state r_2 after the second feature). We believe that this segmentation gives cues about some structures of the speech signal such as formant positions. In fact, in [31] it has been shown that this segmentation information can be used as (additional) features for speech recognition, being (1) discriminant and (2) rather robust in the case of speech degraded by additive noise.

7. Conclusions. In this paper, we have discussed a family of new ASR approaches that have recently been shown to be more robust to noise, without requiring specific adaptation or "multi-style" training.

From all this discussion, and the convergence of independent experiments, we can draw the following preliminary conclusions:

1. Multiband ASR does not seem to be inherently inferior to a full-band approach, although some correlation information is lost due to the division of the frequency space into subbands.[19] Furthermore, it is not clear either that human hearing uses this kind of correlation information.
2. When training subband systems, we should not aim at maximizing the classification performance for every subband. When using the right combination rule, it is better to increase the number of subbands while making sure that at any time at least one subband will be guessing the right answer.[20]

[19]Probably because the advantages of subband based ASR can outweight the slight problem due to independent processing of subbands.

[20]This conclusion is very similar to what is proved mathematically in [4], p. 369, para. 1 (also p. 424).

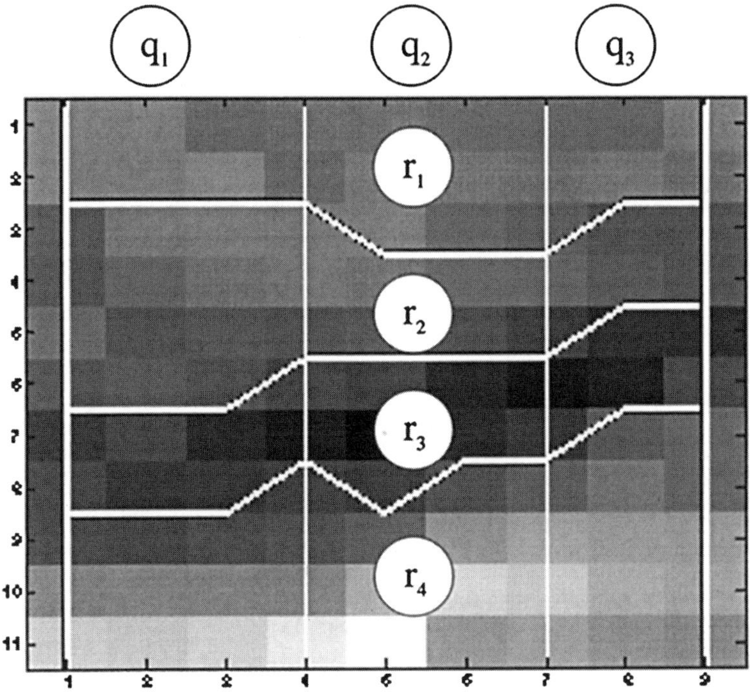

FIG. 7. *Frequency filtered filterbanks and HMM2 resulting (Viterbi) segmentation for a test example of phoneme "w".*

3. Doing this, we should also look at the potential for improvement in subband modeling when combining longer time-scale information streams (trading frequency information for temporal information).
4. The full combination approach discussed here has the potential of providing us with new adaptation schemes in which only the combination weights are automatically adapted (e.g., according to an online EM algorithm).

Finally, it is clear that several key problems remain to be addressed, including:

1. Need for improved expert weighting
2. Need for methods which are robust to noise but still perform well for clean speech.

In subband processing, there is also a need to properly choose the frequency subband, and it is expected that those subbands should be dynamically defined, e.g, following some formant structure. In this respect, the HMM2 formalism also presented here can be considered as a generalization of subband approaches, allowing for optimal (according to a maximum likelihood criterion) subband segmentation and recombination.

Acknowledgments. The content and themes discussed in this paper largely benefited from the collaboration with our colleagues Andrew Morris and Astrid Hagen. This work was partly supported by the Swiss Federal Office for Education and Science (FOES) through the European SPHEAR (TMR, Training and Mobility of Researchers) and RESPITE (ESPRIT Long term Research) projects. Additionally, Katrin Weber is supported by a the Swiss National Science Foundation project MULTICHAN.

REFERENCES

[1] ALLEN J., "How do humans process and recognize speech?," *IEEE Trans. on Speech and Audio Processing*, Vol. **2**, no. 4, pp. 567–577, 1994.

[2] BENGIO S., BOURLARD H., AND WEBER K., "An EM Algorithm for HMMs with Emission Distributions Represented by HMMs," *IDIAP Research Report*, IDIAP-RR-00-11, 2000.

[3] BERTHOMMIER F. AND GLOTIN H., "A new SNR-feature mapping for robust multi-stream speech recognition," *Intl. Conf. of Phonetic Sciences (ICPhS'99)* (San Francisco), to appear, August 1999.

[4] BISHOP C.M., *Neural Networks for Pattern Recognition*, Clarendon Press (Oxford), 1995.

[5] BOURLARD H. AND MORGAN N., *Connectionist Speech Recognition – A Hybrid Approach*, Kluwer Academic Publishers, 1994.

[6] BOURLARD H. AND DUPONT S., "A new ASR approach based on independent processing and combination of partial frequency bands," *Proc. of Intl. Conf. on Spoken Language Processing* (Philadelphia), pp. 422–425, October 1996.

[7] DE VETH J., DE WET F., CRANEN B., AND BOVES L., "Missing feature theory in ASR: make sure you miss the right type of features," *Proceedings of the ESCA Workshop on Robust Speech Recognition* (Tampere, Finland), May 25–26, 1999.

[8] DUDA R.O. AND HART P.E., *Pattern Classification and Scene Analysis*, John Wiley, 1973.

[9] GREENBERG S., "On the origins of speech intelligibility in the real world," *Proc. of the ESCA Workshop on Robust Speech Recognition for Unknown Communication Channels*, pp. 23–32, ESCA, April 1997.

[10] HAGEN A., MORRIS A., AND BOURLARD H., "Subband-based speech recognition in noisy conditions: The full combination approach," *IDIAP Research Report no. IDIAP-RR-98-15*, 1998.

[11] HAGEN A., MORRIS A., AND BOURLARD H., "Different weighting schemes in the full combination subbands approach for noise robust ASR," *Proceedings of the Workshop on Robust Methods for Speech Recognition in Adverse Conditions* (Tampere, Finland), May 25–26, 1999.

[12] HENNEBERT J., RIS C., BOURLARD H., RENALS S., AND MORGAN N. (1997), "Estimation of Global Posteriors and Forward-Backward Training of Hybrid Systems," *Proceedings of EUROSPEECH'97* (Rhodes, Greece, Sep. 1997), pp. 1951–1954.

[13] HERMANSKY H. AND MORGAN N., "RASTA processing of speech," *IEEE Trans. on Speech and Audio Processing*, Vol. **2**, no. 4, pp. 578–589, October 1994.

[14] HERMANSKY H., PAVEL M., AND TRIBEWALA S., "Towards ASR using partially corrupted speech," *Proc. of Intl. Conf. on Spoken Language Processing* (Philadelphia), pp. 458–461, October 1996.

[15] HERMANSKY H. AND SHARMA S., "Temporal patterns (TRAPS) in ASR noisy speech," *Proc. of the IEEE Intl. Conf. on Acoustics, Speech, and Signal Processing* (Phoenix, AZ), pp. 289–292, March 1999.

[16] HOUTGAST T. AND STEENEKEN H.J.M., "A review of the MTF concept in room acoustics and its use for estimating speech intelligibility in auditoria," *J. Acoust. Soc. Am.*, Vol. **77**, no. 3, pp. 1069–1077, March 1985.

[17] IKBAL S., BOURLARD H., BENGIO S., AND WEBER K., "IDIAP HMM/HMM2 System: Theoretical Basis and Software Specifications" *IDIAP Research Report*, IDIAP-RR-01-27, 2001.

[18] KINGSBURY B., MORGAN N., AND GREENBERG S., "Robust speech recognition using the modulation spectrogram," *Speech Communication*, Vol. **25**, nos. 1–3, pp. 117–132, 1998.

[19] LIPPMANN R.P. AND CARLSON B.A., "Using missing feature theory to actively select features for robust speech recognition with interruptions, filtering and noise," *Proc. Eurospeech'97* (Rhodes, Greece, September 1997), pp. KN37-40.

[20] MCGURK H. AND MCDONALD J., "Hearing lips and seeing voices," *Nature*, no. 264, pp.746–748, 1976.

[21] MIRGHAFORI N. AND MORGAN N., "Transmissions and transitions: A study of two common assumptions in multi-band ASR," *Intl. IEEE Conf. on Acoustics, Speech, and Signal Processing* (Seattle, WA, May 1997), pp. 713–716.

[22] MORRIS A.C., COOKE M.P., AND GREEN P.D., "Some solutions to the missing features problem in data classification, with application to noise robust ASR," *Proc. Intl. Conf on Acoustics, Speech, and Signal Processing*, pp. 737–740, 1998.

[23] MORRIS A.C., HAGEN A., AND BOURLARD H., "The full combination subbands approach to noise robust HMM/ANN-based ASR," *Proc. of Eurospeech'99* (Budapest, Sep. 99), to appear.

[24] MOORE B.C.J., *An Introduction to the Psychology of Hearing* (4th edition), Academic Press, 1997.

[25] NADEU C., HERNANDO J., AND GORRICHO M., "On the decorrelation of filter-bank energies in speech recognition," *Proc. of Eurospeech'95* (Madrid, Spain), pp. 1381–1384, 1995.

[26] OKAWA S., BOCCHIERI E., AND POTAMIANOS A., "Multi-band speech recognition in noisy environment," *Proc. IEEE Intl. Conf. on Acoustics, Speech, and Signal Processing*, 1998.

[27] RAO S. AND PEARLMAN W.A., "Analysis of linear prediction, coding, and spectral estimation from subbands," *IEEE Trans. on Information Theory*, Vol. **42**, pp. 1160–1178, July 1996.

[28] TOMLINSON J., RUSSEL M.J., AND BROOKE N.M., "Integrating audio and visual information to provide highly robust speech recognition," *Proc. of IEEE Intl. Conf. on Acoustics, Speech, and Signal Processing* (Atlanta), May 1996.

[29] TOMLINSON M.J., RUSSEL M.J., MOORE R.K., BUCKLAN A.P., AND FAWLEY M.A., "Modelling asynchrony in speech using elementary single-signal decomposition," *Proc. of IEEE Intl. Conf. on Acoustics, Speech, and Signal Processing* (Munich), pp. 1247–1250, April 1997.

[30] VARGA A. AND MOORE R., "Hidden markov model decomposition of speech and noise," *Proc. IEEE Intl. Conf. on Acoustics, Speech and Signal Processing*, pp. 845–848, 1990.

[31] WEBER K., BENGIO S., AND BOURLARD H., "HMM2- Extraction of Formant Features and their Use for Robust ASR", *Proc. of Eurospeech*, pp. 607–610, 2001.

[32] WELLEKENS C.J., KANGASHARJU J., AND MILESI C., "The use of meta-HMM in multistream HMM training for automatic speech recognition," *Proc. of Intl. Conference on Spoken Language Processing* (Sydney), pp. 2991–2994, December 1998.

[33] WU S.-L., KINGSBURY B.E., MORGAN N., AND GREENBERG S., "Performance improvements through combining phone and syllable-scale information in automatic speech recognition," *Proc. Intl. Conf. on Spoken Language Processing* (Sydney), pp. 459–462, Dec. 1998.

GRAPHICAL MODELS AND AUTOMATIC SPEECH RECOGNITION*

JEFFREY A. BILMES†

Abstract. Graphical models provide a promising paradigm to study both existing and novel techniques for automatic speech recognition. This paper first provides a brief overview of graphical models and their uses as statistical models. It is then shown that the statistical assumptions behind many pattern recognition techniques commonly used as part of a speech recognition system can be described by a graph – this includes Gaussian distributions, mixture models, decision trees, factor analysis, principle component analysis, linear discriminant analysis, and hidden Markov models. Moreover, this paper shows that many advanced models for speech recognition and language processing can also be simply described by a graph, including many at the acoustic-, pronunciation-, and language-modeling levels. A number of speech recognition techniques born directly out of the graphical-models paradigm are also surveyed. Additionally, this paper includes a novel graphical analysis regarding why derivative (or delta) features improve hidden Markov model-based speech recognition by improving structural discriminability. It also includes an example where a graph can be used to represent language model smoothing constraints. As will be seen, the space of models describable by a graph is quite large. A thorough exploration of this space should yield techniques that ultimately will supersede the hidden Markov model.

Key words. Graphical Models, Bayesian Networks, Automatic Speech Recognition, Hidden Markov Models, Pattern Recognition, Delta Features, Time-Derivative Features, Structural Discriminability, Language Modeling.

1. Introduction.

Since its inception, the field of automatic speech recognition (ASR) [129, 39, 21, 164, 83, 89, 117, 80] has increasingly come to rely on statistical methodology, moving away from approaches that were initially proposed such as template matching, dynamic time warping, and non-probabilistically motivated distortion measures. While there are still many successful instances of heuristically motivated techniques in ASR, it is becoming increasingly apparent that a statistical understanding of the speech process can only improve the performance of an ASR system. Perhaps the most famous example is the hidden Markov model [129], currently the predominant approach to ASR and a statistical generalization of earlier template-based practices.

A complete state-of-the-art ASR system involves numerous separate components, many of which are statistically motivated. Developing a thorough understanding of a complete ASR system, when it is seen as a collection of such conceptually distinct entities, can take some time. A impressive achievement would be an over-arching and unifying framework within which most statistical ASR methods can be accurately and succinctly de-

*This material is based upon work supported in part by the National Science Foundation under Grant No. 0093430.

†Department of Electrical Engineering, University of Washington, Seattle, Washington 98195-2500.

scribed. Fortunately, a great many of the successful algorithms used by ASR systems can be described in terms of graphical models.

Graphical models (GMs) are a flexible statistical abstraction that have been successfully used to describe problems in a variety of domains ranging from medical diagnosis and decision theory to time series prediction and signal coding. Intuitively, GMs merge probability theory and graph theory. They generalize many techniques used in statistical analysis and signal processing such as Kalman filters [70], auto-regressive models [110], and many information-theoretic coding algorithms [53]. They provide a visual graphical language with which one may observe and reason about some of the most important properties of random processes, and the underlying physical phenomena these processes are meant to represent. They also provide a set of computationally efficient algorithms for probability calculations and decision-making. Overall, GMs encompass an extremely large family of statistical techniques.

GMs provide an excellent formalism within which to study and understand ASR algorithms. With GMs, one may rapidly evaluate and understand a variety of different algorithms, since they often have only minor graphical differences. As we will see in this paper, many of the existing statistical techniques in ASR are representable using GMs — apparently no other known abstraction possesses this property. And even though the set of algorithms currently used in ASR is large, this collection occupies a relatively small volume within GM algorithm space. Because so many existing ASR successes lie within this under-explored space, it is likely that a systematic study of GM-based ASR algorithms could lead to new more successful approaches to ASR.

GMs can also help to reduce programmer time and effort. First, when described by a graph, it is easy to see if a statistical model appropriately represents relevant information contained in a corpus of (speech) data. GMs can help to rule out a statistical model which might otherwise require a large amount of programming effort to evaluate. A GM moreover can be minimally designed so that it has representational power only where needed [10]. This means that a GM-based system might have smaller computational demands than a model designed without the data in mind, further easing programmer effort. Secondly, with the right set of computational tools, many considerably different statistical algorithms can be rapidly evaluated in a speech recognition system. This is because the same underlying graphical computing algorithms are applicable for all graphs, regardless of the algorithm represented by the graph. Section 5 briefly describes the new graphical models toolkit (GMTK)[13], which is one such tool that can be used for this purpose.

Overall, this paper argues that it is both pedagogically and scientifically useful to portray ASR algorithms in the umbrage of GMs. Section 2 provides an overview of GMs showing how they relate to standard statistical procedures. It also surveys a number of GM properties (Section 2.5), such as probabilistic inference and learning. Section 3 casts many of the methods commonly used for automatic speech recognition (ASR) as instances of GMs and their associated algorithms. This includes principle component analysis [44], linear discriminant analysis (and its quadratic and heteroscedastic generalizations) [102], factor analysis, independent component analysis, Gaussian densities, multi-layered perceptrons, mixture models, hidden Markov models, and many language models. This paper further argues that developing novel ASR techniques can benefit from a GM perspective. In doing so, it surveys some recent techniques in speech recognition, some of which have been developed without GMs explicitly in mind (Section 4), and some of which have (Section 5).

In this paper, capital letters will refer to random variables (such as X, Y and Q) and lower-case letters will refer to values they may take on. Sets of variables may be referred to as X_A or Q_B where A and B are sets of indices. Sets may be referred to using a Matlab-like range notation, such as $1{:}N$ which indicates all indices between 1 and N inclusive. Using this notation, one may refer to a length T vector of random variable taking on a vector of values as $P(X_{1:T} = x_{1:T})$.

2. Overview of graphical models. This section briefly reviews graphical models and their associated algorithms — those well-versed in this methodology may wish to skip directly to Section 3.

Broadly speaking, graphical models offer two primary features to those interested in working with statistical systems. First, a GM may be viewed as an abstract, formal, and visual language that can depict important properties (conditional independence) of natural systems and signals when described by multi-variate random processes. There are mathematically precise rules that describe what a given graph means, rules that associate with a graph a family of probability distributions. Natural signals (those that are not purely random) have significant statistical structure, and this can occur at multiple levels of granularity. Graphs can show anything from causal relations between high-level concepts [122] down to the fine-grained dependencies existing within the neural code [5]. Second, along with GMs come a set of algorithms for efficiently performing probabilistic inference and decision making. Typically intractable, the GM inference procedures and their approximations exploit the inherent structure in a graph in a way that can significantly reduce computational and memory demands relative to a naive implementation of probabilistic inference.

Simply put, graphical models describe conditional independence properties amongst collections of random variables. A given GM is identical to a list of conditional independence statements, and a graph represents all distributions for which all these independence statements are true. A random variable X is conditionally independent of a different random variable Y given a third random variable Z under a given probability distribution $p(\cdot)$, if the following relation holds:

$$p(X = x, Y = y|Z = z) = p(X = x|Z = z)p(Y = y|Z = z)$$

for all x, y, and z. This is written $X \perp\!\!\!\perp Y|Z$ (notation first introduced in [37]) and it is said that "X is independent of Y given Z under $p(\cdot)$". This has the following intuitive interpretation: if one has knowledge of Z, then knowledge of Y does not change one's knowledge of X and vice versa. Conditional independence is different from unconditional (or marginal) independence. Therefore, neither $X \perp\!\!\!\perp Y$ implies $X \perp\!\!\!\perp Y|Z$ nor vice versa. Conditional independence is a powerful concept — using conditional independence, a statistical model can undergo enormous changes and simplifications. Moreover, even though conditional independence might not hold for certain signals, making such assumptions might yield vast improvements because of computational, data-sparsity, or task-specific reasons (e.g., consider the hidden Markov model with assumptions that obviously do not hold for speech [10], but that nonetheless empirically appear benign, and actually beneficial as argued in Section 3.9). Formal properties of conditional independence are described in [159, 103, 122, 37].

A GM [103, 34, 159, 122, 84] is a graph $\mathcal{G} = (V, E)$ where V is a set of vertices (also called nodes or random variables) and the set of edges E is a subset of the set $V \times V$. The graph describes an entire *family* of probability distributions over the variables V. A variable can either be scalar- or vector-valued, where in the latter case the vector variable implicitly corresponds to a sub-graphical model over the elements of the vector. The edges E, depending on the graph semantics (see below), encode a set of conditional independence properties over the random variables. The properties specified by the GM are true for all members of its associated family.

Four items must be specified when using a graph to describe a particular probability distribution: the GM *semantics*, *structure*, *implementation*, and *parameterization*. The semantics and the structure of a GM are inherent to the graph itself, while the implementation and parameterization are implicit within the underlying model.

2.1. Semantics. There are many types of GMs, each one with differing semantics. The set of conditional independence assumptions specified by a particular GM, and therefore the family of probability distributions it represents, can be different depending on the GM semantics. The semantics specifies a set of rules about what is or is not a valid graph and what set of distributions correspond to a given graph. Various types of GMs include directed models (or Bayesian networks) [122, 84],[1] undirected networks (or Markov random fields) [27], factor graphs [53, 101], chain graphs [103, 133] which are combinations of directed and undirected GMs, causal models [123], decomposable models (an important sub-family of models [103]),

[1]Note that the name "Bayesian network" does not imply Bayesian statistical inference. In fact, both Bayesian and non-Bayesian Bayesian networks may exist.

dependency networks [76], and many others. In general, different graph semantics will correspond to different families of distributions, but overlap can exist (meaning a particular distribution might be describable by two graphs with different semantics).

A Bayesian network (BN) [122, 84, 75] is one type of directed GM where the graph edges are directed and acyclic. In a BN, edges point from parent to child nodes, and such graphs implicitly portray factorizations that are simplifications of the chain rule of probability, namely:

$$p(X_{1:N}) = \prod_i p(X_i|X_{1:i-1}) = \prod_i p(X_i|X_{\pi_i}).$$

The first equality is the probabilistic chain rule, and the second equality holds under a particular BN, where π_i designates node i's parents according to the BN. A Dynamic Bayesian Network (DBN) [38, 66, 56] has exactly the same semantics as a BN, but is structured to have a sequence of clusters of connected vertices, where edges between clusters point in the direction of increasing time. DBNs are particularly useful to describe time signals such as speech, and as can be seen from Figure 2 many techniques for ASR fall under this or the BN category.

Several equivalent schemata exist that formally define a BN's conditional independence relationships [103, 122, 84]. The idea of d-separation (or directed separation) is perhaps the most widely known: a set of variables A is conditionally independent of a set B given a set C if A is d-separated from B by C. D-separation holds if and only if *all* paths that connect any node in A and any other node in B are *blocked*. A path is blocked if it has a node v with either: 1) the arrows along the path **do not** converge at v (i.e., serial or diverging at v) $v \in C$; or 2) the arrows along the path **do** converge at v, and neither v nor any descendant of v is in C. Note that C can be the empty set in which case d-separation encodes standard statistical independence.

From d-separation, one may compute a list of conditional independence statements made by a graph. This set of probability distributions for which this list of statements is true is precisely the set of distributions represented by the graph. Graph properties equivalent to d-separation include the directed local Markov property [103] (a variable is conditionally independent of its non-descendants given its parents), and the Bayes-ball procedure [143] which is a simple algorithm that one can use to read conditional independence statements from graphs, and which is arguably simpler than d-separation. It is assumed henceforth that the reader is familiar with either d-separation or some equivalent rule.

Conditional independence properties in undirected graphical models (UGMs) are much simpler than for BNs, and are specified using graph separation. For example, assuming that X_A, X_B, and X_C are disjoint sets of nodes in a UGM, $X_A \perp\!\!\!\perp X_B | X_C$ is true when all paths from any node in X_A to any node in X_B intersect some node in X_C. In a UGM, a distribution may be described as the factorization of potential functions where each potential function operates only on collections of nodes that form a clique in the graph. A clique is a set of nodes that are pairwise connected [84].

BNs and DGMs are not the same. Despite the fact that BNs have complicated semantics, they are useful for a variety of reasons. One is that BNs can have a causal interpretation, where if node A is a parent of B, A might be thought of as a cause of B. A second reason is that the family of distributions associated with BNs is not the same as the family associated with UGMs — there are some useful probability models that are concisely representable with BNs but that are not representable at all with UGMs (and vice versa). This issue will arise in Section 3.1 when discussing Gaussian densities. UGMs and BNs do have an overlap, however, and the family of distributions corresponding to this intersection is known as the decomposable models [103]. These models have important properties relating to efficient probabilistic inference (see below).

In general, a lack of an edge between two nodes does not imply that the nodes are independent. The nodes might be able to influence each other indirectly via an indirect path. Moreover, the existence of an edge between two nodes does *not* imply that the two nodes are necessarily dependent — the two nodes could still be independent for certain parameter values or under certain conditions (see later sections). A GM guarantees only that the lack of an edge implies some conditional independence property, determined according to the graph's semantics. It is therefore best, when discussing a given GM, to refer only to its (conditional) independence rather than its dependence properties — it is more accurate to say that there is an edge between A and B than to say that A and B are dependent.

Originally BNs were designed to represent causation, but more recently, models with semantics [123] more precisely representing causality have been developed. Other directed graphical models have been designed as well [76], and can be thought of as the general family of directed graphical models (DGMs).

2.2. Structure. A graph's structure, the set of nodes and edges, determines the set of conditional independence properties for the graph under a given semantics. Note that more than one GM might correspond to exactly the same conditional independence properties even though their structure is entirely different (see the left two models in Figure 1). In this case, multiple graphs will correspond to the same family of probability distributions. In such cases, the various GMs are said to be Markov equivalent [153, 154, 77]. In general, it is not immediately obvious with complicated graphs how to visually determine if Markov equivalence holds, but algorithms are available that can determine the members of an equivalence class [153, 154, 114, 30].

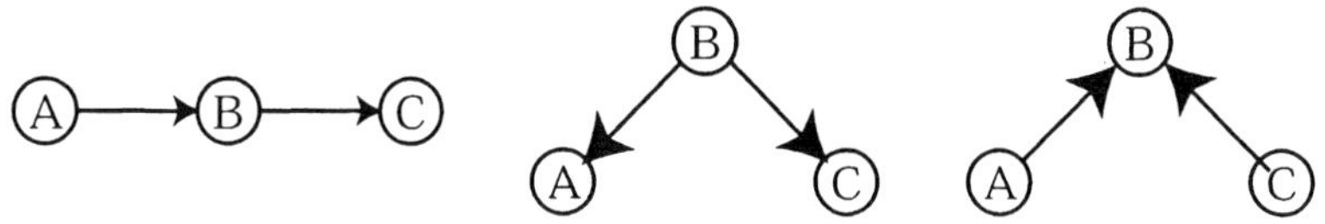

FIG. 1. *This figure shows four BNs with different arrow directions over the same random variables, A, B, and C. On the left side, the variables form a three-variable first-order Markov chain $A \rightarrow B \rightarrow C$. In the middle graph, the same conditional independence statement is realized even though one of the arrow directions has been reversed. Both these networks state that $A \perp\!\!\!\perp C|B$. The right network corresponds to the property $A \perp\!\!\!\perp C$ but* **not** *that $A \perp\!\!\!\perp C|B$.*

Nodes in a graphical model can be either *observed*, or *hidden*. If a variable is observed, it means that its value is known, or that data (or "evidence") is available for that variable. If a variable is hidden, it currently does not have a known value, and all that is available is the conditional distribution of the hidden variables given the observed variables (if any). Hidden nodes are also called confounding, latent, or unobserved variables. Hidden Markov models are so named because they possess a Markov chain that, in some cases, contains only hidden variables. Note that the graphs in GMs do *not* show the zeros that exist in the stochastic transition matrices of a Markov chain — GMs, rather, encode statistical independence properties of a model (see also Section 3.7).

A node in a graph might sometimes be hidden and at other times be observed. With an HMM, for example, the "hidden" chain might be observed during training (because a phonetic or state-level alignment has been provided) and hidden during recognition (because the hidden variable values are not known for test speech data). When making the query "is $A \perp\!\!\!\perp B|C$?", it is implicitly assumed that C is observed. A and B are the nodes being queried, and any other nodes in the network not listed in the query are considered hidden. Also, when a collection of sampled data exists (say as a training set), some of the data samples might have missing values each of which would correspond to a hidden variable. The EM algorithm [40], for example, can be used to train the parameters of hidden variables.

Hidden variables and their edges reflect a belief about the underlying generative process lying behind the phenomenon that is being statistically represented. This is because the data for these hidden variables is either unavailable, is too costly or impossible to obtain, or might not exist since the hidden variables might only be hypothetical (e.g., specified based on human-acquired knowledge about the underlying domain). Hidden variables can be used to indicate the underlying causes behind an information source. In speech, for example, hidden variables can be used to represent the phonetic or articulatory gestures, or more ambitiously, the originating semantic thought behind a speech waveform. One common way of using GMs in ASR, in fact, is to use hidden variables to represent some condition known during training and unknown during recognition (see Section 5).

Certain GMs allow for what are called *switching* dependencies [65, 115, 16]. In this case, edges in a GM can change as a function of other variables in the network. An important advantage of switching dependencies is the reduction in the required number of parameters needed by the model. A related construct allows GMs to have optimized local probability implementations [55] using, for example, decision trees.

It is sometimes the case that certain observed variables are used only as conditional variables. For example, consider the graph $B \rightarrow A$ which implies a factorization of the joint distribution $P(A, B) = P(A|B)P(B)$. In many cases, it is not necessary to represent the marginal distribution over B. In such cases B is a "conditional-only" variable, meaning is always and only to the right of the conditioning bar. In this case, the graph represents $P(A|B)$. This can be useful in a number of applications including classification (or discriminative modeling), where we might only be interested in posterior distributions over the class random variable, or in situations where additional observations (say Z) exist that are marginally independent of a class variable (say C) but that are dependent conditioned on other observations (say X). This can be depicted by the graph $C \rightarrow X \leftarrow Z$, where it is assumed that the distribution over Z is not represented.

Often, the true (or the best) structure for a given task is unknown. This can mean that either some of the edges or nodes (which can be hidden) or both can be unknown. This has motivated research on learning the structure of the model from the data, with the general goal to produce a structure that accurately reflects the important statistical properties in the data set. These can take a Bayesian [75, 77] or frequentist point of view [25, 99, 75]. Structure learning is akin to both statistical model selection [107, 26] and data mining [36]. Several good reviews of structure learning are presented in [25, 99, 75]. Structure learning from a discriminative perspective, thereby producing what is called *discriminative generative models*, was proposed in [10].

Figure 2 depicts a topological hierarchy of both the semantics and structure of GMs, and shows where different models fit in, including several ASR components to be described in Section 3.

2.3. Implementation. When two nodes are connected by a dependency edge, the local conditional probability representation of that dependency may be called its *implementation*. An edge between variable X and Y can represent a lack of independence in a number of ways depending on if the variables are discrete or continuous. For example, one might use discrete conditional probability tables (CPTs) [84], compressed tables [55], decision trees [22], or even a deterministic function (in which case GMs may represent data-flow [1] graphs, or may represent channel coding algorithms [53]). A node in a GM can also depict a constant input parameter since random variables can themselves be constants. Alternatively, the dependence might be linear regression models, mixtures thereof, or non-

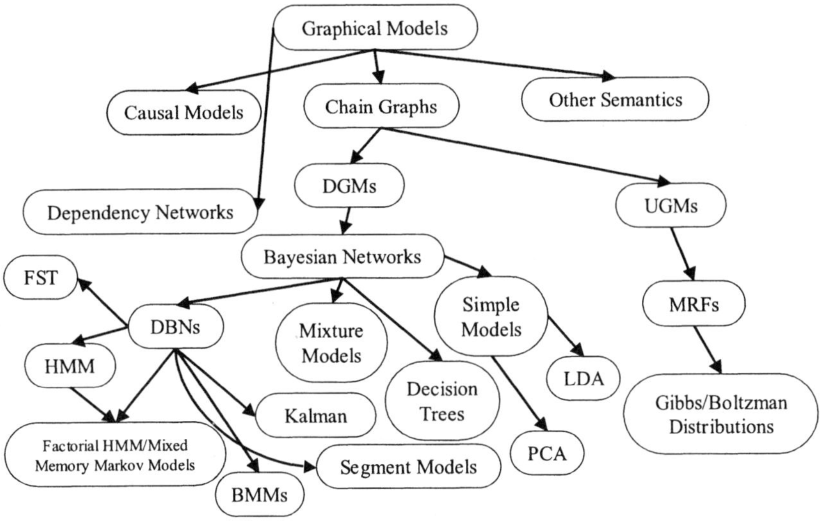

FIG. 2. *A topology of graphical model semantics and structure.*

linear regression (such as a multi-layered perceptron [19], or a STAR [149] or MARS [54] model). In general, different edges in a graph will have different implementations.

In UGMs, conditional distributions are not explicitly represented. Rather a joint distribution over all the variables is constructed using a product of clique potential functions as mentioned in Section 2.1. In general the clique potentials can be arbitrary functions, although certain types are commonly used such as Gibbs or Boltzmann distributions [79]. Many such models fall under what are known as exponential models [44]. The implementation of a dependency in an UGM, therefore, is implicitly specified via these functions in that they specify the way in which subsets of variables, depending on their values, can influence the resulting probability

2.4. Parameterization. The parameterization of a model corresponds to the parameter values of a particular implementation in a particular structure. For example, with linear regression, parameters are simply the regression coefficients; for a discrete probability table the parameters are the table entries. Since parameters of random distributions can themselves be seen as nodes, Bayesian approaches are easily represented [75] with GMs.

Many algorithms exist for training the parameters of a graphical model. These include maximum likelihood [44] such as the EM algorithm [40], discriminative or risk minimization approaches [150], gradient descent [19], sampling approaches [109], or general non-linear optimization [50]. The choice of algorithm depends both on the structure and implementation of the GM. For example, if there are no hidden variables, an EM approach is not required. Certain structural properties of the GM might render certain training procedures less crucial to the performance of the model [16, 47].

2.5. Efficient probabilistic inference. A key application of any statistical model is to compute the probability of one subset of random variables given values for some other subset, a procedure known as probabilistic inference. Inference is essential both to make predictions based on the model and to learn the model parameters using, for example, the EM algorithm [40, 113]. One of the critical advantages of GMs is that they offer procedures for making exact inference as efficient as possible, much more so than if conditional independence is ignored or is used unwisely. And if the resulting savings is not enough, there are GM-inspired approximate inference algorithms that can be used.

Exact inference can in general be quite computationally costly. For example, suppose there is a joint distribution over 6 variables $p(a,b,c,d,e,f)$ and the goal is to compute $p(a|f)$. This requires both $p(a,f)$ and p(f), so the variables b,c,d,e must be "marginalized", or integrated away to form $p(a,f)$. The naive way of performing this computation would entail the following sum:

$$p(a,f) = \sum_{b,c,d,e} p(a,b,c,d,e,f).$$

Supposing that each variable has K possible values, this computation requires $O(K^6)$ operations, a quantity that is exponential in the number of variables in the joint distribution. If, on the other hand, it was possible to factor the joint distribution into factors containing fewer variables, it would be possible to reduce computation significantly. For example, under the graph in Figure 3, the above distribution may be factored as follows:

$$p(a,b,c,d,e,f) = p(a|b)p(b|c)p(c|d,e)p(d|e,f)p(e|f)p(f)$$

so that the sum

$$p(a,f) = p(f)\sum_b p(a|b)\sum_c p(b|c)\sum_d p(c|d,e)\sum_e p(d|e,f)p(e|f)$$

requires only $O(K^3)$ computation. Inference in GMs involves formally defined manipulations of graph data structures and then operations on those data structures. These operations provably correspond to valid operations on probability equations, and they reduce computation essentially by moving sums, as in the above, as far to the right as possible in these equations.

The graph operations and data structures needed for inference are typically described in their own light, without needing to refer back to the original probability equations. One well-known form of inference procedure, for example, is the junction tree (JT) algorithm [122, 84]. In fact,

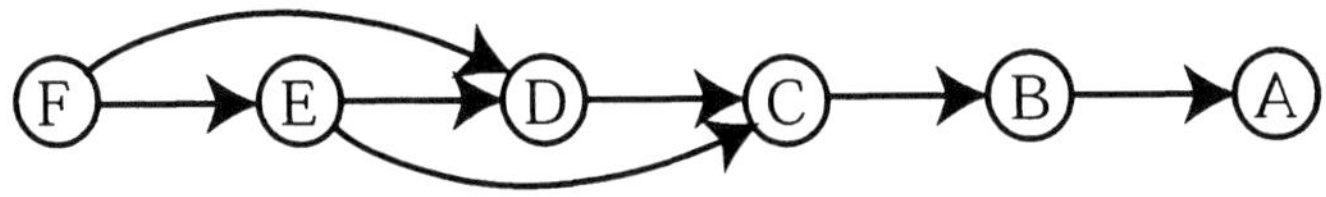

FIG. 3. *The graph's independence properties are used to move sums inside of factors.*

the commonly used forward-backward algorithm [129] for hidden Markov models is just a special case of the junction tree algorithm [144], which is a special case of the generalized distributive law [2].

The JT algorithm requires that the original graph be converted into a junction tree, a tree of cliques with each clique containing nodes from the original graph. A junction tree possesses the running intersection property, where the intersection between any two cliques in the tree is contained in all cliques in the (necessarily) unique path between those two cliques. The junction tree algorithm itself can be viewed as a series of messages passing between the connected cliques of the junction tree. These messages ensure that the neighboring cliques are locally consistent (i.e., that the neighboring cliques have identical marginal distributions on those variables that they have in common). If the messages are passed in a particular order, called the message passing protocol [85], then because of the properties of the junction tree, local consistency guarantees global consistency, meaning that the marginal distributions on all common variables in all cliques are identical, meaning that inference is correct. Because only local operations are required in the procedure, inference can be fast.

For the junction tree algorithm to be valid, however, a decomposable model must first be formed from the original graph. Junction trees exist only for decomposable models, and a message passing algorithm can provably be shown to yield correct probabilistic inference only in that case. It is often the case, however, that a given DGM or UGM is not decomposable. In such cases it is necessary to form a decomposable model from a general GM (directed or otherwise), and in doing so make fewer conditional independence assumptions. Inference is then solved for this larger family of models. Solving inference for a larger family still of course means that inference has been solved for the smaller family corresponding to the original (possibly) non-decomposable model.

Two operations are needed to transform a general DGM into a decomposable model: moralization and triangulation. Moralization joins the unconnected parents of all nodes and then drops all edge directions. This procedure is valid because more edges means fewer conditional independence assumptions or a larger family of probability distributions. Moralization is required to ensure that the resulting UGM does not disobey any of the conditional independence assumptions made by the original DGM. In other words, after moralizing, it is assured that the UGM will make no independence assumption that is not made by the original DGM. Otherwise, inference might not be correct.

After moralization, or if starting from a UGM to begin with, triangulation is necessary to produce a decomposable model. The set of all triangulated graphs corresponds exactly to the set of decomposable models. The triangulation operation [122, 103] adds edges until all cycles in the graph (of length 4 or greater) contain a pair of non-consecutive nodes (along the cycle) that are connected by an edge (i.e., a chord) not part of the cycle edges. Triangulation is valid because more edges enlarge the set of distributions represented by the graph. Triangulation is necessary because only for triangulated (or decomposable) graphs do junction trees exists. A good survey of triangulation techniques is given in [98].

Finally, a junction tree is formed from the triangulated graph by, first, forming all maximum cliques in the graph, next connecting all of the cliques together into a "super" graph, and finally finding a maximum spanning tree [32] amongst that graph of maximum cliques. In this case, the weight of an edge between two cliques is set to the number of variables in the intersection of the two cliques.

For a discrete-node-only network, junction tree complexity is $O(\sum_{c \in C} \prod_{v \in c} |v|)$ where C is the set of cliques in the junction tree, c is the set of variables contained within a clique, and $|v|$ is the number of possible values of variable v — i.e., the algorithm is exponential in the clique sizes, a quantity important to minimize during triangulation. There are many ways to triangulate [98], and unfortunately the operation of finding the optimal triangulation is itself NP-hard. For an HMM, the clique sizes are N^2, where N is the number of HMM states, and there are T cliques leading to the well known $O(TN^2)$ complexity for HMMs. Further information on the junction tree and related algorithms can be found in [84, 122, 34, 85].

Exact inference, such as the above, is useful only for moderately complex networks since inference is NP-hard in general [31]. Approximate inference procedures can, however, be used when exact inference is not feasible. There are several approximation methods including variational techniques [141, 81, 86], Monte Carlo sampling methods [109], and loopy belief propagation [156]. Even approximate inference can be NP-hard however [35]. Therefore, it is always important to use a minimal model, one with least possible complexity that still accurately represents the important aspects of a task.

3. Graphical models and automatic speech recognition. A wide variety of algorithms often used in state-of-the-art ASR systems can easily be described using GMs, and this section surveys a number of them. While many of these approaches were developed without GMs in mind, they turn out to have surprisingly simple and elucidating network structures. Given an understanding of GMs, it is in many cases easier to understand the technique by looking first at the network than at the original algorithmic description.

As is often done, the following sections will separate ASR algorithms into three categories: acoustic, pronunciation, and language modeling. Each of these are essentially statistical models about how the speech data that we observe is generated. Different statistical models, and inference within these models, leads us to the different techniques, but each are essentially special cases of the more general GM techniques described above.

3.1. Acoustic modeling: Gaussians. The most successful and widely used density for acoustic modeling in ASR systems is the multi-dimensional Gaussian. The Gaussian density has a deceptively simple mathematical description that does not disclose many of the useful properties this density possesses (such as that first and second moments completely characterize the distribution). In this section, it will be shown how Gaussians can be viewed as both undirected and directed GMs, and how each of these views describe distinct properties of the Gaussian.

An N-dimensional Gaussian density has the form:

$$p(x) = p(x_{1:N}) = \mathcal{N}(x_{1:N}; \mu, \Sigma) = |2\pi\Sigma|^{-1/2} e^{-\frac{1}{2}(x-\mu)^T \Sigma^{-1}(x-\mu)}$$

where μ is an N-dimensional mean vector, and Σ is an $N \times N$ covariance matrix. Typically, $K = \Sigma^{-1}$ refers to the inverse covariance (or the concentration) matrix of the density.

It will be useful to form partitions of a vector x into a number of parts. For example, a bi-partition of $x = [x_A \; x_B]$ may be formed, where x_A and x_B are sub-vectors of x [68], and where the sum of the dimensions of x_A and x_B equals N. Tri-partitions $x = [x_A \; x_B \; x_C]$ may also be formed. In this way, the mean vector $\mu = [\mu_A \; \mu_B]^T$, and the covariance and concentration matrices can be so partitioned as

$$\Sigma = \begin{pmatrix} \Sigma_{AA} & \Sigma_{AB} \\ \Sigma_{BA} & \Sigma_{BB} \end{pmatrix} \quad \text{and} \quad K = \begin{pmatrix} K_{AA} & K_{AB} \\ K_{BA} & K_{BB} \end{pmatrix}.$$

Conventionally, $\Sigma_{AA}^{-1} = (\Sigma_{AA})^{-1}$, so that the sub-matrix operator takes precedence over the matrix inversion operator. A well known property of Gaussians is that if $\Sigma_{AB} = 0$ then x_A and x_B are marginally independent ($x_A \perp\!\!\!\perp x_B$).

A more interesting and less well-known property of a Gaussian is that for a given tri-partition $x = [x_A \; x_B \; x_C]$ of x, and corresponding tri-partitions of μ and K, then $x_A \perp\!\!\!\perp x_B | x_C$ if and only if, in the corresponding tri-partition of K, $K_{AB} = 0$, a property that may be proven quite readily. For any distribution, the chain rule of probability says

$$p(x) = p(x_A | x_B) p(x_B).$$

When $p(x)$ is a Gaussian density, the marginal distribution $p(x_B)$ is also Gaussian with mean μ_B and covariance Σ_{BB}. Furthermore, $p(x_A | x_B)$ is a

Gaussian having a "conditional" mean and covariance [111, 4]. Specifically, the distribution for x_A given x_B is a conditional Gaussian with an x_B-dependent mean vector

$$\mu_{A|B} = \mu_A + \Sigma_{AB}\Sigma_{BB}^{-1}(x_B - \mu_B)$$

and a fixed covariance matrix

$$\Sigma_{A|B} = \Sigma_{AA} - \Sigma_{AB}\Sigma_{BB}^{-1}\Sigma_{BA}.$$

This means that, if the two vectors X_A and X_B are jointly Gaussian, then given knowledge of one vector, say X_B, the result is a Gaussian distribution over X_A that has a fixed variance for all values of x_B but has a mean that is an affine transformation of the particular value of x_B. Most importantly, it can be shown that K_{AA}, the upper-left partition of the original concentration matrix K, is the inverse of the conditional covariance, specifically $K_{AA} = \Sigma_{A|B}^{-1}$ [103, 159].

Let the partition X_A be further partitioned to form the sub-bi-partition $x_A = [x_{Aa}\ x_{Ab}]$, meaning that $x = [x_{Aa}\ x_{Ab}\ x_B]$. A similar sub-partition is formed of the concentration matrix

$$K_{AA} = \begin{pmatrix} K_{AAaa} & K_{AAab} \\ K_{AAba} & K_{AAbb} \end{pmatrix}.$$

Setting $K_{AAab} = 0$ implies that $x_{Aa} \perp\!\!\!\perp x_{Ab}$, but only when conditioning on x_B. This yields the result desired, but with the matrix and vector partitions renamed. Therefore, zeros in the inverse covariance matrix result in conditional independence properties for a Gaussian, or more specifically if $K_{ij} = 0$ then $X_i \perp\!\!\!\perp X_j | X_{\{1:N\}\setminus\{i,j\}}$.

This property of Gaussians corresponds to their view as an UGM. To see this, first consider a fully connected UGM with N nodes, something that represents all Gaussians. Setting an entry, say K_{ij}, to zero corresponds to the independence property above, which corresponds in the UGM to removing an edge between variable x_i and x_j (see [103] for a formal proof where the pairwise Markov property and the global Markov property are related). This is shown in Figure 4. Therefore, missing edges in a Gaussian UGM correspond exactly to zeros in the inverse covariance matrix.

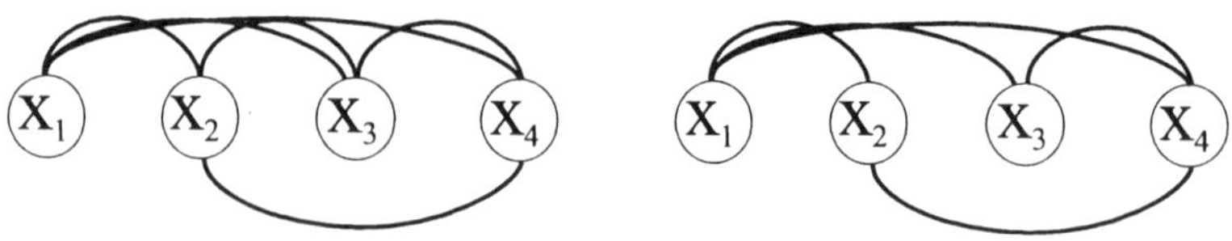

FIG. 4. *A Gaussian viewed as an UGM. On the left, there are no independence assumptions. On the right,* $X_2 \perp\!\!\!\perp X_3 | \{X_1, X_4\}$.

A Gaussian may also be viewed as a BN, and in fact many BNs. Unlike with a UGM, to form a Gaussian BN a specific variable ordering must first be chosen, the same ordering used to factor the joint distribution with the chain rule of probability. A Gaussian can be factored

$$p(x_{1:N}) = \prod_i p(x_i | x_{i+1:N})$$

according to some fixed but arbitrarily chosen variable ordering. Each factor is a Gaussian with conditional mean

$$\mu_{i|i+1:N} = \mu_i + \Sigma_{i,i+1:N}(\Sigma_{i+1:N,i+1:N})^{-1}(x_{i+1:N} - \mu_{i+1:N})$$

and conditional covariance

$$\Sigma_{i|i+1:N} = \Sigma_{ii} - \Sigma_{i,i+1:N}(\Sigma_{i+1:N,i+1:N})^{-1}\Sigma_{i+1:N,i}$$

both of which are unique for a given ordering (these are an application of the conditional Gaussian formulas above, but with A and B set to the specific values $\{i\}$ and $\{(i+1):N\}$ respectively). Therefore, the chain rule expansion can be written:

$$p(x_{1:N}) = \prod_i (2\pi\Sigma_{i|i+1:N})^{-1/2} e^{-\frac{1}{2}(x_i - \mu_{i|i+1:N})^2 \Sigma^{-1}_{i|i+1:N}}. \tag{3.1}$$

An identical decomposition of this Gaussian can be produced in a different way. Every concentration matrix K has a unique factorization $K = U^T D U$ where U is a unit upper-triangular matrix and D is diagonal [111, 73]. A unit triangular matrix is a triangular matrix that has ones on the diagonal, and so has a unity determinant (so is non-singular), therefore, $|K| = |D|$. This corresponds to a form of Cholesky factorization $K = R^T R$, where R is upper triangular, $D^{1/2} = \text{diag}(R)$ is the diagonal portion of R, and $R = D^{1/2}U$. A Gaussian density can therefore be represented as:

$$p(x) = (2\pi)^{-d/2}|D|^{1/2} e^{-\frac{1}{2}(x-\mu)^T U^T D U (x-\mu)}.$$

The unit triangular matrices, however, can be "brought" inside the squared linear terms by considering the argument within the exponential

$$\begin{aligned}
(x-\mu)^T U^T D U (x-\mu) &= (U(x-\mu))^T D (U(x-\mu)) \\
&= (Ux - \tilde{\mu})^T D (Ux - \tilde{\mu}) \\
&= ((I-B)x - \tilde{\mu})^T D((I-B)x - \tilde{\mu}) \\
&= (x - Bx - \tilde{\mu})^T D (x - Bx - \tilde{\mu})
\end{aligned}$$

where $U = I - B$, I is the identity matrix, B is an upper triangular matrix with zeros along the diagonal, and $\tilde{\mu} = U\mu$ is a new mean. Again, this transformation is unique for a given Gaussian and variable ordering. This process exchanges K for a diagonal matrix D, and produces a linear auto-regression of x onto itself, all while not changing the Gaussian normalization factor contained in D. Therefore, a full-covariance Gaussian can be represented as a conditional Gaussian with a regression on x itself, yielding the following:

$$p(x_{1:N}) = (2\pi)^{-d/2}|D|^{1/2}e^{-\frac{1}{2}(x-Bx-\tilde{\mu})^T D(x-Bx-\tilde{\mu})}.$$

In this form the Gaussian can be factored where the ith factor uses only the ith row of B:

$$p(x_{1:N}) = \prod_i (2\pi)^{-1/2} D_{ii}^{1/2} e^{-\frac{1}{2}(x_i - B_{i,i+1:N}x_{i+1:N}-\tilde{\mu}_i)^2 D_{ii}}. \tag{3.2}$$

When this is equated with Equation (3.1), and note is taken of the uniqueness of both transformations, it is the case that

$$B_{i,i+1:N} = \Sigma_{i,i+1:N}(\Sigma_{i+1:N,i+1:N})^{-1}$$

and that $\tilde{\mu}_i = \mu_i - B_{i,i+1:N}\mu_{i+1:N}$. This implies that the regression coefficients within B are a simple function of the original covariance matrix. Since the quantities in the exponents are identical for each factor (which are each an appropriately normalized Gaussian), the variance terms D_{ii} must satisfy:

$$D_{ii} = \Sigma_{i|i+1:N}^{-1}$$

meaning that the D_{ii} values are conditional variances.

Using these equations we can now show how a Gaussian can be viewed as a BN. The directed local Markov property of BNs states that the joint distribution may be factorized as follows:

$$p(x_{1:N}) = \prod_i p(x_i|x_{\pi_i})$$

where $\pi_i \subseteq \{(i+1){:}N\}$ are parents of the variable x_i. When this is considered in terms of Equation (3.2), it implies that the non-zero entries of $B_{i,i+1:N}$ correspond to the set of parents of node i, and the zero entries correspond to missing edges. In other words (under a given variable ordering) the B matrix determines the conditional independence statements for a Gaussians when viewed as a DGM, namely $X_i \perp\!\!\!\perp X_{\{(i+1):N\}\setminus\pi_i}|X_{\pi_i}$ if and only if the entries $B_{i,\{(i+1):N\}\setminus\pi_i}$ are zero.[2]

It is important to realize that these results depend on a particular ordering of the variables $X_{1:N}$. A different ordering might yield a different

[2]Standard notation is used here, where if A and B are sets, $A \setminus B$ is the set of elements in A that are not in B.

B matrix, possibly implying different independence statements (depending on if the graphs are Markov equivalent, see Section 2.2). Moreover, a B matrix can be sparse for one ordering, but for a different ordering the B matrix can be dense, and zeros in B might or might not yield zeros in $K = (I - B)^T D(I - B)$ or $\Sigma = K^{-1}$, and vice versa.

This means that a full covariance Gaussian with $N(N+1)/2$ non-zero covariance parameters might actually employ fewer than $N(N+1)/2$ parameters, since it is in the directed domain where sparse patterns of independence occur. For example, consider a 4-dimensional Gaussian with a B matrix such that $B_{12} = B_{13} = B_{14} = B_{24} = B_{34} = 1$, and along with the other zero B entries, take $B_{23} = 0$. For this B matrix and when $D = I$, neither the concentration nor the covariance matrix has any zeros, although they are both full rank and it is true that $X_2 \perp\!\!\!\perp X_3 | X_4$. It must be that K possesses redundancy in some way, but in the undirected formalism it is impossible to encode this independence statement and one is forced to generalize and to use a model that possesses no independence properties.

The opposite can occur as well, where zeros exist in K or Σ, and less sparsity exists in the B matrix. Take, for example the matrix,

$$K = \begin{pmatrix} 5 & 2 & 0 & 4 \\ 2 & 9 & 1 & 0 \\ 0 & 1 & 5 & 3 \\ 4 & 0 & 3 & 6 \end{pmatrix}.$$

This concentration matrix states that $X_1 \perp\!\!\!\perp X_3 | \{X_2, X_4\}$ and $X_2 \perp\!\!\!\perp X_4 | \{X_1, X_3\}$, but the corresponding B matrix has only a single zero in its upper portion reflecting only the first independence statement.

It was mentioned earlier that UGMs and DGMs represent different families of probability distributions, and this is reflected in the Gaussian case above by a reduction in sparsity when moving between certain B and K matrices. It is interesting to note that Gaussians are able to represent any of the dependency structures captured either in a DGM (via an appropriate order of the variables and zeros in the B matrix) or a UGM (with appropriately placed zeros in the concentration matrix K). Therefore, Gaussians, along with many other interesting and desirable theoretical properties, are quite general in terms of their ability to possess conditional independence relationships.

The question then becomes what form of Gaussian should be used, a DGM or a UGM, and if a DGM, in what variable order. A common goal is to minimize the total number of free parameters. If this is the case, the Gaussian should be represented in a "natural" domain [10], where the least degree of parameter redundancy exists. Sparse matrices often provide the answer, assuming no additional cost exists to represent sparse matrices, since the sparse pattern itself might be considered a parameter needing a representation. This was exploited in [17], where the natural directed Gaussian representation was solicited from data, and where a negligible penalty in WER performance was obtained with a factored sparse covariance matrix having significantly fewer parameters.

Lastly, it is important to realize that while all UGM or DGM dependency structures can be realized by a Gaussian, the implementations in each case are only linear and the random components are only univariate Gaussian. A much greater family of distributions, other than just a Gaussian, can be depicted by a UGM or DGM, as we begin to see in the next sections.

3.2. Acoustic modeling: PCA/FA/ICA. Our second example of GMs for speech consists of techniques commonly used to transform speech feature vectors prior to being used in ASR systems. These include principle component analysis (PCA) (also called the Karhunen-Loéve or KL transform), factor analysis (FA), and independent component analysis (ICA). The PCA technique is often presented without any probabilistic interpretation. Interestingly, when given such an interpretation and seen as a graph, PCA has exactly the same structure as both FA and ICA – the only difference lies in the implementation of the dependencies.

The graphs in this and the next section show nodes both for random variables and their parameters. For example, if X is Gaussian with mean μ, a μ node might be present as a parent of X. Parameter nodes will be indicated as shaded rippled circles. For our purposes, these nodes constitute constant random variables whose probability score is not counted (they are conditional-only variables, always to the right of the conditioning bar in a probability equation). In a more general Bayesian setting [77, 75, 139], however, these nodes would be true random variables with their own distributions and hyper-parameters.

Starting with PCA, observations of a d-dimensional random vector X are assumed to be Gaussian with mean μ and covariance Σ. The goal of PCA is to produce a vector Y that is a zero-mean uncorrelated linear transformation of X. The spectral decomposition theorem [146] yields the factorization $\Sigma = \Gamma\Lambda\Gamma^T$, where Γ is an orthonormal rotation matrix (the columns of Γ are orthogonal eigenvectors, each having unit length), and Λ is a diagonal matrix containing the eigenvalues that correspond to the variances of the elements of X. A transformation achieving PCA's goal is $Y = \Gamma^T(X - \mu)$. This follows since $E[YY^T] = \Gamma^T E[(X-\mu)(X-\mu)^T]\Gamma$ $= \Gamma^T\Sigma\Gamma = \Lambda$. Alternatively, a spherically distributed Y may be obtained by the following transformation: $Y = (\Lambda^{-1/2}\Gamma)^T(X - \mu) = C^T(X - \mu)$ with $C = \Lambda^{-1/2}\Gamma$.

Solving for X as a function of Y yields the following:

$$X = \Gamma Y + \mu.$$

Slightly abusing notation, one can say that $X \sim \mathcal{N}(\Gamma Y + \mu, 0)$, meaning that X, conditioned on Y, is a linear-conditional constant "Gaussian" — i.e., a conditional-Gaussian random variable with mean $\Gamma Y + \mu$ and zero variance.[3] In this view of PCA, Y consists of the latent or hidden "causes" of the observed vector X, where $Y \sim \mathcal{N}(0, \Lambda)$, or if the C-transformation is used above, $Y \sim \mathcal{N}(0, I)$ where I is the identity matrix. In either case, the variance in X is entirely explained by the variance within Y, as X is simply a linear transformation of these underlying causes. PCA transforms a given X to the most likely values of the hidden causes. This is equal to the conditional mean $E[Y|X] = \Gamma^T(X - \mu)$ since $p(y|x) \sim \mathcal{N}(\Gamma^T(x - \mu), 0)$.

The two left graphs of Figure 5 show the probabilistic interpretations of PCA as a GM, where the dependency implementations are all linear. The left graph corresponds to the case where Y is spherically distributed. The hidden causes Y are called the "principle components" of X. It is often the case that only the components (i.e., elements of Y) corresponding to the largest eigenvalues of Σ are used in the model, the other elements of Y are removed, so that Y is k-dimensional with $k < d$. There are many properties of PCA [111] – for example, using the principle k elements of Y leads to the smallest reconstruction error of X in a mean-squared sense. Another notable property (which motivates factor analysis below) is that PCA is not scale invariant — if the scale of X changes (say by converting from inches to centimeters), both Γ and Λ will also change, leading to different components Y. In this sense, PCA explains the variance in X using only variances found in the hidden causes Y.

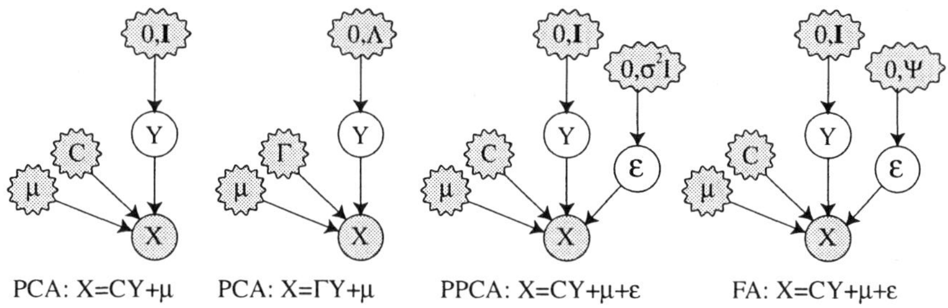

FIG. 5. *Left two graphs: two views of principle components analysis (PCA); middle: probabilistic PCA; right; factor analysis (FA). In general, the graph corresponds to the equation* $X = CY + \mu + \epsilon$, *where* $Y \sim \mathcal{N}(0, \Lambda)$ *and* $\epsilon \sim \mathcal{N}(0, \Psi)$. X *is a random conditional Gaussian with mean* $CY + \mu$ *and variance* $C\Lambda C^T + \Psi$. *With PCA,* $\Psi = 0$ *so that* $\epsilon = 0$ *with probability 1. Also, either (far left)* $\Lambda = I$ *is the identity matrix and* C *is general, or (second from left)* Λ *is diagonal and* $C = \Gamma$ *is orthonormal. With PPCA,* $\Psi = \sigma^2 I$ *is a spherical covariance matrix, with diagonal terms* σ^2. *With FA,* Ψ *is diagonal. Other generalizations are of possible, but they can lead to an indeterminacy of the parameters.*

Factor analysis (the right-most graph in Figure 5) is only a simple modification of PCA — a single random variable is added onto the PCA equation above, yielding:

[3]This of course corresponds to a degenerate Gaussian, as the covariance matrix is singular.

$$X = CY + \mu + \epsilon$$

where $Y \sim \mathcal{N}(0, I)$, and $\epsilon \sim \mathcal{N}(0, \Psi)$ with Ψ a non-negative diagonal matrix. In factor analysis, C is the *factor loading* matrix and Y the *common factor* vector. Elements of the residual term $\epsilon = X - CY - \mu$, are called the *specific factors*, and account both for noise in the model and for the underlying variance in X. In other words, X possesses a non-zero variance, even conditional on Y, and Y is constrained to be unable to explain the variance in X since Y is forced to have I as a covariance matrix. C, on the other hand, is compelled to represent just the correlation between elements of X irrespective of its individual variance terms, since correlation can not be represented by ϵ. Therefore, unlike PCA, if the scale of an element of X changes, the resulting Y will not change as it is ϵ that will absorb the change in X's variance. As in PCA, it can be seen that in FA X is being explained by underlying hidden causes Y, and the same graph (Figure 5) can describe both PCA and FA.

Probabilistic PCA (PPCA) (second from the right in Figure 5) [147, 140] while not widely used in ASR is only a simple modification to FA, where $\Psi = \sigma^2 I$ is constrained so that ϵ is a spherically-distributed Gaussian.

In all of the models above, the hidden causes Y are uncorrelated Gaussians, and therefore are marginally independent. Any statistical dependence between elements of X exist only in how they are jointly dependent on one or more of the hidden causes Y. It is possible to use a GM to make this marginal Y dependence explicit, as is provided on the left in Figure 6 where all nodes are now scalars. In this case, $Y_j \sim \mathcal{N}(0, 1)$, $\epsilon_j \sim \mathcal{N}(0, \psi_j)$, and $p(x_i) \sim \mathcal{N}(\sum_j C_{ij} y_j + \mu_i, \psi_i)$ where $\psi_j = 0$ for PCA.

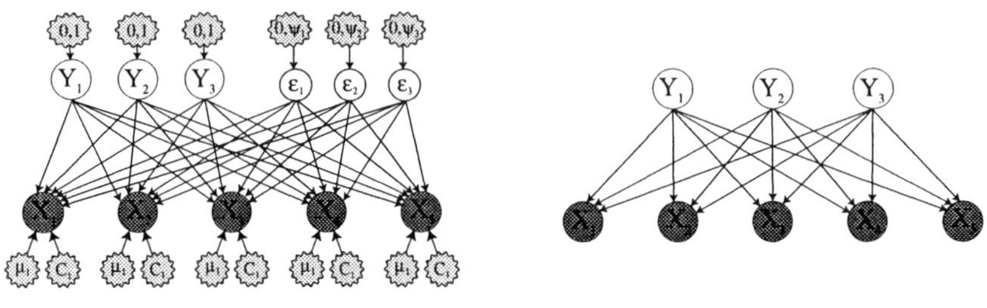

FIG. 6. *Left: A graph showing the explicit scalar variables (and therefore their statistical dependencies) for PCA, PPCA, and FA. The graph also shows the parameters for these models. In this case, the dependencies are linear and the random variables are all Gaussian. Right: The graph for PCA/PPCA/FA (parameters not shown) which is the same as the graph for ICA. For ICA, the implementation of the dependencies and the random variable distributions can be arbitrary, different implementations lead to different ICA algorithms. The key goal in all cases is to explain the observed vector X with a set of statistically independent causes Y.*

The PCA/PPCA/FA models can be viewed without the parameter and noise nodes, as shown on the right in Figure 6. This, however, is the general model for independent component analysis (ICA) [7, 92], another method that explains data vectors X with independent hidden causes. Like PCA and FA, a goal of ICA is to first learn the parameters of the model that explain X. Once done, it is possible to find Y, the causes of X, that are as statistically independent as possible. Unlike PCA and FA, however, dependency implementations in ICA neither need to be linear nor Gaussian. Since the graph on the right in Figure 6 does not depict implementations, the vector Y can be any non-linear and/or non-Gaussian causes of X. The graph insists only that the elements of Y are marginally independent, leaving alone the operations needed to compute $E[Y|X]$. Therefore, ICA can be seen simply as supplying the mechanism for different implementation of the dependencies used to infer $E[Y|X]$. Inference can still be done using the standard graphical-model inference machinery, described in Section 2.5.

Further generalizations of PCA/FA/ICA can be obtained simply by using different implementations of the basic graph given in Figure 6. Independent factor analysis [6] occurs when the hidden causes Y are described by a mixture of Gaussians. Moreover, a multi-level factor analysis algorithm, shown in Figure 7, can easily be described where the middle hidden layer is a possibly non-independent explanation for the final marginally independent components. The goal again is to train parameters to explain X, and to compute $E[Z|X]$. With graphs it is therefore easy to understand all of these techniques, and simple structural or implementation changes can lead to dramatically different statistical procedures.

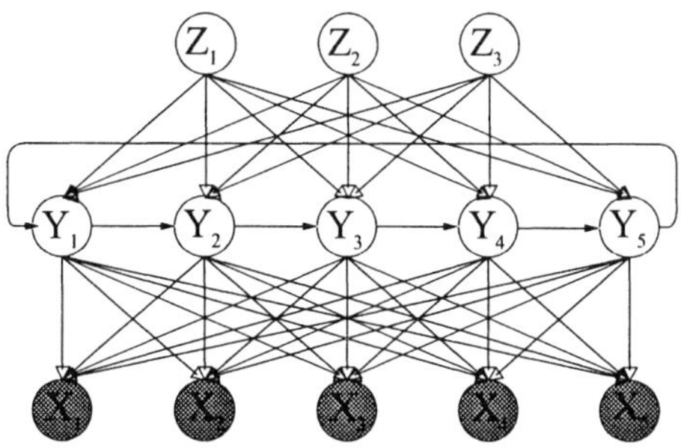

FIG. 7. *Multi-level ICA.*

3.3. Acoustic modeling: LDA/QDA/MDA/QMDA. When the goal is pattern classification [44] (deciding amongst a set of classes for X), it is often beneficial to first transform X to a space spanned neither by the principle nor the independent components, but rather to a space that best discriminatively represents the classes. Let C be a variable that indicates the class of X, $|C|$ the cardinality of C. As above, a linear transformation can be used, but in this case it is created to maximize the between-class covariance while minimizing the within-class covariance in the transformed space. Specifically, the goal is to find the linear transformation matrix A to form $Y = AX$ that maximizes $\text{tr}(BW^{-1})$ [57] where

$$W = \sum_i p(C = i) E_{p(y|C=i)}[(Y - \mu_y^i)(Y - \mu_y^i)^T]$$

and

$$B = \sum_i p(C = i)(\mu_y^i - \mu_y)(\mu_y^i - \mu_y)^T$$

where μ_y^i is the class conditional mean and μ_y is the global mean in the transformed space. This is a multi-dimensional generalization of Fisher's original linear discriminant analysis (LDA) [49].

LDA can also be seen as a particular statistical modeling assumption about the way in which observation samples X are generated. In this case, it is assumed that the class conditional distributions in the transformed space $P(Y|C = i)$ are Gaussians having priors $P(C = i)$. Therefore, Y is a mixture model $p(y) = \sum_i p(C = i)p(y|C = i)$, and classification of y is optimally performed using the posterior:

$$p(C = i|y) = \frac{p(y|i)p(i)}{\sum_j p(y|j)p(j)}.$$

For standard LDA, it is assumed that the Gaussian components $p(y|j) = \mathcal{N}(y; \mu_j, \Sigma)$ all have the same covariance matrix, and are distinguished only by their different means. Finally, it is assumed that there is a linear transform relating X to Y. The goal of LDA is to find the linear transformation that maximizes the likelihood $P(X)$ under the assumptions given by the model above. The statistical model behind LDA can therefore be graphically described as shown on the far left in Figure 8.

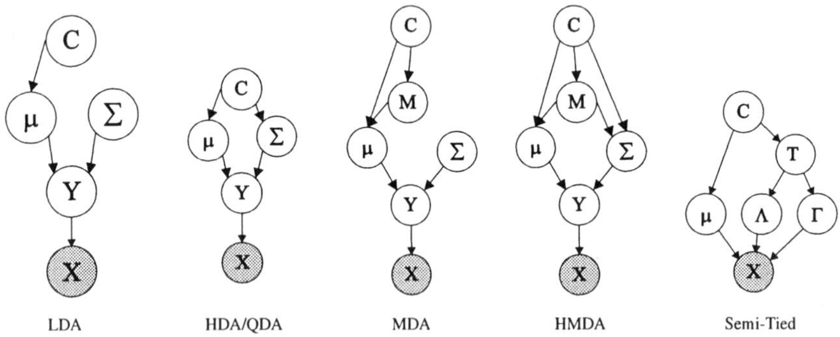

FIG. 8. *Linear discriminant analysis (left), and its generalizations.*

There is a intuitive way in which these two views of LDA (a statistical model or simply an optimizing linear transform) can be seen as identical. Consider two class-conditional Gaussian distributions with identical covariance matrices. In this case, the discriminant functions are linear, and effectively project any unknown sample down to an affine set[4], in this case a line, that points in the direction of the difference between the two means [44]. It is possible to discriminate as well as possible by choosing a threshold along this line — the class of X is determined by the side of the threshold X's projection lies.

More generally, consider the affine set spanned by the means of $|C|$ class-conditional Gaussians with identical covariance matrices. Assuming the means are distinct, this affine set has dimensionality $\min\{|C| - 1, \dim(X)\}$. Discriminability is captured entirely within this set since the decision regions are hyperplanes orthogonal to the lines containing pairs of means [44]. The linear projection of X onto the $|C| - 1$ dimensional affine set Y spanned by the means leads to no loss in classification accuracy, assuming Y indeed is perfectly described with such a mixture. If fewer than $C - 1$ dimensions are used for the projected space (as is often the case with LDA), this can lead to a dimensionality reduction algorithm that has a minimum loss in discriminative information. It is shown in [102] that the original formulation of LDA ($Y = AX$ above) is identical to the maximum likelihood linear transformation from the observations X to Y under the model described by the graph shown on the left in Figure 8.

When LDA is viewed as graphical model, it is easy to extend it to more general techniques. The simplest extension allows for different covariance matrices so that $p(x|i) = \mathcal{N}(x; \mu_i, \Sigma_i)$, leading to the GM second from the left in Figure 8. This has been called quadratic discriminant analysis (QDA) [44, 113], because decision boundaries are quadratic rather than linear, or heteroscedastic discriminant analysis (HDA) [102], because covariances are not identical. In the latter case, it is assumed that only a portion of the mean vectors and covariance matrices are class specific — the remainder corresponds in the projected space to the dimensions that do not carry discriminative information.

Further generalizations to LDA are immediate. For example, if the class conditional distributions are Gaussians mixtures, every component sharing the same covariance matrix, then mixture discriminant analysis (MDA) [74] is obtained (3rd from the left in Figure 8). A further generalization yields what could be called be called heteroscedastic MDA, as described 2nd from the right in Figure 8. If non-linear dependencies are allowed between the hidden causes and the observed variables, then one may obtain non-linear discriminant analysis methods, similar to the neural-network feature preprocessing techniques [51, 95, 78] that have recently been used.

Taking note of the various factorizations one may perform on a positive-definite matrix [73], a concentration matrix K within a Gaussian distribution can be factored as $K = \Lambda^T \Gamma \Lambda$. Using such a factorization, each

[4]An affine set is simply a translated subspace [135].

Gaussian component in a Gaussian mixture can use one each from a shared pool of Λs and Γs, leading to what are called semi-tied covariance matrices [62, 165]. Once again, this form of tying can be described by a GM as shown by the far right graph in Figure 8.

3.4. Acoustic modeling: Mixture models. In speech recognition, hidden Markov model observation distributions rarely use only single component Gaussian distributions. Much more commonly, mixtures of such Gaussians are used. A general mixture distribution for $p(x)$ assumes the existence of a hidden variable C that determines the active mixture component as in:

$$p(x) = \sum_i p(x, C = i) = \sum_i p(C = i)p(x|C = i)$$

where $p(x|C = i)$ is a component of the mixture. A GM may simply describe a general mixture distribution as shown in the graph $C \to X$. Conditional mixture generalizations, where X requires Z, are quite easy to obtain using the graph $Z \to C \to X$, leading to the equation:

$$p(x|z) = \sum_i p(x, C = i, z) = \sum_i p(C = i|z)p(x|C = i).$$

Many texts such as [148, 112] describe the properties of mixture distributions, most of which can be described using graphs in this way.

3.5. Acoustic modeling: Acoustic classifier combination. It has often been found that when multiple separately trained classifiers are used to make a classification decision in tandem, the resulting classification error rate often decreases. This has been found in many instances both empirically and theoretically [82, 100, 124, 160, 19]. The theoretical results often make assumptions about the statistical dependencies amongst of the various classifiers, such as that their errors are assumed to be statistically independent. The empirical results for ASR have found that combination is useful at the acoustic feature level [12, 94, 72, 95], the HMM state level [96], the sub-word or word level [163], and even at the utterance level [48].

Assume that $p_i(c|x)$ is a probability distribution corresponding to the ith classifier, where c is the class for feature set x. A number of classification combination rules exist such as the sum rule [97] where $p(c|x) = \sum_i p(i)p_i(c|x)$, or the product rule where $p(c|x) \propto \prod_i p_i(c|x)$. Each of these schemes can be explained statistically, by assuming a statistical model that leads to the particular combination rule. Ideally, the combination rule that performs best will correspond to the model that best matches the data. For example, the sum rule corresponds to a mixture model described above, and the product rule can be derived by the independence assumptions corresponding to a naive Bayes classifier [18]. Additional combination schemes, moreover, can be defined under the assumption of different models, some of which might not require the errors to be statistically independent.

More advanced combination schemes can be defined by, of course, assuming more sophisticated models. One such example is shown in Figure 9, where a true class variable C drives several error-full (noisy) versions of the class C_i, each of which generates a (possibly quite dependent) set of feature vectors. By viewing the process of classifier combination as a graph, and by choosing the right graph, one may quickly derive combination schemes that best match the data available and that need not make assumptions which might not be true.

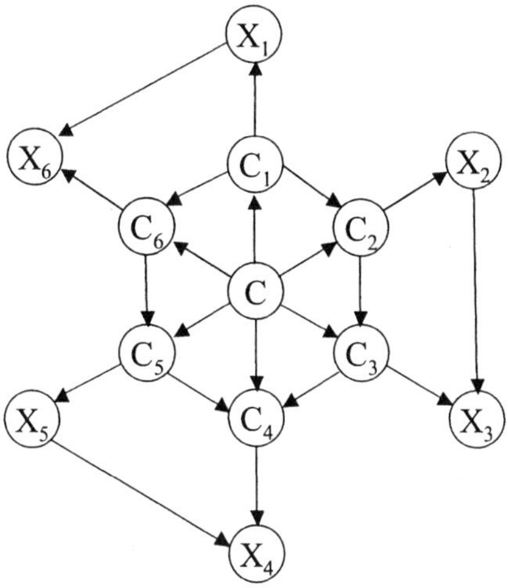

FIG. 9. *A GM to describe the process of classifier combination. The model does not require in all cases that the errors are statistically independent.*

3.6. Acoustic modeling: Adaptation. It is typically the case that additional ASR WER improvements can be obtained by additional adaptation of the model parameters after training has occurred, but before the final utterance hypothesis is decided upon. Broadly, these take the form of vocal-tract length normalization (VTLN) [91], and explicit parameter adaptation such as maximum-likelihood linear regression (MLLR) [104]. It turns out that these procedures may also be described with a GM.

VTLN corresponds to augmenting an HMM model with an additional global hidden variable that indicates the vocal tract length. This variable determines the transformation on the acoustic feature vectors that should be performed to "normalize" the affect of vocal-tract length on these features. It is common in VTLN to perform all such transformations, and the one yielding the highest likelihood of the data is ultimately chosen to produce a probability score. The graph in Figure 10 shows this model, where A indicates vocal-tract length and that can potentially affect the entire model as shown (in this case, the figure shows a hidden Markov model which will be described in Section 3.9). In a "Viterbi" approach, only the most probable assignment of A is used to form a probability score. Also, A is often a conditional-only variable (see Section 2.2) so the prior $P(A)$ is not counted. If a prior is available, it is also possible to integrate over all values to produce the final probability score.

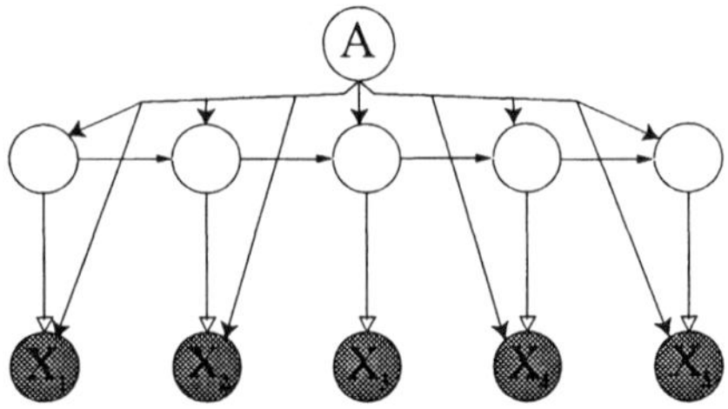

FIG. 10. *A GM to describe various adaptation and global parameter transformation methods, such as VTLN, MLLR, and SAT. The variable A indicates that the parameters of the entire model can be adapted.*

MLLR [104], or more generally speaker adaptation [164], corresponds to adjusting parameters of a model at test time using adaptation data that is not available at training time. In an ASR system, this takes the form of training on a speaker or an acoustic environment that is not (necessarily) encountered in training data. Since supervised training requires supervisory information, either that is available (supervised speaker adaptation), or an initial recognition pass is performed to acquire hypothesized answers for an unknown utterance (unsupervised speaker adaptation) — in either case, these hypotheses are used as the supervisory information to adjust the model. After this is done, a second recognition pass is performed. The entire procedure may also be repeated. The amount of novel adaptation information is often limited (typically a single utterance), so rather than adjust all the parameters of the model directly, typically a simple global transformation of those parameters is learned (e.g., a linear transformation of all of the means in a Gaussian-mixture HMM system). This procedure is also described in Figure 10, where A in this case indicates the global transformation. During adaptation, all of the model parameters are held fixed except for A which is adjusted to maximize the likelihood of the adaptation data.

Finally, speaker adaptive training (SAT) [3] is the dual of speaker adaptation. Rather than learn a transformation that maps the parameters from being speaker-independent to being speaker-dependent and doing so at recognition time, in SAT such a transformation is learned at training time. With SAT, the speaker-independent parameters of a model along with speaker-specific transformations are learned simultaneously. This procedure corresponds to a model that possesses a variable that identifies the speaker, is observed during training, and is hidden during testing. The speaker variable is the parent of the transformation mapping from speaker-independent to speaker-dependent space, and the transformation could potentially affect all the remaining parameters in the system. Figure 10 once again describes the basic structure, with A the speaker variable. During recognition, either the most likely transformation can be used (a Viterbi approach), or all speaker transformations can be used to form an integrative score.

In the cases above, novel forms of VTLN, MLLR, or SAT would arise simply by using different implementations of the edges between A and the rest of the model.

3.7. Pronunciation modeling. Pronunciation modeling in ASR systems involves examining each word in a lexicon, and finding sets of phone strings each of which describes a valid instance of the corresponding word [28, 134, 45, 52, 89]. Often these strings are specified probabilistically, where the probability of a given phone depends on the preceding phone (as in a Markov chain), thus producing probabilities of pronunciation variants of a word. The pronunciation may also depend on the acoustics [52].

Using the chain rule, the probability of a string of T phones $V_{1:T}$, where V_i is a phone, can be written as:

$$P(V_{1:T}) = \prod_i p(V_t|V_{1:t-1}).$$

If it is assumed that only the previous K phones are relevant for determining the current phone probability, this yields a Kth-order Markov chain. Typically, only a first-order model is used for pronunciation modeling, as is depicted in Figure 11.

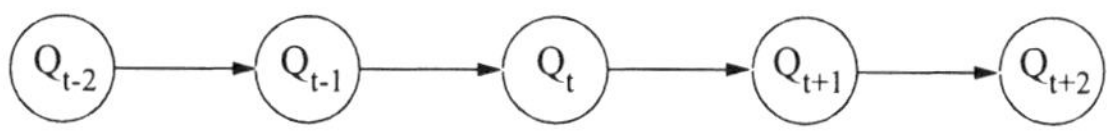

FIG. 11. *A simple first-order Markov chain. This graph encodes the relationship* $Q_t \perp\!\!\!\perp Q_{1:t-2}|Q_{t-1}$.

Phones are typically shared across multiple words. For example, in the two words "bat" and "bag", the middle phone /ae/ is the same. Therefore, it is advantageous in the acoustic Gaussian model for /ae/ to be shared between these two words. With a first-order model, however, it is possible only to select the distribution over the next state given the current one. This seems to present a problem since $P(V_t|/ae/)$ should choose a /t/ for "bat" and a /g/ for "bag". Clearly, then, there must be a mechanism, even in a first order case, to specify that the following V_t might need to depend on more than just the current phone.

Fortunately, there are several ways of achieving this issue. The easiest way is to expand the cardinality of V_t (i.e., increase the state space in the Markov chain). That is, the set of values of V_t represents not only the different phones, but also different positions of different words. Different values of v_t, corresponding to the same phone in different words, would then correspond to the same acoustic Gaussians, but the distribution of v_{t+1} given v_t would be appropriate for the word containing v_t and the position within that word. This procedure is equivalent to turning a Kth-order Markov chain into a first-order chain [83].

Another way to achieve this effect is rather than condition on the previous phone, condition instead on the word W_t and the sequential position of a phone in the word S_t, as in $P(V_t|W_t, S_t)$. The position variable is needed to select the current phone. A Markov chain can be used over the two variables W_t and S_t. This approach corresponds to expanding the graph in Figure 11 to one that explicitly mentions the variables needed to keep track of and use each phone, as further described in Section 5.

A GM view of a pronunciation model does not explicitly mention the non-zero entries in the stochastic matrices in the Markov chain. Stochastic finite state automata (SFSA) [130] diagrams are ideally suited for that purpose. A GM, rather, explains only the independence structure of a model. It is important to realize that while SFSAs are often described using graphs (circles and arrows), SFSA graphs describe entirely different properties of a Markov chain than do the graphs that are studied in this text.

Pronunciation modeling often involves a mapping from base-forms (isolated word dictionary-based pronunciations) to surface forms (context-dependent and data-derived pronunciations more likely to correspond to what someone might say). Decision trees are often used for this purpose [28, 134], so it is elucidative at this point to see how they may be described using GMs [86][5]. Figure 12 shows a standard decision tree on the right, and a stochastic GM version of the same decision tree on the left. In the graphical view, there is an input node I, an output node O, and a series of decision random variables D_i. The cardinality of the decision variables D_i is equal to the arity of the corresponding decision tree node (at roughly the same horizontal level in the figure) — the figure shows that all nodes have an arity of two (i.e., correspond to binary random variables).

In the GM view, the answer to a question at each level of the tree is made with a certain probability. All possible questions are considered, and a series of answers from the top to the bottom of the tree provides the probability of one of the possible outputs of the decision tree. The probability of an answer at a node is conditioned on the set of answers higher in the tree that lead to that node. For example, $D_1 = 0$ means the answer is 0 to the first question asked by the tree. This answer occurs with probability $P(D_1 = i|I)$. The next answer is provided with probability $P(D_2 = j|D_1, I)$ based on the first decision, leading to the graph on the left of the figure.

[5]These GMs also describe hierarchical mixtures of experts [87].

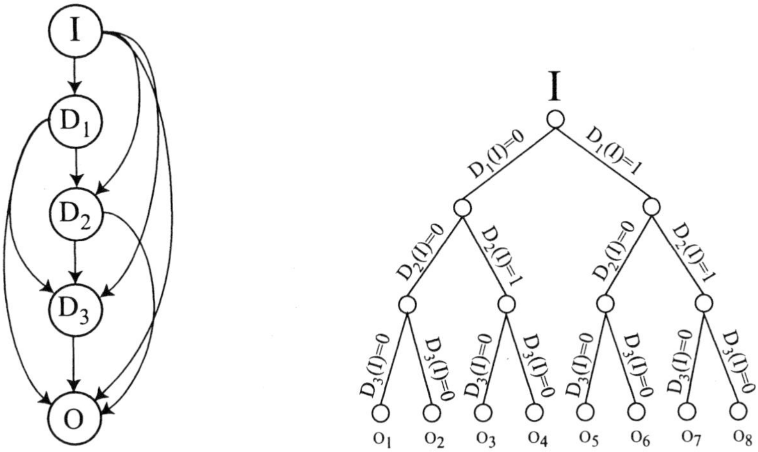

FIG. 12. *Left: A GM view of a decision tree, which is a probabilistic generalization of the more familiar decision tree on the right.*

In a normal decision tree, only one decision is made at each level in the tree. A GM can represent such a "crisp" tree by insisting that the distributions at each level D_ℓ (and the final decision O) of the tree are Dirac-delta functions, such as $P(D_\ell = i|I) = \delta_{i,f_\ell(I,d_{1:\ell-1})}$ where $f_\ell(I, d_{1:\ell-1})$ is a deterministic function of the input I and previously made decisions $d_{1:\ell-1}$, where $d_\ell = f_\ell(I, d_{1:\ell-1})$. Therefore, with the appropriate implementation of dependencies, it can be seen that the GM-view is a probabilistic generalization of normal decision trees.

3.8. Language modeling. Similar to pronunciation modeling, the goal of language modeling is to provide a probability for any possible string of words $W_{1:T}$ in a language. There are many varieties of language models (LMs) [83, 136, 118, 29], and it is beyond the scope of this paper to describe them all. Nevertheless, the following section uses GMs to portray some of the more commonly and successfully used LMs.

At this time, the most common and successful language model is the n-gram. Similar to pronunciation modeling, the chain rule is applied to a joint distribution over words $p(W_{1:T})$. Within each conditional factor $p(W_t|W_{1:t-1})$, the most distant parent variables are dropped until an $(n-1)^{th}$ order Markov chain results $p(W_t|W_{t-n+1:t-1}) = p(W_t|H_t)$, where H_t is the length $n-1$ word history. For a bi-gram ($n = 2$), this leads to a graph identical to the one shown in Figure 11. In general, tri-grams (i.e., 2nd-order Markov chains) have so far been most successful for language modeling among all values n [29].

While a graphical model showing an $(n-1)^{th}$-order Markov chain accurately depicts the statistical independence assumptions made by an

n-gram, it does not portray how the parameters of such a model are typically obtained, a procedure that can be quite involved [29]. In fact, much research regarding n-grams involves methods to cope with data-sparsity — because of insufficient training data, "smoothing" methodology must be employed, whereby a Kth order model is forced to provide probability for length $K+1$ strings of words that did not occur in training data. If a purely maximum-likelihood procedure was used, these strings would be given zero probability.

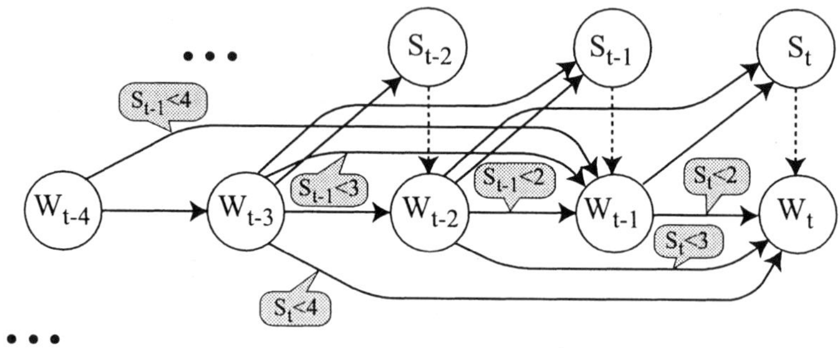

FIG. 13. *A GM view of a LM. The dashed arcs indicate that the parents are switching. The hidden switching parents S_t switch between the word variables W_t forming either a zeroth ($S_t = 1$), first ($S_t = 2$), or second ($S_t = 3$) order Markov chain. The switching parents also possess previous words as parents so that the probability of the Markov-chain order is itself context dependent.*

Often, smoothing takes the form of mixing together higher- and lower-order sub-models, with mixing weights determined from data not used for training any of the sub-models [83, 29]. In such a case, a language model mixture can be described by the following equation:

$$\begin{aligned} p(w_t|w_{t-1}, w_{t-2}) = \alpha_3(w_{t-1}, w_{t-2}) f(w_t|w_{t-1}, w_{t-2}) \\ + \alpha_2(w_{t-1}, w_{t-2}) f(w_t|w_{t-1}) \\ + \alpha_1(w_{t-1}, w_{t-2}) f(w_t) \end{aligned}$$

where $\sum_i \alpha_i = 1$ for all word histories, and where the α coefficients are some (possibly) history-dependent mixing values that determine how much each sub-model should contribute to the total probability score. Figure 13 shows this mixture using a graph with switching parents (see Section 2.2). The variables S_t correspond to the α coefficients, and the edges annotated with values for S_t exist only in the case that S_t has those values. The dashed edges between S_t and W_t indicate that the S_t variables are switching rather than normal parents. The graph describes the statistical underpinnings of many commonly used techniques such as deleted interpolation [83], which is a form of parameter training for the S_t variables. Of course, much of the success of a language model depends on the form of smoothing that is used [29], and such methods are not depicted by Figure 13 (but see Figure 15).

A common extension to the above LM is to cluster words together and then form a Markov chain over the word group clusters, generally called a class-based LM [24]. There are a number of ways that these clusters can be formed, such as by grammatical category or by data-driven approaches that might use decision trees (as discussed in [83, 24]). Whatever the method, the underlying statistical model can also be described by a GM, as shown in Figure 14. In this figure, a Markov chain exists over a (presumably) much lower dimensional class variables C rather than the high-dimensional word variables. This representation can therefore considerably decrease model complexity and therefore lower parameter estimation variance.

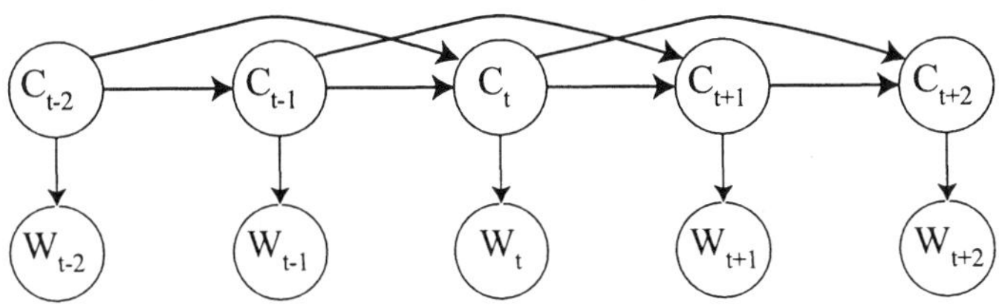

FIG. 14. *A class-based language model. Here, a Markov chain is used to model the dynamics of word classes rather than the words themselves.*

The class-based language model can be further extended to impose certain desirable constraints with respect to words that do not occur in training material, or the so-called unknown words. It is common in a language model to have a special token called **unk** indicating the unknown word. Whenever a word is encountered in a test set that has not occurred in the training set, the probability of **unk** should be given as the probability of the unknown word. The problem, however, is that if maximum likelihood estimates of the parameters of the language model are obtained using the training set, the probability of **unk** will be zero. It is therefore typical to force this token to have a certain non-zero probability and in doing so, essentially "steal" probability mass away from some of the tokens that do indeed occur in the training set. There are many ways of implementing such a feature, generally called language model back-off [83]. For our purposes here, it will be sufficient to provide a simple model, and show how it can be enforced by an explicit graph structure.[6]

Suppose that the vocabulary of words $\mathcal{W}$ can be divided into three disjoint sets: $\mathcal{W} = \{\textbf{unk}\} \bigcup \mathcal{S} \bigcup \mathcal{M}$, where **unk** is the token representing the unknown word, $\mathcal{S}$ is the set of items that have occurred only one time in the training set (the singletons), and $\mathcal{M}$ is the set of all other lexical items. Let

[6]Thanks to John Henderson who first posed the problem to me of how to represent this construct using a graphical model, in the context of building a word-tagger.

us suppose also that we have a maximum-likelihood distribution p_{ml} over words in $\mathcal{S}$ and $\mathcal{M}$, such that $\sum_w p_{ml}(w) = 1$, $p_{ml}(\texttt{unk}) = 0$, and in general

$$p_{ml}(w) = \frac{N(w)}{N},$$

where $N(w)$ is the number of times word w occurs in the training set, and N is the total number of words in the training set. This means, for example, that $N(w) = 1, w \in \mathcal{S}$.

One possible assumption is to force the probability of unk to be 0.5 times the probability of the entire singleton set, i.e., $p(\text{unk}) = 0.5 * p_{ml}(\mathcal{S}) = 0.5 * \sum_{w\in\mathcal{S}} p_{ml}(w)$. This requires that probability be taken away from tokens that do occur in the training set. In this case probability is removed from the singleton words, leading to the following desired probability model $p_d(w)$:

$$p_d(w) = \begin{cases} 0.5p_{ml}(\mathcal{S}) & \text{if } w = \texttt{unk} \\ 0.5p_{ml}(w) & \text{if } w \in \mathcal{S} \\ p_{ml}(w) & \text{otherwise.} \end{cases} \tag{3.3}$$

This model, of course, can easily be modified so that it is conditioned on the current class $p_d(w|c)$ and so that it uses the conditional maximum likelihood distribution $p_{ml}(w|c)$. Note that this is still a valid probability model, as $\sum_w p_d(w|c) = 1$.

The question next becomes how to represent the constraints imposed by this model using a directed graph. One can of course produce a large conditional probability table that stores all values as appropriate, but the goal here is to produce a graph that explicitly represents the constraints above, and that can be used to train such a model. It might seem at first not to be possible because the variable W must be used both to switch in different distributions and, once a given distribution has been selected, as a probabilistic query variable. The variable W can not exist both on the left of the conditioning bar (where it is possible to produce a probability of $W = w$) and also on the right of the conditioning bar (where it can be used to select the current distribution).

Surprisingly, there exists a solution within the space of directed graphical models shown in Figure 15. This graph, a modified class-based language model, includes a child variable V_t at each time that is always observed to be $V_t = 1$. This means, rather than compute $p(w_t|c_t)$ at each time step, we compute $p(w_t, V_t = 1|c_t)$. The goal is to show that it is possible for $p(w_t, V_t = 1|c_t) = p_d(w_t|c_t)$. The variables B_t and K_t are both binary valued for all t, and these variables are also switching parents (see Section 2.2) of W_t. From the graph, we see that there are a number of conditional distributions that need to be defined. Before doing that, two auxiliary distributions are defined so as to make the definitions that follow simpler:

$$p_M(w|c) \triangleq \begin{cases} \dfrac{p_{ml}(w|c)}{p(\mathcal{M}|c)} & \text{if } w \in \mathcal{M} \\ 0 & \text{else} \end{cases} \tag{3.4}$$

where $p(\mathcal{M}|c) \triangleq \sum_{w\in\mathcal{M}} p_{ml}(w|c)$, and

$$p_S(w|c) \triangleq \begin{cases} \dfrac{p_{ml}(w|c)}{p(\mathcal{S}|c)} & \text{if } w \in \mathcal{S} \\ 0 & \text{else} \end{cases} \tag{3.5}$$

where $p(\mathcal{S}|c) \triangleq \sum_{w\in\mathcal{S}} p_{ml}(w|c)$. Note that both p_M and p_S are valid normalized distributions over all words. Also, $p(\mathcal{S}|c) + p(\mathcal{M}|c) = 1$ since these two quantities together use up all the probability mass contained in the maximum likelihood distribution.

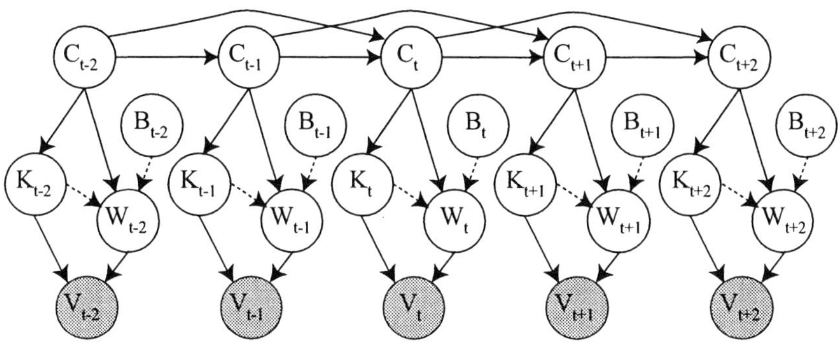

FIG. 15. *A class-based language model that forces the probability of* unk *to be* α *times the probability of all singleton words. The* V_t *variables are shaded, indicating that they are observed, and have value* $V_t = 1, \forall t$. *The* K_t *and* B_t *variables are switching parents of* W_t.

The remaining distributions are as follows. First, B_t has a binary uniform distribution:

$$p(B_t = 0) = p(B_t = 1) = 0.5. \tag{3.6}$$

The observation variable $V_t = 1$ simply acts as in indicator, and has a distribution that produces probability one only if certain conditions are met, and otherwise produces probability zero:

$$p(V_t = 1|w_t, k_t) = 1_{\{(w_t\in\mathcal{S},k_t=1) \text{ or } (w_t\in\mathcal{M},k_t=0) \text{ or } (w_t=\textbf{unk},k_t=1)\}} \tag{3.7}$$

where 1_A is a binary indicator function that is unity only when the event A is true, and is zero otherwise. Next, the word distribution will switch between one of three distributions depending on the values of the switching parents K_t and B_t, as follows:

$$p(w_t|k_t, b_t, c_t) = \begin{cases} p_M(w_t|c_t) & \text{if } k_t = 0 \\ p_S(w_t|c_t) & \text{if } k_t = 1 \text{ and } b_t = 1 \\ \delta_{\{w_t=\mathtt{unk}\}} & \text{if } k_t = 1 \text{ and } b_t = 0. \end{cases} \tag{3.8}$$

Note that the third distribution is simply a Dirac-delta distribution, giving probability one only when w_t is the unknown word. Last, the distribution for K_t is as follows:

$$p(k_t|c_t) = \begin{cases} p(\mathcal{S}|c_t) & \text{if } k_t = 1 \\ 1 - p(\mathcal{S}|c_t) & \text{otherwise.} \end{cases} \tag{3.9}$$

This model correctly produces the probabilities that are given in Equation 3.3. First, when $w_t = \mathtt{unk}$:

$$\begin{aligned} p(w_t = \mathtt{unk}, v_t = 1) &= p(v_t = 1|k_t = 1, w_t = \mathtt{unk}) \\ &\quad \times p(w_t = \mathtt{unk}|k_t = 1, b_t = 0, c_t) \\ &\quad \times p(b_t = 0) \times p(k_t = 1|c_t) \\ &= 1 \times 1 \times 0.5 \times p(\mathcal{S}|c_t) \\ &= 0.5p(\mathcal{S}|c_t) \end{aligned}$$

as desired. This follows because the other terms, for different values of the hidden variables, are all zero. Next, when $w_t \in \mathcal{S}$,

$$\begin{aligned} p(w_t, v_t = 1) &= p(v_t = 1|k_t = 1, w_t \in \mathcal{S}) \times p(w_t|k_t = 1, b_t = 1, c_t) \\ &\quad \times p(b_t = 1) \times p(k_t = 1|c_t) \\ &= 1 \times p_S(w_t|c_t) \times 0.5 \times p(\mathcal{S}|c_t) \\ &= 0.5 * p_{ml}(w_t|c_t) \end{aligned}$$

again as desired. Lastly, when $w_t \in \mathcal{M}$,

$$\begin{aligned} p(w_t, v_t = 1) &= p(v_t = 1|k_t = 0, w_t \in \mathcal{M}) \times \left(\sum_{b_t \in \{0,1\}} p(w_t|k_t = 0, b_t, c_t) p(b_t) \right) \\ &\quad \times p(k_t = 0|c_t) \\ &= 1 \times p_M(w_t|c_t) \times p(\mathcal{M}|c_t) \\ &= p_{ml}(w_t|c_t). \end{aligned}$$

In this last case, B_t has no influence as it is marginalized away — this is because the event $K_t = 0$ removes the parent B_t from W_t. Once the graph structures and implementations are set up, standard GM learning algorithms can be used to obtain smoothed parameters for this distribution.

Many other such models can be described by directed graphs in a similar way. Moreover, many language models are members of the family of exponential models[44]. These include those models whose parameters are learned by maximum entropy methods [126, 83, 9, 137], and are derived by establishing a number of constraints that the underlying probability distribution must possess. The goal is to find a distribution satisfying these constraints and that otherwise has maximum entropy (or minimum KL-divergence with some desired distribution [126]). Note that such an approach can also be used to describe the distribution over an entire sentence[138] at a time, rather than a conditional distribution of the current word w_t given the current history h_t. Such maximum entropy models can be described by UGMs, where the edges between words indicate that there is some dependency induced by the constraint functions. In many cases, the resulting graphs can become quite interesting.

Overall, however, it is clear that there are a multitude of ways to depict language models with GMs, and this section has only begun to touch upon this topic.

3.9. GMs for basic speech models. The Hidden Markov model (HMM) is still the most successful statistical technique used in ASR. The HMM encompasses standard acoustic, pronunciation, and most language modeling into a single unified framework. This is because pronunciation and language modeling can be seen as a large finite-state automata that can be "flattened" down to a single first-order Markov chain [116, 83]. This Markov chain consists of a sequence of serially connected discrete hidden variables during recognition, thus the name HMM.

Most generally, a hidden Markov model (HMM) is collection of T discrete scalar random variables $Q_{1:T}$ and T other variables $X_{1:T}$ that may be either discrete or continuous (and either scalar- or vector-valued). These variables, collectively, possess the following conditional independence properties:

$$Q_{t:T} \perp\!\!\!\perp Q_{1:t-2} | Q_{t-1} \tag{3.10}$$

and

$$X_t \perp\!\!\!\perp \{Q_{\neg t}, X_{\neg t}\} | Q_t \tag{3.11}$$

for each $t \in 1 : T$. $Q_{\neg t}$ refers to all variables Q_τ except for the one at time $\tau = t$. The length T of these sequences is itself an integer-valued random variable having a complex distribution. An HMM consists of a hidden Markov chain of random variables (the unshaded nodes) and a collection of nodes corresponding to the speech utterance (the shaded nodes). In most ASR systems, the hidden chain corresponds to sequences of words, phones, and sub-phones.

This set of properties can be concisely described using the GM shown in Figure 16. The figure shows two equivalent representations of an HMM, one as a BN and another as an UGM. They are equivalent because moralizing the BN introduces no edges, and because the moralized HMM graph is already triangulated and therefore decomposable. The UGM on the right is the result of moralizing the BN on the left. Interestingly, the same graph describes the structure of a Kalman filter [70] where all variables are continuous and Gaussian and all dependency implementations are linear. Kalman filter operations are simply applications of the formulas for conditional Gaussians (Section 3.1), used in order to infer conditional means and covariances (the sufficient statistics for Gaussians).

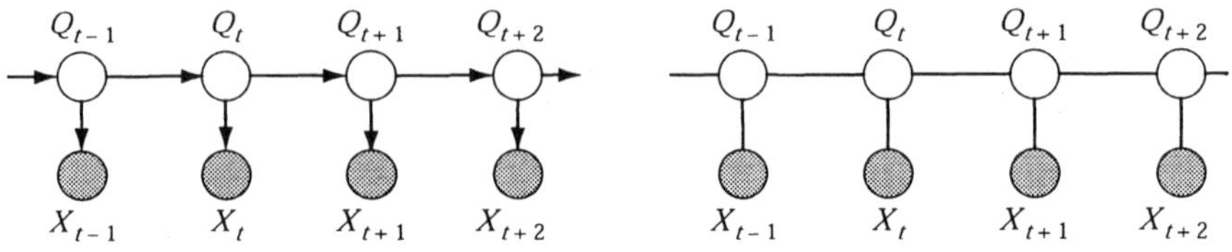

FIG. 16. *A hidden Markov model (HMM), viewed as a graphical model. Note that an HMM may be equivalently viewed either as a directed (left) or an undirected (right) model, as in this case the conditional independence properties are the same.*

In a standard HMM with Gaussian mixture observation densities, each value of the hidden variable (i.e., each state) corresponds to a separate (possibly tied) mixture distribution (Figure 17). Other forms of HMM also exist, such as when there is a single global pool of Gaussians, and each state corresponds to a particular mixture over this global pool. This is often called a semi-continuous HMM (similar to vector quantization [69]), and corresponds to the state-conditional observation equation:

$$p(x|Q = q) = \sum_i p(C = i|Q = q)p(x|C = i).$$

In other words, each state uses a mixture with components from this globally shared set of distributions. The GM for such an HMM loses an edge between Q and X as shown on the right in Figure 17. In this case, all of the represented dependence occurs via the hidden mixture variable at each time.

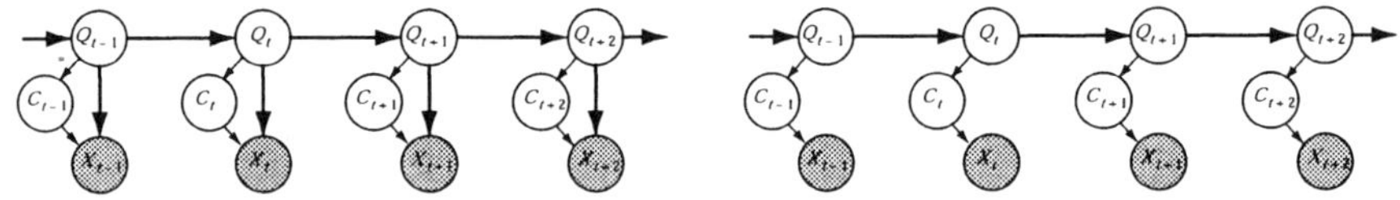

FIG. 17. *An HMM with mixture observation distributions (left) and a semi-continuous HMM (right).*

Still another modification of HMMs relaxes one of the HMM conditional independence statements, namely that successive feature vectors are conditionally independent given the state. Auto-regressive, or correlation HMMs [157, 23, 120], place an additional edges between successive observation vectors. In other words, the variable X_t might have as a parent not only the variable Q_t but also the variables X_{t-l} for $l = 1, 2, \ldots, K$ for some K. The case where $K = 1$ is shown in Figure 18. When the additional dependencies are linear and Gaussian, these are sometimes called conditional Gaussian HMMs [120].

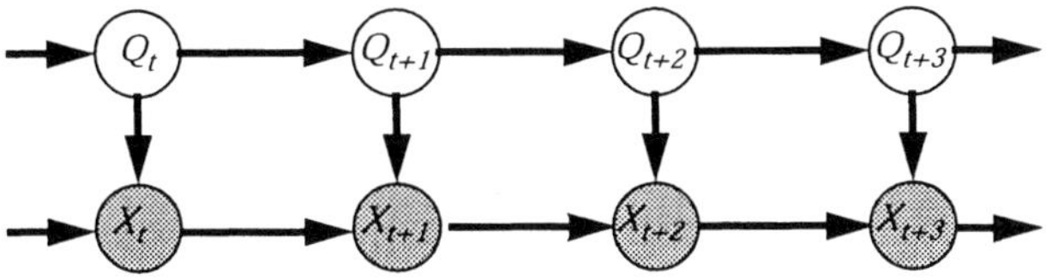

FIG. 18. *An Auto-regressive HMM as a GM.*

Note that although these models are sometimes called vector-valued auto-regressive HMMs, they are not to be confused with auto-regressive, linear predictive, or hidden filter HMMs [127, 128, 88, 129]. These latter models are HMMs that have been inspired from the use of linear-predictive coefficients for speech [129]. They use the observation distribution that arises from random Gaussian noise sources passed through a hidden-state dependent auto-regressive filter. The filtering occurs at the raw acoustic (signal) level rather than on the observation feature vector (frame) level. These earlier models can also be described by an GM that depicts state-conditioned auto-regressive models at the speech sample level.

Our last example of an augmented HMM is something often called an input-output HMM [8] (See Figure 20). In this case, there are variables at each time frame corresponding both to the input and the output. The output variables are to be inferred. Given a complete input feature stream $X_{1:T}$, one might want to find $E[Y|X]$, the most likely values for the output. These HMMs can therefore be used to map from a continuous variable length input feature streams to output stream. Such a model shows promise for speech enhancement.

While HMMs account for much of the technology behind existing ASR, GMs include a much larger space of models. It seems quite improbable that within this space, it is the HMM alone that is somehow intrinsically superior to all other models. While there are of course no guarantees to the following, it seems reasonable to assume that because the space of GMs is large and diverse, and because it includes HMMs, that there exists some model within this space that will greatly outperform the HMM. Section 4 begins to explore more advanced speech models as viewed from a GM perspective.

3.10. Why delta features work. State-of-the-art ASR systems augment HMM feature vectors X_t with approximations to their first and second order time-derivatives (called delta- and delta-delta- features [46, 58–60], or just "dynamic" features). Most often, estimates of the derivative are obtained using linear regression [129], namely:

$$\dot{x}_t = \frac{\sum_{k=-K}^{K} k x_t}{\sum_{k=-K}^{K} k^2}$$

where K in this case is the number of points used to fit the regression. This can be viewed as a regression because

$$\dot{x}_t = \sum_{k=-K}^{K} a_k x_{t-k} + \epsilon$$

where a_k are defined accordingly, and ϵ can be seen as a Gaussian error term. A new feature vector is then produced that consists of x_t and $\dot{x}_t$ appended together.

It is elucidating to expand the joint distribution of the features and the deltas, namely $p(x_{1:T}, \dot{x}_{1:T}) = \sum_{q_{1:T}} p(x_{1:T}, \dot{x}_{1:T}|q_{1:T})p(q_{1:T})$. The state conditioned joint distribution within the sum can be expanded as:

$$p(x_{1:T}, \dot{x}_{1:T}|q_{1:T}) = p(\dot{x}_{1:T}|x_{1:T}, q_{1:T})p(x_{1:T}|q_{1:T}).$$

The conditional distribution $p(x_{1:T}|q_{1:T})$ can be expanded as is normal for an HMM [129, 11], but

$$p(\dot{x}_{1:T}|x_{1:T}, q_{1:T}) = \prod_t p(\dot{x}_t|\text{parents}(\dot{x}_t)).$$

This last equation follows because, observing the process to generate delta features, $\dot{X}_t$ is independent of everything else given its parents. The parents of $\dot{X}_t$ are a subset of $X_{1:T}$ and they do not include the hidden variables Q_t. This leads to the GM on the left in Figure 19, a generative model for HMMs augmented with delta features. Note that the edges between the feature stream X_t, and the delta feature stream $\dot{X}_t$ correspond to deterministic linear implementations. In this view, delta-features appear to be similar to fixed-dependency auto-regressive HMMs (Figure 18), where each child feature has additional parents both from the past and from the future. In this figure, however, there are no edges between $\dot{X}_t$ and Q_t, because $\dot{X}_t \perp\!\!\!\perp Q_t|\text{parents}(\dot{X}_t)$. This means that parents$(\dot{X}_t)$ contain all the information about $\dot{X}_t$, and Q_t is irrelevant.

It is often asked why delta features help ASR performance as much as they do. The left of Figure 19 does not portray the model typically used with delta features. A goal of speech recognition is for the features to contain as much information as possible about the underlying word sequence as represented via the vector $Q_{1:T}$. The generative model on the left in Figure 19 shows, however, that there is zero information between the $\dot{X}_t$ and Q_t. When the edges between $\dot{X}_t$ and its parents parents$(\dot{X}_t)$ are

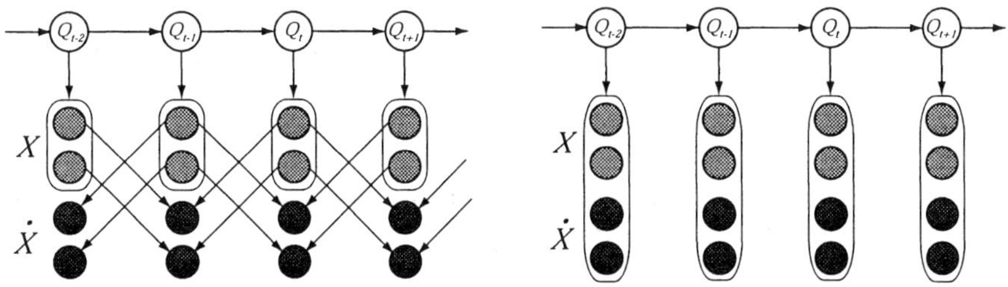

FIG. 19. *An GM-based explanation of why delta features work in HMM-based ASR systems. The left figure gives a GM that shows the generative process of HMMs with delta features. The right figure shows how delta features are typically used in an HMM system, where the information between $\dot{X}_t$ and Q_t is greatly increased relative to the left figure.*

removed, the mutual information [33] between $\dot{X}_t$ and Q_t can only *increase* (from zero to something greater) relative to the generative model. The right of Figure 19 thus shows the standard model used with deltas, where it is not the case that $\dot{X}_t \perp\!\!\!\perp Q_t$. Since in the right model, it is the case that more information about $\dot{X}_t$ and Q_t exist, it might be said that this model has a structure that is inherently more discriminative (see Section 5).

Interestingly, the above analysis demonstrates that additional conditional independence assumptions (i.e., fewer edges) in a model can increase the amount of mutual information that exists between random variables. When edges are added between the delta features and the generative parents X_t, the delta features become less useful since there is less (or zero) mutual information between them and Q_t.

Therefore, the very conditional independence assumptions that are commonly seen as a flaw of the HMM provide a benefit when using delta features. More strongly put, the *incorrect* statistical independence properties made by the HMM model on the right of Figure 19 (relative to truth, as shown by the generative model on the left) are the very thing that enable delta features to decrease recognition error. The standard HMM model with delta features seem to be an instance of a model with an inherently discriminative structure [16, 47] (see also Section 5).

In general, can the removal of edges or additional processing lead to and overall increase in the information between the entire random vectors $X_{1:T}$ and $Q_{1:T}$? The data processing inequality [33] says it can not. In the above, each feature vector $(\dot{X}_t, X_t)$ will have more information about the temporally local hidden variable Q_t — this can sometimes lead to better word error scores. This same analysis can be used to better understand other feature processing strategies derived from multiple frames of speech, such as PCA or LDA preprocessing over multiple windows [71] and other non-linear generalizations [51, 95, 78].

It has often been found that conditionally Gaussian HMMs (as in Figure 18) often do not provide an improvement when delta features are included in the feature stream [20, 23, 93, 158]. The above provides one possible explanation, namely that by having a delta feature $\dot{X}_t$ include as its parent say X_{t-1}, the mutual information between $\dot{X}_t$ and Q_t decreases (perhaps to zero). Note, however, that improvements were reported with the use of delta features in [161, 162] where discriminative output distributions were used. In [105, 106], successful results were obtained using delta features but where the conditional mean, rather than being linear, was non-linear and was implemented using a neural network. Furthermore, Buried Markov models [16] (to be described below) also found an improvement with delta features and additional dependencies, but only when the edges were added discriminatively.

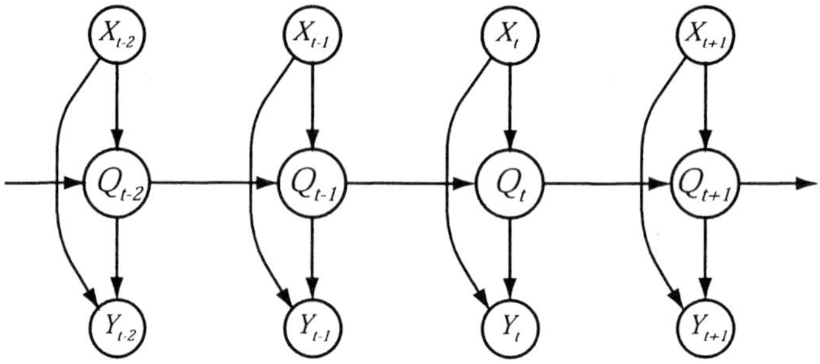

FIG. 20. *An input-output HMM.* $X_{1:T}$ *the input is transformed via integration over a Markov chain* $Q_{1:T}$ *into the output* $Y_{1:T}$.

4. GMs for advanced speech models. Many non-HMM models for speech have been developed outside the GM paradigm but turn out to be describable fairly easily as GMs — this section to describe some of them. While each of these models are quite different from one another, they can all be described with only simple modifications of an underlying graph structure.

The first example presented is a factorial HMM [67]. In this case, rather than a single Markov chain, multiple Markov chains are used to guide the temporal evolution of the probabilities over observation distributions (see Figure 21). The multiple hidden chains can be used to represent a number of real-world phenomena. For example, one chain might represent speech and another could represent an independent and dynamic noise source [90]. Alternatively, one chain could represent the speech to be recognized and the other chain could represent confounding background speech [151, 152][7], or the two chains might each represent two underlying

[7]A related method to estimate the parameters of a composite HMM given a collection of separate, independent, and already trained HMMs is called parallel model combination [64].

concurrent and independent sub-processes governing the realization of the observation vectors [61, 155, 108]. Such factored hidden state representations have also been called HMM decomposition [151, 152] in the past.

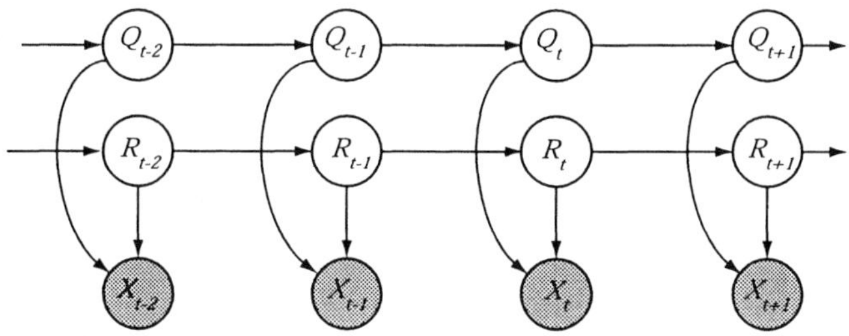

FIG. 21. *A factorial HMM where there are multiple hidden Markov chains.*

One can imagine many modifications of this basic structure, where edges are added between variables at each time step. Often, these separate Markov chains have been used for modeling separate loosely coupled streams of hidden articulatory information [131, 132] or to represent a coupling between phonetic and articulatory information [167, 145].

It is interesting to note that the factorial HMMs described above are all special cases of HMMs. That is, they are HMMs with tied parameters and state transition restrictions made according to the factorization. Starting with a factorial HMM consisting of two hidden chains Q_t and R_t, an equivalent HMM may be constructed by using $|\mathcal{Q}||\mathcal{R}|$ states and by restricting the set of state transitions and parameter assignments to be those only allowed by the factorial model. A factorial HMM using M hidden Markov chains each with K states that all span T time steps can have time complexity $O(TMK^{M+1})$ [67]. If one translates the factorial HMM into an HMM having K^M states, the complexity becomes $O(TK^{2M})$ which is significantly larger. An unrestricted HMM with K^M states will, however, have more expressive power than a factorial HMM with M chains each with K states because in the HMM there are no required state transition restrictions and any form of correlation may be represented between the separate chains. It is possible, however, that such an expanded state space would be more flexible than needed for a given task. Consider, as an example, the fact that many HMMs used for ASR have only simple left-to-right Markov chain structures.

As mentioned earlier, the GM for an HMM is identical to that of a Kalman filter — it is only the nodes and the dependency implementations that differ. Adding a discrete hidden Markov chain to a Kalman filter allows it to behave in much more complex ways than just a large joint Gaussian. This has been called a switching Kalman filter, as shown in Figure 22. A version of this structure, applied to ASR, has been called a hidden dynamic model [125]. In this case, the implementations of the dependences are such that the variables are non-linearly related.

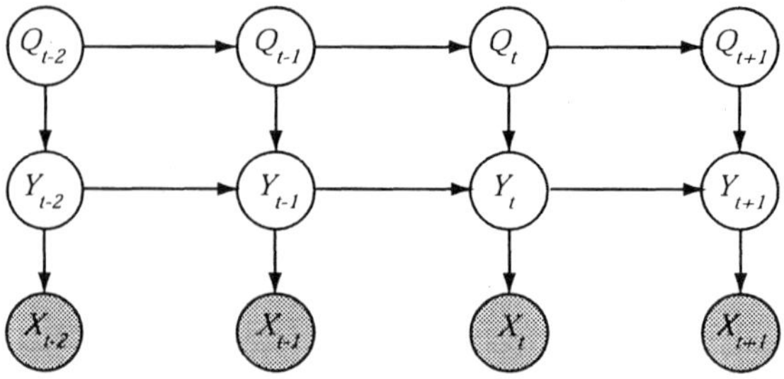

FIG. 22. *The GM corresponding to a switching Kalman filter (SKM). The Q variables are discrete, but the Y and X variables are continuous. In the standard SKM, the implementations between continuous variables are linear Gaussian, other implementations can be used as well and have been applied to the ASR problem.*

Another class of models well beyond the boundaries of HMMs are called segment or trajectory models [120]. In such cases, the underlying hidden Markov chain governs the evolution not of the statistics of individual observation vectors. Instead, the Markov chain determines the allowable sequence of observation segments, where each segment may be described using an arbitrary distribution. Specifically, a segment model uses the joint distribution over a variable length segment of observations conditioned on the hidden state for that segment. In the most general form, the joint distribution for a segment model is as follows:

$$\begin{aligned} p(X_{1:T} = x_{1:T}) &= \\ \sum_{\tau} \sum_{q_{1:\tau}} \sum_{\ell_{1:\tau}} \prod_{i=1}^{\tau} p(x_{t(i,1)}, x_{t(i,2)}, \ldots, x_{t(i,\ell_i)}, \ell_i | q_i, \tau) p(q_i | q_{i-1}, \tau) p(\tau). \end{aligned} \tag{4.1}$$

There are T time frames and τ segments where the ith segment is of a hypothesized length ℓ_i. The collection of lengths are constrained such that $\sum_{i=1}^{\tau} \ell_i = T$. For a particular segmentation and set of lengths, the ith segment starts at time frame $t(i,1) = f(q_{1:\tau}, \ell_{1:\tau}, i, 1)$ and ends at time frame $t(i, \ell_i) = f(q_{1:\tau}, \ell_{1:\tau}, i, \ell_i)$. In this general case, the time variable t could be a general function $f()$ of the complete Markov chain assignment $q_{1:\tau}$, the complete set of currently hypothesized segment lengths $\ell_{1:\tau}$, the segment number i, and the frame position within that segment 1 through ℓ_i. It is assumed that $f(q_{1:\tau}, \ell_{1:\tau}, i, \ell_i) = f(q_{1:\tau}, \ell_{1:\tau}, i+1, 1) - 1$ for all values of all quantities.

Renumbering the time sequence for a segment starting at one, an observation segment distribution is given by:

$$p(x_1, x_2, \ldots, x_\ell, \ell | q) = p(x_1, x_2, \ldots, x_\ell | \ell, q) p(\ell | q)$$

where $p(x_1, x_2, \ldots, x_\ell | \ell, q)$ is the length ℓ segment distribution under hidden Markov state q, and $p(\ell|q)$ is the explicit duration model for state q.

A plain HMM may be represented using this framework if $p(\ell|q)$ is a geometric distribution in ℓ and if

$$p(x_1, x_2, \ldots, x_\ell|\ell, q) = \prod_{j=1}^{\ell} p(x_j|q)$$

for a state specific distribution $p(x|q)$. One of the first segment models [121] is a generalization that allows observations in a segment to be additionally dependent on a region within a segment

$$p(x_1, x_2, \ldots, x_\ell|\ell, q) = \prod_{j=1}^{\ell} p(x_j|r_j, q)$$

where r_j is one of a set of fixed regions within the segment. A more general model is called a segmental hidden Markov model [63]

$$p(x_1, x_2, \ldots, x_\ell|\ell, q) = \int p(\mu|q) \prod_{j=1}^{\ell} p(x_j|\mu, q) d\mu$$

where μ is the multi-dimensional conditional mean of the segment and where the resulting distribution is obtained by integrating over all possible state-conditioned means in a Bayesian setting. More general still, in trended hidden Markov models [41, 42], the mean trajectory within a segment is described by a polynomial function over time. Equation 4.1 generalizes many models including the conditional Gaussian methods discussed above. A summary of segment models, their learning equations, and a complete bibliography is given in [120].

One can view a segment model as a GM as shown in Figure 23. A single hidden variable τ is shown that determines the number of segments. Within each segment, additional dependencies exist. The segment model allows for the set of dependencies within a segment to be arbitrary, so it is likely that many of the dependencies shown in the figure would not exist in practice. Moreover, there may be additional dependencies not shown in the figure, since it is the case that there must be constraints on the segment lengths. Nevertheless, this figure quickly details the essential structure behind a segment model.

5. GM-motivated speech recognition. There have been several cases where graphical models have themselves been used as the cruxes of speech recognition systems — this section explores several of them.

Perhaps the easiest way to use a graphical model for speech recognition is to start with the HMM graph given in Figure 16, and extend it with either additional edges or additional variables. In the former case, edges can be added between the hidden variables [43, 13] or between observed variables [157, 23, 14]. A crucial issue is how should the edges be added, as

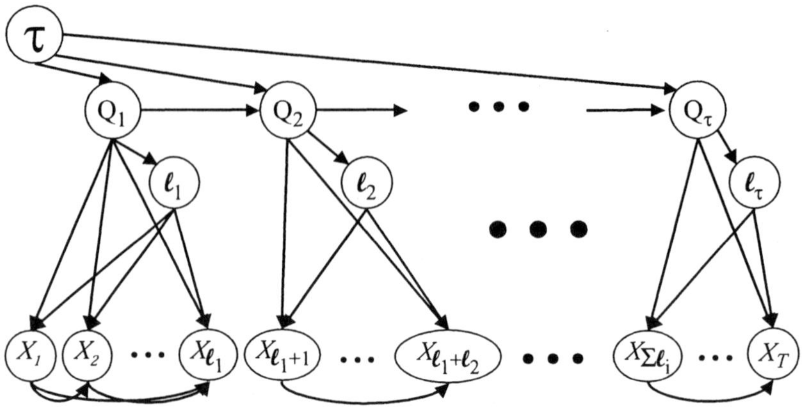

FIG. 23. *A Segment model viewed as a GM.*

mentioned below. In the latter case, a variable might indicate a condition such as noise level or quality, gender, vocal tract length, speaking mode, prosody, pitch, pronunciation, channel quality, microphone type, and so on. The variables might be observed during training (when the condition is known), and hidden during testing (when the condition can be unknown). In each case, the number of parameters of the system will typically increase — in the worst of cases, the number of parameters will increase by a factor equal to the number of different conditions.

In Section 3.7 it was mentioned that for an HMM to keep track of the differences that exist between a phone that occurs in multiple contexts, it must expand the state space so that multiple HMM states share the same acoustic Gaussian mixture corresponding to a particular phone. It turns out that a directed graph itself may be used to keep track of the necessary parameter tying and to control the sequencing needed in this case [167]. The simplest of cases is shown in Figure 24, which shows a sequence of connected triangles — for each time frame a sequence variable S_t, a phone variable Q_t, and a transition variable R_t is used. The observation variable X_t has as its parent only Q_t since it is only the phone that determines the observation distribution. The other variables are used together to appropriately sequence through valid phones for a given utterance.

In this particular figure, straight lines are used to indicate that the implementations of the dependencies are strictly deterministic, and rippled lines are used to indicate that the implementations correspond to true random dependencies. This means, for example, that $p(S_{t+1} = i | R_t, S_t) = \delta_{i,f(R_t,S_t)}$ is a Dirac-delta function having unity probability for only one possible value of S_{t+1} given a particular pair of values for R_t and S_t.

In the figure, S_t is the current sequence number (i.e., 1, 2, 3, etc.) and indicates the sub-word position in a word (e.g., the first, second, or third

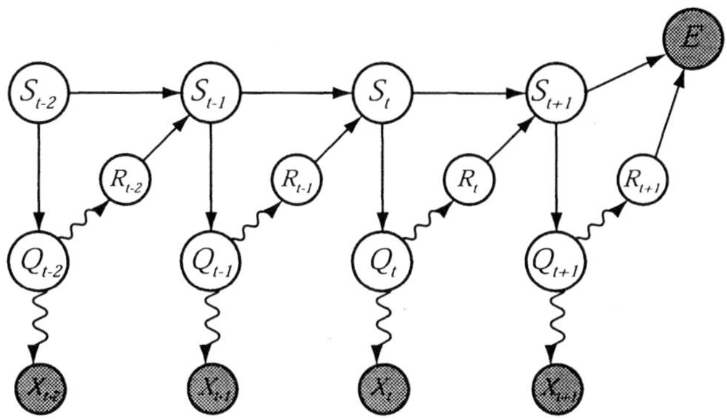

FIG. 24. *A BN used to explicitly represent parameter tying. In this figure, the straight edges correspond to deterministic implementations and the rippled edges correspond to stochastic implementations.*

phone). S_t does not determine the identity of the phone. Often, S_t will be a monotonically increasing sequence of successive integers, where either $S_{t+1} = S_t$ (the value stays the same) or $S_{t+1} = S_t + 1$ (an increment occurs). An increment occurs only if $R_t = 1$. R_t is a binary indicator variable that has unity value only when a transition between successive phone positions occurs. R_t is a true random variable and depending on the phone (Q_t), R_t will have a different binary distribution, thereby yielding the normal geometric duration distributions found in HMMs. Q_t is a deterministic function of the position S_t. A particular word might use a phone multiple times (consider the phone /aa/ in the word "yamaha"). The variable S_t sequences, say, from 1 through to 6 (the number of phones in "yamaha"), and Q_t then gets the identity of the phone via a deterministic mapping from S_t to Q_t for each position in the word (e.g., 1 maps to /y/, 2 maps to /aa/, 3 maps to /m/, and so on). This general approach can be extended to multiple hidden Markov chains, and to continuous speech recognition to provide graph structures that explicitly represent the control structures needed for an ASR system [167, 13, 47].

As mentioned above, factorial HMMs require a large expansion of the state space and therefore a large number of parameters. A recently proposed system that can model dependencies in a factorial HMM using many fewer parameters are called mixed memory Markov models [142]. Viewed as a GM as in Figure 25, this model uses an additional hidden variable for each time frame and chain. Each normal hidden variables possesses an additional switching parent (as depicted by dotted edges in the figure, and as described in Section 2.2). The switching conditional independence assumptions for one time slice are that $Q_t \perp\!\!\!\perp R_{t-1} | S_t = 0$, $Q_t \perp\!\!\!\perp Q_{t-1} | S_t = 1$ and the symmetric relations for R_t. This leads to the following distributional simplification:

$$p(Q_t|Q_{t-1}, R_{t-1}) = p(Q_t|Q_{t-1}, S_t = 0)P(S_t = 0) + p(Q_t|R_{t-1}, S_t = 1)P(S_t = 1)$$

which means that, rather than needing a single three-dimensional table for the dependencies, only two two-dimensional tables are required. These models have been used for ASR in [119].

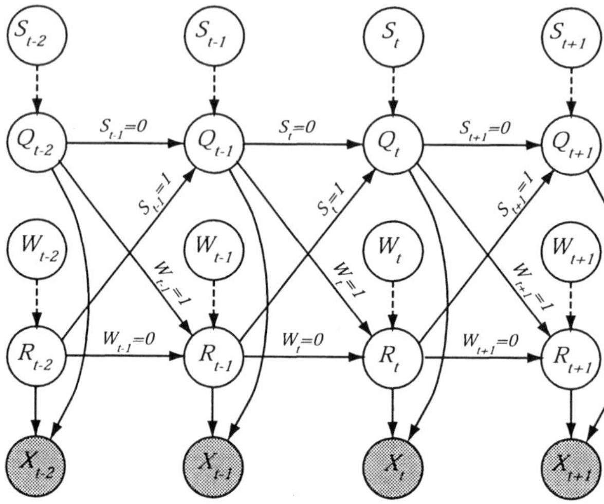

FIG. 25. *A mixed-memory hidden Markov model. The dashed edges indicate that the S and the W nodes are switching parents.*

A Buried Markov model (BMM) [16, 15, 14] is another recently proposed GM-based approach to speech recognition. A BMM is based on the idea that one can quantitatively measure where the conditional independence properties of a particular HMM are poorly representing a corpus of data. Wherever the model is found to be most lacking, additional edges are added (i.e., conditional independence properties are removed) relative to the original HMM. The BMM is formed to include only those data-derived, sparse, hidden-variable specific, and discriminative dependencies (between observation vectors) that are most lacking in the original model. In general, the degree to which $X_{t-1} \perp\!\!\!\perp X_t|Q_t$ is true can be measured using conditional mutual information $I(X_{t-1}; X_t|Q_t)$ [33]. If this quantity is zero, the model needs no extension, but if it is greater than zero, there is a modeling inaccuracy. Ideally, however, edges should be added discriminatively, to produce a discriminative generative model, and when the structure is formed discriminatively, the notion has been termed structural discriminability [16, 47, 166, 47]. For this purpose, the "EAR" (explaining away residual) measure has been defined that measures the discriminative mutual information between a variable X and its potential set of parents Z as follows:

$$\text{EAR}(X, Z) \triangleq I(X; Z|Q) - I(X; Z).$$

It can be shown that choosing Z to optimize the EAR measure can be equivalent to optimizing the posterior probability of the class Q [16]. Since it attempts to minimally correct only those measured deficiencies in a particular HMM, and since it does so discriminatively, this approach has the potential to produce better performing and more parsimonious models for speech recognition.

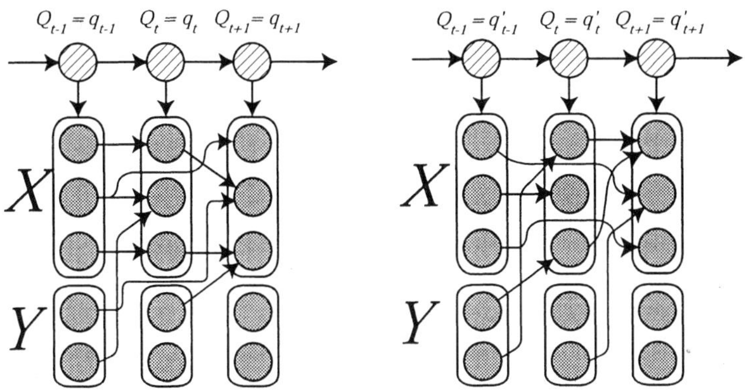

FIG. 26. *A Buried Markov Model (BMM) with two hidden Markov chain assignments, $Q_{1:T} = q_{1:T}$ on the left, and $Q_{1:T} = q'_{1:T}$ on the right.*

It seems apparent at this point that the set of models that can be described using a graph is enormous. With the options that are available in choosing hidden variables, the different sets of dependencies between those hidden variables, the dependencies between observations, choosing switching dependencies, and considering the variety of different possible implementations of those dependencies and the various learning techniques, it is obvious that the space of possible models is practically unlimited. Moreover, each of these modeling possibilities, if seen outside of the GM paradigm, requires a large software development effort before evaluation is possible with a large ASR system. This effort must be spent without having any guarantees as to the model's success.

In answer to these issues, a new flexible GM-based software toolkit has been developed (GMTK) [13]. GMTK is a graphical models toolkit that has been optimized for ASR and other time-series processing tasks. It supports EM and GEM parameter training, sparse linear and non-linear dependencies between observations, arbitrary parameter sharing, Gaussian vanishing and splitting, decision-tree implementations of dependencies, sampling, switching parent functionality, exact and log-space inference, multi-rate and multi-stream processing, and a textual graph programming language. The toolkit supports structural discriminability and arbitrary model selection, and makes it much easier to begin to experiment with GM-based ASR systems.

6. Conclusion. This paper has provided an introductory survey of graphical models, and then has provided a number of examples of how many existing ASR techniques can be viewed as instances of GMs. It is hoped that this paper will help to fuel the use of GMs for further speech recognition research. While the number of ASR models described in this document is large, it is of course the case that many existing ASR techniques have not even been given a mention. Nevertheless, it is apparent that ASR collectively occupies a relatively minor portion of the space of models representable by a graph. It therefore seems quite improbable that a thorough exploration of the space of graphical models would not ultimately yield a model that performs better than the HMM. The search for such a novel model should ideally occur on multiple fronts: on the one hand guided by our high-level domain knowledge about speech and thereby utilize phonetics, linguistics, psycho-acoustics, and so on. On the other hand, the data should have a strong say, so there should be significant data-driven model selection procedures to determine the appropriate natural graph structure [10]. And since ASR is inherently an instance of pattern classification, the notion of discriminability (parameter training) and structural discriminability (structure learning) might play a key role in this search. All in all, graphical models opens many doors to novel speech recognition research.

REFERENCES

[1] A.V. AHO, R. SETHI, AND J.D. ULLMAN. *Compilers: Principles, Techniques and Tools.* Addison-Wesley, Inc., Reading, Mass., 1986.

[2] S.M. AJI AND R.J. MCELIECE. The generalized distributive law. *IEEE Transactions in Information Theory*, **46**:325–343, March 2000.

[3] T. ANASTASAKOS, J. MCDONOUGH, R. SCHWARTZ, AND J MAKHOUL. A compact model for speaker adaptive training. In *Proc. Int. Conf. on Spoken Language Processing*, pp. 1137–1140, 1996.

[4] T.W. ANDERSON. *An Introduction to Multivariate Statistical Analysis.* Wiley Series in Probability and Statistics, 1974.

[5] J.J. ATICK. Could information theory provide an ecological theory of sensory processing? *Network*, **3**:213–251, 1992.

[6] H. ATTIAS. Independent Factor Analysis. *Neural Computation*, **11**(4):803–851, 1999.

[7] A.J. BELL AND T.J. SEJNOWSKI. An information maximisation approach to blind separation and blind deconvolution. *Neural Computation*, **7**(6):1129–1159, 1995.

[8] Y. BENGIO. Markovian models for sequential data. *Neural Computing Surveys*, **2**:129–162, 1999.

[9] A. L. BERGER, S.A. DELLA PIETRA, AND V.J. DELLA PIETRA. A maximum entropy approach to natural language processing. *Computational Linguistics*, **22**(1):39–71, 1996.

[10] J. BILMES. *Natural Statistical Models for Automatic Speech Recognition.* PhD thesis, U.C. Berkeley, Dept. of EECS, CS Division, 1999.

[11] J. Bilmes. What HMMs can do. Technical Report UWEETR-2002-003, University of Washington, Dept. of EE, 2002.

[12] J. Bilmes, N. Morgan, S.-L. Wu, and H. Bourlard. Stochastic perceptual speech models with durational dependence. *Intl. Conference on Spoken Language Processing*, November 1996.

[13] J. Bilmes and G. Zweig. The Graphical Models Toolkit: An open source software system for speech and time-series processing. *Proc. IEEE Intl. Conf. on Acoustics, Speech, and Signal Processing*, 2002.

[14] J.A. Bilmes. Data-driven extensions to HMM statistical dependencies. In *Proc. Int. Conf. on Spoken Language Processing*, Sidney, Australia, December 1998.

[15] J.A. Bilmes. Buried Markov models for speech recognition. In *Proc. IEEE Intl. Conf. on Acoustics, Speech, and Signal Processing*, Phoenix, AZ, March 1999.

[16] J.A. Bilmes. Dynamic Bayesian Multinets. In *Proceedings of the 16th conf. on Uncertainty in Artificial Intelligence.* Morgan Kaufmann, 2000.

[17] J.A. Bilmes. Factored sparse inverse covariance matrices. In *Proc. IEEE Intl. Conf. on Acoustics, Speech, and Signal Processing*, Istanbul, Turkey, 2000.

[18] J.A. Bilmes and K. Kirchhoff. Directed graphical models of classifier combination: Application to phone recognition. In *Proc. Int. Conf. on Spoken Language Processing*, Beijing, China, 2000.

[19] C. Bishop. *Neural Networks for Pattern Recognition.* Clarendon Press, Oxford, 1995.

[20] H. Bourlard. Personal communication, 1999.

[21] H. Bourlard and N. Morgan. *Connectionist Speech Recognition: A Hybrid Approach.* Kluwer Academic Publishers, 1994.

[22] L. Breiman, J.H. Friedman, R.A. Olshen, and C.J. Stone. *Classification and Regression Trees.* Wadsworth and Brooks, 1984.

[23] P.F. Brown. *The Acoustic Modeling Problem in Automatic Speech Recognition.* PhD thesis, Carnegie Mellon University, 1987.

[24] P.F. Brown, V.J. Delle Pietra, P.V. deSouza, J.C. Lai, and R.L. Mercer. Class-based n-gram models of natural language. *Computational Linguistics*, **18**(4):467–479, 1992.

[25] W. Buntine. A guide to the literature on learning probabilistic networks from data. *IEEE Trans. on Knowledge and Data Engineering*, **8**:195–210, 1994.

[26] K.P. Burnham and D.R. Anderson. *Model Selection and Inference : A Practical Information-Theoretic Approach.* Springer-Verlag, 1998.

[27] R. Chellappa and A. Jain, eds. *Markov Random Fields: Theory and Application.* Academic Press, 1993.

[28] Francine R. Chen. Identification of contextual factors for pronunciation networks. *Proc. IEEE Intl. Conf. on Acoustics, Speech, and Signal Processing*, pp. 753–756, 1990.

[29] S.F. Chen and J. Goodman. An empirical study of smoothing techniques for language modeling. In Arivind Joshi and Martha Palmer, editors, *Proceedings of the Thirty-Fourth Annual Meeting of the Association for Computational Linguistics*, pp. 310–318, San Francisco, 1996. Association for Computational Linguistics, Morgan Kaufmann Publishers.

[30] D.M. Chickering. *Learning from Data: Artificial Intelligence and Statistics*, chapter Learning Bayesian networks is NP-complete, pp. 121–130. Springer-Verlag, 1996.

[31] G. Cooper and E. Herskovits. Computational complexity of probabilistic inference using Bayesian belief networks. *Artificial Intelligence*, **42**:393–405, 1990.

[32] T.H. Cormen, C.E. Leiserson, and R.L. Rivest. *Introduction to Algorithms.* McGraw Hill, 1990.

[33] T.M. Cover and J.A. Thomas. *Elements of Information Theory.* Wiley, 1991.

[34] R.G. COWELL, A.P. DAWID, S.L. LAURITZEN, AND D.J. SPIEGELHALTER. *Probabilistic Networks and Expert Systems.* Springer-Verlag, 1999.

[35] P. DAGUM AND M. LUBY. Approximating probabilistic inference in Bayesian belief networks is NP-hard. *Artificial Intelligence*, **60**(141–153), 1993.

[36] Data mining and knowledge discovery. Kluwer Academic Publishers. Maritime Institute of Technology, Maryland.

[37] A.P. DAWID. Conditional independence in statistical theory. *Journal of the Royal Statistical Society B*, **41**(1):1–31, 1989.

[38] T. DEAN AND K. KANAZAWA. Probabilistic temporal reasoning. *AAAI*, pp. 524–528, 1988.

[39] J.R. DELLER, J.G. PROAKIS, AND J.H.L. HANSEN. *Discrete-time Processing of Speech Signals.* MacMillan, 1993.

[40] A.P. DEMPSTER, N.M. LAIRD, AND D.B. RUBIN. Maximum-likelihood from incomplete data via the EM algorithm. *J. Royal Statist. Soc. Ser. B.*, **39**, 1977.

[41] L. DENG, M. AKSMANOVIC, D. SUN, AND J. WU. Speech recognition using hidden Markov models with polynomial regression functions as non-stationary states. *IEEE Trans. on Speech and Audio Proc.*, **2**(4):101–119, 1994.

[42] L. DENG AND C. RATHINAVELU. A Markov model containing state-conditioned second-order non-stationarity: application to speech recognition. *Computer Speech and Language*, **9**(1):63–86, January 1995.

[43] M. DEVIREN AND K. DAOUDI. Structure learning of dynamic bayesian networks in speech recognition. In *European Conf. on Speech Communication and Technology (Eurospeech)*, 2001.

[44] R.O. DUDA, P.E. HART, AND D.G. STORK. *Pattern Classification.* John Wiley and Sons, Inc., 2000.

[45] E. EIDE. Automatic modeling of pronunciation variations. In *European Conf. on Speech Communication and Technology (Eurospeech)*, 6th, 1999.

[46] K. ELENIUS AND M. BLOMBERG. Effects of emphasizing transitional or stationary parts of the speech signal in a discrete utterance recognition system. In *Proc. IEEE Intl. Conf. on Acoustics, Speech, and Signal Processing*, pp. 535–538, 1982.

[47] J. BILMES et al. Discriminatively structured graphical models for speech recognition: JHU-WS-2001 final workshop report. Technical report, CLSP, Johns Hopkins University, Baltimore MD, 2001.

[48] J.G. FISCUS. A post-processing system to yield reduced word error rates: Recognizer output voting error reduction (ROVER). In *Proceedings of IEEE Workshop on Automatic Speech Recognition and Understanding*, Santa Barbara, California, 1997.

[49] R.A. FISHER. The use of multiple measurements in taxonomic problems. *Ann. Eugen.*, **7**:179–188, 1936.

[50] R. FLETCHER. *Practical Methods of Optimization.* John Wiley & Sons, New York, NY, 1980.

[51] V. FONTAINE, C. RIS, AND J.M.BOITE. Nonlinear discriminant analysis for improved speech recognition. In *European Conf. on Speech Communication and Technology (Eurospeech)*, 5th, pp. 2071–2074, 1997.

[52] E. FOSLER-LUSSIER. *Dynamic Pronunciation Models for Automatic Speech Recognition.* PhD thesis, University of California, Berkeley., 1999.

[53] B. FREY. *Graphical Models for Machine Learning and Digital Communication.* MIT Press, 1998.

[54] J.H. FRIEDMAN. Multivariate adaptive regression splines. *The Annals of Statistics*, **19**(1):1–141, 1991.

[55] N. FRIEDMAN AND M. GOLDSZMIDT. *Learning in Graphical Models*, chapter Learning Bayesian Networks with Local Structure. Kluwer Academic Publishers, 1998.

[56] N. FRIEDMAN, K. MURPHY, AND S. RUSSELL. Learning the structure of dynamic probabilistic networks. *14th Conf. on Uncertainty in Artificial Intelligence*, 1998.

[57] K. FUKUNAGA. *Introduction to Statistical Pattern Recognition, 2nd Ed.* Academic Press, 1990.

[58] S. FURUI. Cepstral analysis technique for automatic speaker verification. *IEEE Transactions on Acoustics, Speech, and Signal Processing*, **29**(2):254–272, April 1981.

[59] S. FURUI. Speaker-independent isolated word recognition using dynamic features of speech spectrum. *IEEE Transactions on Acoustics, Speech, and Signal Processing*, **34**(1):52–59, February 1986.

[60] S. FURUI. On the role of spectral transition for speech perception. *Journal of the Acoustical Society of America*, **80**(4):1016–1025, October 1986.

[61] M.J.F. GALES AND S. YOUNG. An improved approach to the hidden Markov model decomposition of speech and noise. In *Proc. IEEE Intl. Conf. on Acoustics, Speech, and Signal Processing*, pp. I–233–236, 1992.

[62] M.J.F. GALES. Semi-tied covariance matrices for hidden Markov models. *IEEE Transactions on Speech and Audio Processing*, **7**(3):272–281, May 1999.

[63] M.J.F. GALES AND S.J. YOUNG. Segmental hidden Markov models. In *European Conf. on Speech Communication and Technology (Eurospeech)*, 3rd, pp. 1579–1582, 1993.

[64] M.J.F. GALES AND S.J. YOUNG. Robust speech recognition in additive and convolutional noise using parallel model combination. *Computer Speech and Language*, **9**:289–307, 1995.

[65] D. GEIGER AND D. HECKERMAN. Knowledge representation and inference in similarity networks and Bayesian multinets. *Artificial Intelligence*, **82**:45–74, 1996.

[66] Z. GHAHRAMANI. *Lecture Notes in Artificial Intelligence*, Chapter Learning Dynamic Bayesian Networks. Springer-Verlag, 1998.

[67] Z. GHAHRAMANI AND M. JORDAN. Factorial hidden Markov models. *Machine Learning*, **29**, 1997.

[68] G.H. GOLUB AND C.F. VAN LOAN. *Matrix Computations.* Johns Hopkins, 1996.

[69] R.M. GRAY AND A. GERSHO. *Vector Quantization and Signal Compression.* Kluwer, 1991.

[70] M.S. GREWAL AND A.P. ANDREWS. *Kalman Filtering: Theory and Practice.* Prentice Hall, 1993.

[71] X.F. GUO, W.B. ZHU, Q. SHI, S. CHEN, AND R. GOPINATH. The IBM LVCSR system used for 1998 mandarin broadcast news transcription evaluation. In *The 1999 DARPA Broadcast News Workshop*, 1999.

[72] A.K. HALBERSTADT AND J.R. GLASS. Heterogeneous measurements and multiple classifiers for speech recognition. In *Proc. Int. Conf. on Spoken Language Processing*, pp. 995–998, 1998.

[73] D.A. HARVILLE. *Matrix Algebra from a Statistician's Perspective.* Springer-Verlag, 1997.

[74] T. HASTIE AND R. TIBSHIRANI. Discriminant analysis by Gaussian mixtures. *Journal of the Royal Statistical Society series B*, **58**:158–176, 1996.

[75] D. HECKERMAN. A tutorial on learning with Bayesian networks. Technical Report MSR-TR-95-06, Microsoft, 1995.

[76] D. HECKERMAN, MAX CHICKERING, CHRIS MEEK, ROBERT ROUNTHWAITE, AND CARL KADIE. Dependency networks for density estimation, collaborative filtering, and data visualization. In *Proceedings of the 16th conf. on Uncertainty in Artificial Intelligence.* Morgan Kaufmann, 2000.

[77] D. HECKERMAN, D. GEIGER, AND D.M. CHICKERING. Learning Bayesian networks: The combination of knowledge and statistical data. Technical Report MSR-TR-94-09, Microsoft, 1994.

[78] H. HERMANSKY, D. ELLIS, AND S. SHARMA. Tandem connectionist feature stream extraction for conventional HMM systems. In *Proc. IEEE Intl. Conf. on Acoustics, Speech, and Signal Processing*, Istanbul, Turkey, 2000.

[79] J. HERTZ, A. KROGH, AND R.G. PALMER. *Introduction to the Theory of Neural Computation.* Allan M. Wylde, 1991.

[80] X.D. HUANG, A. ACERO, AND H.-W. HON. *Spoken Language Processing: A Guide to Theory, Algorithm, and System Development.* Prentice Hall, 2001.

[81] T.S. JAAKKOLA AND M.I. JORDAN. *Learning in Graphical Models*, chapter Improving the Mean Field Approximations via the use of Mixture Distributions. Kluwer Academic Publishers, 1998.

[82] R.A. JACOBS. Methods for combining experts' probability assessments. *Neural Computation*, **7**:867–888, 1995.

[83] F. JELINEK. *Statistical Methods for Speech Recognition.* MIT Press, 1997.

[84] F.V. JENSEN. *An Introduction to Bayesian Networks.* Springer-Verlag, 1996.

[85] M.I. JORDAN AND C.M. BISHOP, eds. *An Introduction to Graphical Models.* to be published, 200x.

[86] M.I. JORDAN, Z. GHAHRAMANI, T.S. JAAKKOLA, AND L.K. SAUL. *Learning in Graphical Models*, chapter An Introduction to Variational Methods for Graphical Models. Kluwer Academic Publishers, 1998.

[87] M.I. JORDAN AND R. JACOBS. Hierarchical mixtures of experts and the EM algorithm. *Neural Computation*, **6**:181–214, 1994.

[88] B.-H. JUANG AND L.R. RABINER. Mixture autoregressive hidden Markov models for speech signals. *IEEE Trans. Acoustics, Speech, and Signal Processing*, **33**(6):1404–1413, December 1985.

[89] D. JURAFSKY AND J.H. MARTIN. *Speech and Language Processing.* Prentice Hall, 2000.

[90] M. KADIRKAMANATHAN AND A.P. VARGA. Simultaneous model re-estimation from contaminated data by composed hidden Markov modeling. In *Proc. IEEE Intl. Conf. on Acoustics, Speech, and Signal Processing*, pp. 897–900, 1991.

[91] T. KAMM, G. ANDREOU, AND J. COHEN. Vocal tract normalization in speech recognition compensating for systematic speaker variability. In *Proc. of the 15th Annual speech research symposium*, pp. 175–178. CLSP, Johns Hopkins University, 1995.

[92] J. KARHUNEN. Neural approaches to independent component analysis and source separation. In *Proc 4th European Symposium on Artificial Neural Networks (ESANN '96)*, 1996.

[93] P. KENNY, M. LENNIG, AND P. MERMELSTEIN. A linear predictive HMM for vector-valued observations with applications to speech recognition. *IEEE Transactions on Acoustics, Speech, and Signal Processing*, **38**(2):220–225, February 1990.

[94] B.E.D. KINGSBURY AND N. MORGAN. Recognizing reverberant speech with RASTA-PLP. *Proceedings ICASSP-97*, 1997.

[95] K. KIRCHHOFF. Combining acoustic and articulatory information for speech recognition in noisy and reverberant environments. In *Proceedings of the International Conference on Spoken Language Processing*, 1998.

[96] K. KIRCHHOFF AND J. BILMES. Dynamic classifier combination in hybrid speech recognition systems using utterance-level confidence values. *Proceedings ICASSP-99*, pp. 693–696, 1999.

[97] J. KITTLER, M. HATAF, R.P.W. DUIN, AND J. MATAS. On combining classifiers. *IEEE Transactions on Pattern Analysis and Machine Intelligence*, **20**(3):226–239, 1998.

[98] K. KJAERULFF. Triangulation of graphs - algorithms giving small total space. Technical Report R90-09, Department of Mathematics and Computer Science. Aalborg University, 1990.

[99] P. KRAUSE. Learning probabilistic networks. *Philips Research Labs Tech. Report*, 1998.

[100] A. KROGH AND J. VEDELSBY. Neural network ensembles, cross validation, and active learning. In *Advances in Neural Information Processing Systems 7*. MIT Press, 1995.

[101] F.R. KSCHISCHANG, B. FREY, AND H.-A. LOELIGER. Factor graphs and the sum-product algorithm. *IEEE Trans. Inform. Theory*, **47**(2):498–519, 2001.

[102] N. KUMAR. *Investigation of Silicon Auditory Models and Generalization of Linear Discriminant Analysis for Improved Speech Recognition.* PhD thesis, Johns Hopkins University, 1997.

[103] S.L. LAURITZEN. *Graphical Models.* Oxford Science Publications, 1996.

[104] C.J. LEGGETTER AND P.C. WOODLAND. Maximum likelihood linear regression for speaker adaptation of continuous density hidden Markov models. *Computer Speech and Language*, **9**:171–185, 1995.

[105] E. LEVIN. Word recognition using hidden control neural architecture. In *Proc. IEEE Intl. Conf. on Acoustics, Speech, and Signal Processing*, pp. 433–436. IEEE, 1990.

[106] E. LEVIN. Hidden control neural architecture modeling of nonlinear time varying systems and its applications. *IEEE Trans. on Neural Networks*, **4**(1):109–116, January 1992.

[107] H. LINHART AND W. ZUCCHINI. *Model Selection.* Wiley, 1986.

[108] B.T. LOGAN AND P.J. MORENO. Factorial HMMs for acoustic modeling. *Proc. IEEE Intl. Conf. on Acoustics, Speech, and Signal Processing*, 1998.

[109] D.J.C. MACKAY. *Learning in Graphical Models*, chapter Introduction to Monte Carlo Methods. Kluwer Academic Publishers, 1998.

[110] J. MAKHOUL. Linear prediction: A tutorial review. *Proc. IEEE*, **63**:561–580, April 1975.

[111] K.V. MARDIA, J.T. KENT, AND J.M. BIBBY. *Multivariate Analysis.* Academic Press, 1979.

[112] G.J. MCLACHLAN. *Finite Mixture Models.* Wiley Series in Probability and Statistics, 2000.

[113] G.J. MCLACHLAN AND T. KRISHNAN. *The EM Algorithm and Extensions.* Wiley Series in Probability and Statistics, 1997.

[114] C. MEEK. Causal inference and causal explanation with background knowledge. In Besnard, Philippe and Steve Hanks, editors, *Proceedings of the 11th Conference on Uncertainty in Artificial Intelligence (UAI'95)*, pp. 403–410, San Francisco, CA, USA, August 1995. Morgan Kaufmann Publishers.

[115] M. MEILĂ. *Learning with Mixtures of Trees.* PhD thesis, MIT, 1999.

[116] M. MOHRI, F.C.N. PEREIRA, AND M. RILEY. The design principles of a weighted finite-state transducer library. *Theoretical Computer Science*, **231**(1):17–32, 2000.

[117] N. MORGAN AND B. GOLD. *Speech and Audio Signal Processing.* John Wiley and Sons, 1999.

[118] H. NEY, U. ESSEN, AND R. KNESER. On structuring probabilistic dependencies in stochastic language modelling. *Computer Speech and Language*, **8**:1–38, 1994.

[119] H.J. NOCK AND S.J. YOUNG. Loosely-coupled HMMs for ASR. In *Proc. Int. Conf. on Spoken Language Processing*, Beijing, China, 2000.

[120] M. OSTENDORF, V. DIGALAKIS, AND O. KIMBALL. From HMM's to segment models: A unified view of stochastic modeling for speech recognition. *IEEE Trans. Speech and Audio Proc.*, **4**(5), September 1996.

[121] M. OSTENDORF, A. KANNAN, O. KIMBALL, AND J. ROHLICEK. Continuous word recognition based on the stochastic segment model. *Proc. DARPA Workshop CSR*, 1992.

[122] J. PEARL. *Probabilistic Reasoning in Intelligent Systems: Networks of Plausible Inference.* Morgan Kaufmann, 2nd printing edition, 1988.

[123] J. PEARL. *Causality.* Cambridge, 2000.

[124] M.P. Perrone and L.N. Cooper. When networks disagree: ensemble methods for hybrid neural networks. In R.J. Mammone, editor, *Neural Networks for Speech and Image Processing*, page Chapter 10, 1993.

[125] J. Picone, S. Pike, R. Regan, T. Kamm, J. Bridle, L. Deng, Z. Ma, H. Richards, and M. Schuster. Initial evaluation of hidden dynamic models on conversational speech. In *Proc. IEEE Intl. Conf. on Acoustics, Speech, and Signal Processing*, 1999.

[126] S.D. Pietra, V.D. Pietra, and J. Lafferty. Inducing features of random fields. Technical Report CMU-CS-95-144, CMU, May 1995.

[127] A.B. Poritz. Linear predictive hidden Markov models and the speech signal. *Proc. IEEE Intl. Conf. on Acoustics, Speech, and Signal Processing*, pp. 1291–1294, 1982.

[128] A.B. Poritz. Hidden Markov models: A guided tour. *Proc. IEEE Intl. Conf. on Acoustics, Speech, and Signal Processing*, pp. 7–13, 1988.

[129] L.R. Rabiner and B.-H. Juang. *Fundamentals of Speech Recognition*. Prentice Hall Signal Processing Series, 1993.

[130] L.R. Rabiner and B.H. Juang. An introduction to hidden Markov models. *IEEE ASSP Magazine*, 1986.

[131] M. Richardson, J. Bilmes, and C. Diorio. Hidden-articulator markov models for speech recognition. In *Proc. of the ISCA ITRW ASR2000 Workshop*, Paris, France, 2000. LIMSI-CNRS.

[132] M. Richardson, J. Bilmes, and C. Diorio. Hidden-articulator markov models: Performance improvements and robustness to noise. In *Proc. Int. Conf. on Spoken Language Processing*, Beijing, China, 2000.

[133] T.S. Richardson. *Learning in Graphical Models*, chapter Chain Graphs and Symmetric Associations. Kluwer Academic Publishers, 1998.

[134] M.D. Riley. A statistical model for generating pronunciation networks. *Proc. IEEE Intl. Conf. on Acoustics, Speech, and Signal Processing*, pp. 737–740, 1991.

[135] R.T. Rockafellar. *Convex Analysis*. Princeton, 1970.

[136] R. Rosenfeld. *Adaptive Statistical Language Modeling: A Maximum Entropy Approach*. PhD thesis, School of Computer Science, CMU, Pittsburgh, PA, April 1994.

[137] R. Rosenfeld. Two decades of statistical language modeling: Where do we go from here? *Proceedings of the IEEE*, **88**(8), 2000.

[138] R. Rosenfeld, S.F. Chen, and X. Zhu. Whole-sentence exponential language models: a vehicle for linguistic-statistical integration. *Computer Speech and Language*, **15**(1), 2001.

[139] D.B. Rowe. *Multivariate Bayesian Statistics: Models for Source Separation adn Signal Unmixing*. CRC Press, Boca Raton, FL, 2002.

[140] S. Roweis and Z. Ghahramani. A unifying review of linear gaussian models. *Neural Computation*, **11**:305–345, 1999.

[141] L.K. Saul, T. Jaakkola, and M.I. Jordan. Mean field theory for sigmoid belief networks. *JAIR*, **4**:61–76, 1996.

[142] L.K. Saul and M.I. Jordan. Mixed memory markov models: Decomposing complex stochastic processes as mixtures of simpler ones. *Machine Learning*, 1999.

[143] R.D. Shachter. Bayes-ball: The rational pastime for determining irrelevance and requisite information in belief networks and influence diagrams. In *Uncertainty in Artificial Intelligence*, 1998.

[144] P. Smyth, D. Heckerman, and M.I. Jordan. Probabilistic independence networks for hidden Markov probability models. Technical Report A.I. Memo No. 1565, C.B.C.L. Memo No. 132, MIT AI Lab and CBCL, 1996.

[145] T. Stephenson, H. Bourlard, S. Bengio, and A. Morris. Automatic speech recognition using dynamic bayesian networks with both acoustic and artic-

ulatory variables. In *Proc. Int. Conf. on Spoken Language Processing*, pp. 951–954, Beijing, China, 2000.

[146] G. Strang. *Linear Algebra and its applications, 3rd Edition.* Saunders College Publishing, 1988.

[147] M.E. Tipping and C.M. Bishop. Probabilistic principal component analysis. *Journal of the Royal Statistical Society, Series B*, **61**(3):611–622, 1999.

[148] D.M. Titterington, A.F.M. Smith, and U.E. Makov. *Statistical Analysis of Finite Mixture Distributions.* John Wiley and Sons, 1985.

[149] H. Tong. *Non-linear Time Series: A Dynamical System Approach.* Oxford Statistical Science Series 6. Oxford University Press, 1990.

[150] V. Vapnik. *Statistical Learning Theory.* Wiley, 1998.

[151] A.P. Varga and R.K. Moore. Hidden Markov model decomposition of speech and noise. In *Proc. IEEE Intl. Conf. on Acoustics, Speech, and Signal Processing*, pp. 845–848, Alburquerque, April 1990.

[152] A.P. Varga and R.K. Moore. Simultaneous recognition of concurrent speech signals using hidden makov model decomposition. In *European Conf. on Speech Communication and Technology (Eurospeech)*, 2nd, 1991.

[153] T. Verma and J. Pearl. Equivalence and synthesis of causal models. In *Uncertainty in Artificial Intelligence.* Morgan Kaufmann, 1990.

[154] T. Verma and J. Pearl. An algorithm for deciding if a set of observed independencies has a causal explanation. In *Uncertainty in Artificial Intelligence.* Morgan Kaufmann, 1992.

[155] M.Q. Wang and S.J. Young. Speech recognition using hidden Markov model decomposition and a general background speech model. In *Proc. IEEE Intl. Conf. on Acoustics, Speech, and Signal Processing*, pp. I–253–256, 1992.

[156] Y. Weiss. Correctness of local probability propagation in graphical models with loops. *Neural Computation*, **12**(1):1–41, 2000.

[157] C.J. Wellekens. Explicit time correlation in hidden Markov models for speech recognition. *Proc. IEEE Intl. Conf. on Acoustics, Speech, and Signal Processing*, pp. 384–386, 1987.

[158] C.J. Wellekens. Personal communication, 2001.

[159] J. Whittaker. *Graphical Models in Applied Multivariate Statistics.* John Wiley and Son Ltd., 1990.

[160] D.H. Wolpert. Stacked generalization. *Neural Networks*, **5**:241–259, 1992.

[161] P.C. Woodland. Optimizing hidden Markov models using discriminative output distributions. In *Proc. IEEE Intl. Conf. on Acoustics, Speech, and Signal Processing*, 1991.

[162] P.C. Woodland. Hidden Markov models using vector linear prediction and discriminative output distributions. In *Proc. IEEE Intl. Conf. on Acoustics, Speech, and Signal Processing*, pp. I–509–512, 1992.

[163] Su-Lin Wu, Michael L. Shire, Steven Greenberg, and Nelson Morgan. Integrating syllable boundary information into speech recognition. In *Proc. IEEE Intl. Conf. on Acoustics, Speech, and Signal Processing*, Vol. 1, Munich, Germany, April 1997. IEEE.

[164] S. Young. A review of large-vocabulary continuous-speech recognition. *IEEE Signal Processing Magazine*, **13**(5):45–56, September 1996.

[165] K.H. Yuo and H.C. Wang. Joint estimation of feature transformation parameters and gaussian mixture model for speaker identification. *Speech Communications*, **3**(1), 1999.

[166] G. Zweig, J. Bilmes, T. Richardson, K. Filali, K. Livescu, P. Xu, K. Jackson, Y. Brandman, E. Sandness, E. Holtz, J. Torres, and B. Byrne. Structurally discriminative graphical models for automatic speech recognition — results from the 2001 Johns Hopkins summer workshop. *Proc. IEEE Intl. Conf. on Acoustics, Speech, and Signal Processing*, 2002.

[167] G. Zweig and S. Russell. Speech recognition with dynamic Bayesian networks. *AAAI-98*, 1998.

AN INTRODUCTION TO MARKOV CHAIN MONTE CARLO METHODS

JULIAN BESAG*

Abstract. This article provides an introduction to Markov chain Monte Carlo methods in statistical inference. Over the past twelve years or so, these have revolutionized what can be achieved computationally, especially in the Bayesian paradigm. Markov chain Monte Carlo has exactly the same goals as ordinary Monte Carlo and both are intended to exploit the fact that one can learn about a complex probability distribution if one can sample from it. Although the ordinary version can only rarely be implemented, it is convenient initially to presume otherwise and to focus on the rationale of the sampling approach, rather than computational details. The article then moves on to describe implementation via Markov chains, especially the Hastings algorithm, including the Metropolis method and the Gibbs sampler as special cases. Hidden Markov models and the autologistic distribution receive some emphasis, with the noisy binary channel used in some toy examples. A brief description of perfect simulation is also given. The account concludes with some discussion.

Key words. Autologistic distribution; Bayesian computation; Gibbs sampler; Hastings algorithm; Hidden Markov models; Importance sampling; Ising model; Markov chain Monte Carlo; Markov random fields; Maximum likelihood estimation; Metropolis method; Noisy binary channel; Perfect simulation; Reversibility; Simulated annealing.

1. The computational challenge.

1.1. Introduction. Markov chain Monte Carlo (MCMC) methods have had a profound influence on computational statistics over the past twelve years or so, especially in the Bayesian paradigm. The intention here is to cover the basic ideas and to provide references to some more specialized topics. Other descriptions include the books by (or edited by) Fishman (1996), Gilks *et al.* (1996), Newman and Barkema (1999), Robert and Casella (1999), Chen *et al.* (2000), Doucet *et al.* (2001), Liu (2001) and MacCormick (2002). Although none of these addresses speech *per se*, the last three include descriptions of sequential Monte Carlo methods and particle filters, with applications to on–line signal processing and target tracking, for example. These books may therefore be of particular interest to readers of this volume.

In the remainder of this section, we introduce the basic computational task in MCMC. In Section 2, we discuss ordinary Monte Carlo methods and their conceptual relevance to Bayesian inference, especially hidden Markov models, to maximum likelihood estimation and to function optimization. Unfortunately, ordinary Monte Carlo is rarely practicable for high–dimensional problems, even for minor enhancements of hidden Markov models. However, the underlying ideas transfer quite smoothly to MCMC, with random samples replaced by dependent samples from a

*Department of Statistics, University of Washington, Box 354322, Seattle, WA 98195, USA (julian@stat.washington.edu).

Markov chain, as we discuss in Section 3. We also describe the Hastings algorithm, including the Metropolis method and the Gibbs sampler, and perfect MCMC simulation via monotone coupling from the past. The paper includes a few toy examples based on the noisy binary channel. Finally, Section 4 provides some discussion. The paper is mostly a distillation of Besag (2001), where some applications and more specialized topics, such as MCMC p–values (Besag and Clifford, 1989, 1991), cluster algorithms (Swendsen and Wang, 1987), Langevin–Hastings algorithms (Besag, 1994a) and reversible jump MCMC (Green, 1995), can be found. For the most recent developments in a rapidly expanding field, the interested reader should consult the MCMC website at

`http://www.statslab.cam.ac.uk/~mcmc/`

1.2. The main task. Let X denote a random quantity: in practice, X will have many components and might represent, for example, a random vector or a multi–way contingency table or a grey–level pixel image (perhaps augmented by other variables). Also, some components of X might be discrete and others continuous. However, it is most convenient for the moment to think of X as a single random variable (r.v.), having a finite but huge sample space. Indeed, in a sense, such a formulation is perfectly general because ultimately all our calculations are made on a finite machine. It is only in considering specific MCMC algorithms, such as the Gibbs sampler, or applications, such as hidden Markov models, that we need to address the individual components of X.

Thus, let $\{\pi(x) : x \in S\}$ denote the probability mass function (p.m.f.) of X, where S is the corresponding support of X; that is, $S = \{x : \pi(x) > 0\}$. We assume that $\pi(x)$ is known up to scale, so that

$$\pi(x) \;=\; h(x)/c, \qquad x \in S, \tag{1}$$

where $h(x)$ is completely specified, but that the normalizing constant

$$c \;=\; \sum_{x\in S} h(x) \tag{2}$$

is not known in closed form and that S is too unwieldy for c to be found numerically from the sum in (2). Nevertheless, our goal is to compute expectations of particular functions g under π; that is, we require

$$\mathrm{E}_{\pi} g \;=\; \sum_{x\in S} g(x)\pi(x), \tag{3}$$

for any relevant g, where again the summation in equation (3) cannot be evaluated directly.

As an especially important special case, note that (3) includes the *probability* of any particular event concerning X. Explicitly, for any relevant subset B of S,

$$\Pr(X \in B) = \sum_{x \in S} 1[x \in B]\, \pi(x), \tag{4}$$

where $1[\,.\,]$ is the usual indicator function; that is, $1[x \in B] = 1$ if the outcome x implies that the event B occurs and $1[x \in B] = 0$ otherwise. Indeed, one of the major strengths of MCMC is that it can focus directly on probabilities, in contrast to the usual tradition in statistics of indirect calculations based on large sample asymptotics.

2. Ordinary Monte Carlo calculations.

2.1. Monte Carlo estimation. Suppose that, despite the complexity of S, we can generate random draws from the target p.m.f. $\pi(x)$. If we produce m such draws, $x^{(1)}, \ldots, x^{(m)}$, then the natural estimate of $\mathrm{E}_{\pi} g$ is the empirical mean,

$$\bar{g} = \frac{1}{m} \sum_{t=1}^{m} g(x^{(t)}). \tag{5}$$

This is unbiased for $\mathrm{E}_{\pi} g$ and its sampling variance can be assessed in the usual way.

Thinking ahead, we remark that (5) may provide an approximation to $\mathrm{E}_{\pi} g$ even when $(x^{(1)}, \ldots, x^{(m)})$ is not a random sample from π. In particular, this occurs when m is sufficiently large and $x^{(1)}, x^{(2)}, \ldots$, seeded by some $x^{(0)} \in S$, are successive observations from a Markov chain with (finite) state space S and limiting distribution π. This extension provides the basis of MCMC when random sampling from π is no longer feasible. It requires that useful general recipes exist for constructing appropriate Markov chains, as in Section 3.

2.2. Bayesian computation. For a description of parametric Bayesian inference, see e.g. Gelman *et al.* (1995). Here we begin in the simplest possible context. Thus, let x now denote a meaningful constant whose value we wish to estimate. Suppose we know that this *parameter* x lies in a finite space S and that our initial beliefs about its value can be represented by a *prior* p.m.f. $\{\rho(x) : x \in S\}$. If y denotes relevant discrete data, then the probability of y given x, viewed as a function of x, is called the *likelihood* $L(y|x)$. In the Bayesian paradigm, the prior information and the likelihood are combined via Bayes theorem to produce the *posterior* p.m.f.

$$\pi(x|y) \propto L(y|x)\rho(x), \qquad x \in S, \tag{6}$$

for x given y. All inferences are based on (6) and require the evaluation of corresponding expectations. In terms of (1), (2), (3) and (4), we replace $\pi(x)$ by $\pi(x|y)$ and let $h(x)$ be proportional to $L(y|x)\rho(x)$. Note that $L(y|x)$ and $\rho(x)$ need be known only up to scale.

Unfortunately, it is rarely possible to calculate the expectations directly. A potential remedy is to generate a large random sample $x^{(1)}, \ldots,$

$x^{(m)}$ from the p.m.f. (6) and then use (5) to approximate the corresponding posterior mean and variance, to evaluate posterior probabilities about x and to construct corresponding credible intervals. In principle, this approach extends to multicomponent parameters but it is generally impossible to implement because of the difficulty of sampling directly from complex multivariate p.m.f.'s. This hindered applied Bayesian inference until it was recognized that ordinary Monte Carlo could be replaced by MCMC.

As an example of how the availability of random samples from $\pi(x|y)$ would permit trivial solutions to ostensibly very complicated problems, consider a clinical, industrial or agricultural trial in which the aim is to compare different treatment effects θ_i. Then $x = (\theta, \phi)$, where θ is the vector of θ_i's and ϕ is a vector of other, possibly uninteresting, parameters in the posterior distribution. A natural quantity of interest from a Bayesian perspective is the posterior probability that any particular treatment effect is best or is among the best three, say, where here we suppose best to mean having the largest effect. Such calculations are usually far beyond the capabilities of conventional numerical methods, because they involve sums (or integrals) of non–standard functions over awkward regions of the parameter space S. However, in a sampling approach, we can closely approximate the probability that treatment i is best, simply by the proportion of simulated $\theta^{(t)}$'s among which $\theta_i^{(t)}$ is the largest component; and the probability that treatment i is one of the best three by the proportion of $\theta^{(t)}$'s for which $\theta_i^{(t)}$ is one of the largest three components. Note that the values obtained for components that are not of immediate interest are simply ignored and that this procedure is entirely rigorous.

Ranking and selection is just one area in which the availability of random samples from posterior distributions would have had a profound influence on applied Bayesian inference. Not only does MCMC deliver what is beyond ordinary Monte Carlo methods but also it encourages the investigator to build and analyze more realistic statistical models. Indeed, one must sometimes resist the temptation to build representations whose complexity cannot be justified by the underlying scientific problem or by the available data.

2.2.1. Hidden Markov models. Basic hidden Markov models (HMM's) are among the most complex formulations for which ordinary Monte Carlo can be implemented. In addition to their popularity in speech recognition (e.g. Rabiner, 1989; Juang and Rabiner, 1991), HMM's also occur in neurophysiology (e.g. Fredkin and Rice, 1992), in computational biology (e.g. Haussler *et al.*, 1993; Eddie *et al.*, 1995; Liu *et al.*, 1995), in climatology (e.g. Hughes *et al.*, 1999), in epidemiologic surveillance (Le Strat and Carrat, 1999), and elsewhere. For a useful survey, see MacDonald and Zucchini (1997). Here, we briefly describe HMM's and the corresponding role of the Baum *et al.* (1970) algorithm.

Let $x_1, x_2, \ldots$, where $x_i \in \{0, 1, \ldots, s\}$, denote successive states of a process. These states are unknown but, for each $i = 1, \ldots, n$ independently, x_i generates an observation y_i with probability $f(x_i, y_i)$, so that the data $y = (y_1, \ldots, y_n)$ has probability

$$L(y|x) = \prod_{i=1}^{n} f(x_i, y_i), \tag{7}$$

given x. Now suppose $x_1, x_2, \ldots$ is (modeled as) the output of a Markov chain, with transition probability $q(x_i, x_{i+1})$ of the ith component x_i being followed by x_{i+1}, so that the prior probability for $x = (x_1, \ldots, x_n)$ is

$$\rho(x) = q(x_1) \prod_{i=1}^{n-1} q(x_i, x_{i+1}), \qquad x \in S = \{0, 1, \ldots, s\}^n, \tag{8}$$

where $q(.)$ is the p.m.f. of x_1. Then the posterior probability of x, given y, is

$$\pi(x|y) \propto q(x_1) f(x_1, y_1) \prod_{i=2}^{n} q(x_{i-1}, x_i) f(x_i, y_i), \qquad x \in S. \tag{9}$$

The goal is to make inferences about the true x and perhaps about some future x_i's. For example, we might require the *marginal posterior modes* (MPM) estimate x^* of x, defined by

$$x_i^* = \arg\max_{x_i} \pi(x_i|y), \tag{10}$$

where $\pi(x_i|y)$ is the posterior marginal p.m.f. for x_i. Note here that x^* is generally distinct from the *maximum a posteriori* (MAP) estimate $\breve{x}$, defined by

$$\breve{x} = \arg\max_{x} \pi(x|y), \tag{11}$$

as we discuss later for the noisy binary channel.

We remark in passing that, if the above Markov formulation provides a physically correct model, then there is nothing intrinsically Bayesian in the use of Bayes theorem to obtain the p.m.f. (9). Furthermore, if the three p.m.f.'s on the right–hand side are known, then the Baum algorithm can be used directly to evaluate many expectations (3) of interest, without any need for simulation; and, if the p.m.f.'s are unknown, then Baum can be preceded by the Baum–Welch algorithm, which determines their nonparametric maximum likelihood estimates. Thus, it may seem that a Bayesian formulation and a sampling approach to the analysis of HMM's are both somewhat irrelevant. However, the Bayesian paradigm has distinct advantages when, as is usual, the HMM is more a conceptual rather than a physical model of the underlying process; and more important here, a sampling approach is viable under almost any modification of the basic HMM formulation, but with ordinary Monte Carlo replaced by MCMC, for which the Baum algorithm retains its relevance if an HMM lives within the overall formulation. We comment further at the end of Section 3.7; see also Robert *et al.* (2000).

The Baum *et al.* (1970) recursions for (9) exploit the fact that $\pi(x) \equiv \pi(x|y)$ inherits the Markov property, though its transition probabilities are functions of y and therefore nonhomogeneous. Specifically,

$$\pi(x|y) = \pi(x_1|y) \prod_{i=2}^{n} \pi(x_i|x_{i-1}, y_{\geq i}), \tag{12}$$

where $y_{\geq i} = (y_i, \ldots, y_n)$, a form of notation we use freely below. The factorization (12) provides access to the calculation of expected values and to sampling from $\{\pi(x|y) : x \in S\}$, except that the conditional probabilities in (12) are not immediately available. It can easily be shown that

$$\pi(x_{\geq i}|x_{i-1}, y_{\geq i}) \propto \prod_{k=i}^{n} q(x_{k-1}, x_k) f(x_k, y_k), \qquad i = 2, \ldots, n, \tag{13}$$

which determines $\pi(x_i|x_{i-1}, y_{\geq i})$ by summing over $x_{>i}$ but the summations are impracticable unless n is tiny. The Baum algorithm avoids the problem by using the results,

$$\pi(x_1|y) \propto f(x_1, y_1)\, q(x_1)\, \Pr(y_{>1}|x_1), \tag{14}$$

$$\pi(x_i|x_{i-1}, y_{\geq i}) \propto f(x_i, y_i)\, q(x_{i-1}, x_i)\, \Pr(y_{>i}|x_i), \qquad i = 2, \ldots, n. \tag{15}$$

Here, $\Pr(y_{>n}|x_n) \equiv 1$ and the other $\Pr(y_{>i}|x_i)$'s for $x_i = 0, 1, \ldots, s$ can be evaluated successively for $i = n-1, \ldots, 1$ from the backward recursion,

$$\Pr(y_{>i}|x_i) = \sum_{x_{i+1}} \Pr(y_{>i+1}|x_{i+1})\, f(x_{i+1}, y_{i+1})\, q(x_i, x_{i+1}). \tag{16}$$

Then (14) and (15) are used forwards to calculate expectations or to sample from $\pi(x|y)$. Some care is needed in using (16) because the probabilities quickly become vanishingly small. However, as they are required only up to scale in (14) and (15), a dummy normalization can be carried out at each stage to remedy the problem.

Example. Noisy binary channel. The noisy binary channel provides a convenient numerical illustration of sampling via the Baum algorithm. We additionally impose symmetry and stationarity, merely to ease the notation. The noisy binary channel is not only the simplest HMM but is also a rather special case of the autologistic distribution, which we consider in Section 3.4 and which is not generally amenable to the Baum algorithm.

Thus, suppose that both the hidden x_i's and the observed y_i's are binary and that the posterior probability (9) of a true signal $x \in S$ given data $y \in S$, where $S = \{0,1\}^n$, is

$$\pi(x|y) \propto \exp\left(\alpha \sum_{i=1}^{n} 1[x_i = y_i] + \beta \sum_{i=1}^{n-1} 1[x_i = x_{i+1}]\right), \tag{17}$$

where $1[\,.\,]$ again denotes the usual indicator function. Here α is the log–odds of correct to incorrect transmission of each x_i and β is the log–odds in favor of $x_{i+1} = x_i$. In particular, we set $\alpha = \ln 4$, corresponding to a corruption probability of 0.2, and $\beta = \ln 3$, so that x_{i+1} is a repeat of x_i with probability 0.75.

As a trite example, suppose $y = 11101100000100010111$, so that $|S| = 2^{20} = 1048576$. For such a tiny space, we can calculate expected values simply by enumeration but we also applied the Baum algorithm to generate a random sample of size 10000 from $\pi(x|y)$. First, consider the posterior marginal probabilities for the x_i's. We obtained $x_1 = 1$ in 8989 of the samples $x^{(t)}$ and hence our estimate of the corresponding probability is 0.899, versus the exact value 0.896; for $x_2 = 1$, we obtained 0.927 versus 0.924; and so on. Hence, the MPM estimate x^* of x, defined by (10), is correctly identified as $x^* = 11111100000000010111$. Clearly, x^* is a smoothed version of the data, with two fewer isolated bits. The x_i^*'s for positions $i = 4$, 12, 16 and 17 are the most doubtful, with estimated (exact) probabilities of $x_i = 1$ equal to 0.530 (0.541), 0.421 (0.425), 0.570 (0.570) and 0.434 (0.432). Note that, neither component 16 nor 17 flips in the MPM estimate but that, if we examine them as a single unit, the posterior probabilities of 00, 10, 01 and 11 are 0.362 (0.360), 0.203 (0.207), 0.068 (0.070) and 0.366 (0.362), respectively. Thus, there is a preference for 00 or 11, rather than the 10 obtained in x^*.

The previous point illustrates how an estimate may be affected by choosing either a marginal or a multivariate criterion. Indeed, at the opposite extreme to MPM is the MAP estimate (11), which here is equally 11111100000000011111 or 11111100000000000111, both of which are easily seen to have the same posterior probability. In our random sample, they were indeed the two most frequent configurations, occurring 288 and 323 times, respectively, compared with the exact probability 0.0304. Note that x^* and y itself occurred 138 and 25 times, compared with the exact probabilities 0.0135 and 0.0027. If one requires a single–shot estimate of x, then the choice of a particular criterion, ultimately in the form of a loss function, should depend on the practical goals of the analysis. For example, the MAP estimate corresponds to zero loss for the correct x and unit loss for any incorrect estimate, regardless of the number of errors among its components; whereas MPM arises from a componentwise loss function and minimizes the expected total number of errors among all the components. The writer's own view is that a major benefit of a sampling approach is that it enables one to investigate various aspects of the posterior distribution, rather than forcing one to concentrate on a single criterion; but note that sampling from the posterior is not generally suitable for finding the MAP estimate, which we address in Section 2.5 on simulated annealing.

As a more taxing toy example, we applied the Baum algorithm to generate a 100 realizations x from a noisy binary channel, again with $\alpha = \ln 4$ and $\beta = \ln 3$ but now with $y = 111001110011100\ldots$, a vector of length 100000, so that $|S| = 2^{100000}$. The MPM and MAP estimates of x, obtained in the conventional manner from Baum and Viterbi algorithms, both coincide with the data y in this case. The majority vote classification from the 100 random samples was correct for all 100000 components, although the average success rate for a single sample was only 77.7%, with a maximum of 78.1%. For a sample of size 10000, these figures were 77.7% and 78.2%, respectively. We return to this example in Sections 2.5 and 3.8.

Finally, we mention some modifications of HMM's that one might want to make in practice. For instance, the three p.m.f.'s on the right–hand side of (9) could be partially unknown, with their own (hyper)priors; the degradation mechanism forming the data y might be more complex, with each y_i depending on several components of x; there could be multiple y's for each x; the Markov formulation x might be inappropriate in known or unknown segments of x; and so on. In such cases, it is very likely that standard methods for HMM's break down but a sampling approach via MCMC can still be adopted. Slowing down of the algorithm can be countered by sequential Monte Carlo; see below. An interesting and largely unexplored further possibility is to cater for complications by incorporating MCMC in an otherwise deterministic algorithm.

2.3. Importance sampling. The notion of learning about an otherwise intractable fixed probability distribution π via Monte Carlo simulation is of course quite natural. However, we now describe a more daunting task in which the goal is to approximate $\mathrm{E}_{\pi^*} g$ for distributions π^* that are close to a baseline distribution π from which we have a random sample. For example, in Bayesian sensitivity analysis, we need to assess how changes in the basic formulation affect our conclusions. This may involve posterior distributions that have different functional forms and yet are not far apart. An analogous problem arises in difficult maximum likelihood estimation, as we discuss in the next section. Importance sampling also drives sequential Monte Carlo methods and particle filters, in which observations on a process arrive as a single or multiple time series and the goal is to update inference as each new piece of information is received, without the need to run a whole new simulation; see especially Doucet *et al.* (2001), Liu (2001) and MacCormick (2002). Particle filters provide the most relevant MCMC methods for problems in speech recognition, though the writer is not aware of any specific references. We now describe how ordinary importance sampling works.

Suppose we have a random sample $x^{(1)}, \ldots, x^{(m)}$ from $\pi(x) = h(x)/c > 0$ for $x \in S$ but that our real interest lies in $\mathrm{E}_{\pi^*} g$, for some specific g, where

$$\pi^*(x) = h^*(x)/c^* > 0, \qquad x \in S^*,$$

with h^* known and crucially $S^* \subseteq S$. Now

$$\text{(18)} \quad \mathrm{E}_{\pi} \frac{gh^*}{h} = \sum_{x \in S} \frac{g(x)h^*(x)}{h(x)} \frac{h(x)}{c} = \frac{c^*}{c} \sum_{x \in S^*} g(x) \frac{h^*(x)}{c^*} = \frac{c^*}{c} \mathrm{E}_{\pi^*} g,$$

so we can estimate the right–hand side of (18) by the mean value of $g(x^{(t)})h^*(x^{(t)})/h(x^{(t)})$. Usually, c^*/c is unknown but, as a special case of (18),

$$\mathrm{E}_{\pi}(h^*/h) = c^*/c,$$

so that, as our eventual approximation to $\mathrm{E}_{\pi^*} g$, we adopt the ratio estimate,

$$\text{(19)} \qquad \sum_{t=1}^{m} w(x^{(t)})\, g(x^{(t)}),$$

where

$$w(x^{(t)}) = \frac{h^*(x^{(t)})/h(x^{(t)})}{\sum_{t=1}^{m} \{h^*(x^{(t)})/h(x^{(t)})\}}.$$

Note that the $w(x^{(t)})$'s are independent of g and are well defined because $S^* \subseteq S$. The estimate (19) is satisfactory if (5) is adequate for $\mathrm{E}_{\pi} g$ and there are no large weights among the $w(x^{(t)})$'s. In practice, the latter condition requires that h and h^* are not too far apart. There are modifications of the basic method described here that can extend its range (e.g. umbrella sampling).

2.4. Monte Carlo maximum likelihood estimation. Let $x^{(0)}$ denote an observation, generally a vector, from a p.m.f.

$$\pi(x;\theta) = h(x;\theta)/c(\theta), \qquad x \in S, \quad \theta \in \Theta,$$

where $c(\theta) = \sum_{x \in S} h(x;\theta)$. Suppose we require the maximum likelihood estimate,

$$\hat{\theta} = \arg\max_{\theta \in \Theta} \pi(x^{(0)};\theta)$$

of θ but that, although h is quite manageable, $c(\theta)$ and its derivatives cannot be calculated directly, even for particular values of θ.

Instead, suppose that we can generate a random sample from $\pi(x;\theta)$ for any given θ, and let $(x^{(1)},\ldots,x^{(m)})$ denote such a sample for $\theta=\breve{\theta}$, a current approximation to $\hat{\theta}$. Then, trivially, we can always write

$$(20)\quad \hat{\theta} \;=\; \arg\max_{\theta\in\Theta} \ln\frac{\pi(x^{(0)};\theta)}{\pi(x^{(0)};\breve{\theta})} \;=\; \arg\max_{\theta\in\Theta}\left\{\ln\frac{h(x^{(0)};\theta)}{h(x^{(0)};\breve{\theta})}-\ln\frac{c(\theta)}{c(\breve{\theta})}\right\}.$$

The first quotient on the right–hand side of (20) is known and the second can be approximated using (18), where $c(\theta)$, $c(\breve{\theta})$, $h(x^{(0)};\theta)$ and $h(x^{(0)};\breve{\theta})$ play the roles of c^*, c, h^* and h, respectively. That is,

$$\frac{c(\theta)}{c(\breve{\theta})} \;=\; \sum_{x\in S}\frac{h(x;\theta)}{c(\breve{\theta})} \;=\; \sum_{x\in S}\frac{h(x;\theta)}{h(x;\breve{\theta})}\,\pi(x;\breve{\theta})$$

can be approximated by the empirical average,

$$\frac{1}{m}\sum_{t=1}^{m}\frac{h(x^{(t)};\theta)}{h(x^{(t)};\breve{\theta})}\,,$$

for any θ in the neighborhood of $\breve{\theta}$. It follows that, at least when θ is one– or two–dimensional, an improved approximation to $\hat{\theta}$ can be found by direct search, though, in higher dimensions, it is necessary to implement a more sophisticated approach, usually involving derivatives and corresponding approximations. In practice, several stages of Monte Carlo sampling may be required to reach an acceptable approximation to $\hat{\theta}$.

Unfortunately, in most applications where standard maximum likelihood estimation is problematical, so too is the task of producing a random sample from π. The above approach must then be replaced by an MCMC version, as introduced by Penttinen (1984), in spatial statistics, and by Geyer (1991) and Geyer and Thompson (1992), in more general settings. For an exception to this rule, see Besag (2003), though in fact this is a swindle because it uses perfect MCMC to generate the random samples! For a quite complicated example of genuine MCMC maximum likelihood, see Tjelmeland and Besag (1998).

2.5. Simulated annealing. Simulated annealing (Kirkpatrick *et al.*, 1983) is a general purpose MCMC algorithm for the optimization of discrete high–dimensional functions. Here we describe a toy version, based on ordinary Monte Carlo sampling, and comment briefly on the MCMC implementation that is required in practice.

Let $\{h(x) : x \in S\}$, with S finite, denote a bounded non–negative function, specified at least up to scale. Let $\breve{x} = \arg\max_x h(x)$. We assume for the moment that $\breve{x}$ is unique but that S is too complicated for $\breve{x}$ to be found by complete enumeration and that h does not have a sufficiently nice structure for $\breve{x}$ to be determined by simple hill–climbing methods. In operations research, where such problems abound, h is sometimes amenable to mathematical programming techniques; for example, the simplex method applied to the traveling salesman problem. However, here we make no such assumption.

Let $\{\pi(x) : x \in S\}$ denote the corresponding finite p.m.f. defined by (1) and (2), with c generally unknown. Clearly, $\breve{x} = \arg\max_x \pi(x)$ and, indeed, the original task may have been to locate the global mode of π, as in our example below. The goal in simulated annealing is not to produce a random draw from π but to bias the selection overwhelmingly in favor of the most probable value $\breve{x}$.

We begin by defining a sequence of distributions $\{\pi_k(x)\}$ for $k = 1, 2, \ldots$, where

$$\pi_k(x) \;\propto\; \{h(x)\}^{m_k}, \qquad\qquad x \in S, \tag{21}$$

and the m_k's form a specified increasing sequence. Then, each of the distributions has its mode at $\breve{x}$ and, as k increases, the mode becomes more and more prominent. Thus, if we make a random draw from each successive $\pi_k(x)$, eventually we shall only produce $\breve{x}$, with the proviso that, if there are multiple global maxima, observations are eventually drawn uniformly from among the corresponding $\breve{x}$'s.

Example. Noisy binary channel. We return to the second case of the noisy binary channel in Section 2.2.1, with $y = 111001110011100\ldots$, a vector of length 100000. The ordinary Viterbi algorithm identifies y itself as the mode of $\pi(x|y)$ but we also deduced this by sampling from $\pi_k(x) \propto \{\pi(x|y)\}^k$, which requires a trivial amendment of the original sampling algorithm. Thus, we generated x's from $\pi_k(x)$ for $m_k = 1$ (done already), 2, $\ldots, 25$ and noted the number of disagreements with y. For $m_k = 1, 2, 3, 4, 8, 12, 16, 20, 21, 22, 23, 24, 25$, there were 22290, 11928, 6791, 3826, 442, 30, 14, 0, 0, 2, 0, 0, 0 discrepancies, respectively. Although still a toy example, $\pi(y|y) \approx 5 \times 10^{-324}$, so the task was not entirely trivial from a sampling perspective.

Of course, in the real world, it is typical that, if $\breve{x}$ cannot be found directly, then nor can we generate draws from $\pi_k(x)$. In that case, we must implement an MCMC version in which successive π_k's in a single run of the algorithm are sampled approximately rather than exactly. This requires some care in selecting a "schedule" for how the m_k's in (21) should increase, because the observation attributed to π_k must also serve as an approximate draw from π_{k+1}. It is typical that eventually the m_k's must increase extremely slowly at a rate closer to logarithmic than to linear. Simulated annealing can also be extended to continuous functions via Langevin diffusion; see Geman and Hwang (1986).

3. Markov chain Monte Carlo calculations.

3.1. Markov chains, stationary distributions and ergodicity. In ordinary Monte Carlo calculations, we require perfect draws from the target distribution $\{\pi(x) : x \in S\}$. We now assume that this is impracticable but that we can construct a Markov transition probability matrix (t.p.m.) P with state space S and limiting distribution π and that we can generate a very long realization from the corresponding Markov chain. In Section 3.2, we discuss some general issues in the construction and implementation of suitable t.p.m.'s. At present, this may all seem bizarre: generally S is astronomically large, π is an arbitrary probability distribution on S and, even if we can find a suitable P, we cannot possibly store it! Nevertheless, in Sections 3.3 to 3.7, we describe a general recipe for any π, due to Hastings (1970), with the Gibbs sampler and the Metropolis algorithm as special cases. Section 3.8 considers the more specialized topic of perfect MCMC.

We begin by recalling some useful definitions and results for Markov chains with finite or countable state spaces. Our notation differs from that for the Markov chains in Section 2.2.1 but is chosen for consistency with Section 2.1. Thus, let $X^{(0)}, X^{(1)}, \ldots$ denote a Markov chain with state space S and t.p.m. P, whose (x, x') element $P(x, x')$ is the probability of a one–step transition from $x \in S$ to $x' \in S$. Define p_0 to be the row vector representing the p.m.f. of the initial state $x^{(0)}$. Then the marginal p.m.f. p_t of $X^{(t)}$ is given by

$$p_t = p_0 P^t, \qquad t = 0, 1, \ldots, \tag{22}$$

and, if π is a probability vector satisfying *general balance*

$$\pi P = \pi, \tag{23}$$

then π is called a *stationary distribution* for P. That is, P maintains π: if $p_0 = \pi$, then $p_t = \pi$ for all $t = 1, 2, \ldots$. What we require is something more: that, given π (up to scale), we can always find a P for which $p_t \to \pi$ as $t \to \infty$, *irrespective* of p_0. The additional condition is that P should be *ergodic*; that is, *irreducible* and *aperiodic*, in which case π in (23) is unique. Irreducible means that there exists a finite path between any pair of states $x, x' \in S$ that has non–zero probability. Aperiodic means that there is no state that can recur only after a multiple of d steps, where $d \geq 2$. A sufficient condition for an irreducible P to be aperiodic is that at least one diagonal element $P(x, x)$ of P is non–zero, which is automatically satisfied by almost any P in MCMC. More succinctly, P is ergodic if and only if all elements of P^m are positive for some positive integer m. It then follows that $\bar{g}$, defined in (5) or, more correctly, the corresponding sequence of r.v.'s, also converges almost surely to $\mathrm{E}_{\pi} g$ as $m \to \infty$. Furthermore, as in ordinary Monte Carlo, the sampling variance of $\bar{g}$ can be assessed and is of order $1/m$. For details, see almost any textbook covering Markov chains.

Stationarity and irreducibility are somewhat separate issues in MCMC. Usually, one uses the Hastings recipe in Section 3.3 to identify a whole collection of t.p.m.'s P_k, each of which maintains π and is simple to apply but is not individually irreducible with respect to S. One then combines these P_k's appropriately to achieve irreducibility. In particular, note that, if $P_1, \dots, P_n$ maintain π, then so do

$$P = P_1 P_2 \dots P_n, \tag{24}$$

equivalent to applying $P_1, \dots, P_n$ in turn, and

$$P = \frac{1}{n}(P_1 + \dots + P_n), \tag{25}$$

equivalent to choosing one of the P_k's at random. Amalgamations such as (24) or (25) are very common in practice. For example, (25) ensures that, if a transition from x to x' is possible using any single P_k, then this is inherited by P. In applications of MCMC, where $x \in S$ has many individual components, $x = (x_1, \dots, x_n)$, it is typical to specify a P_i for each i, where P_i allows change only in x_i. Then P in (24) allows change in each component in turn and (25) in any single component of x, so that, in either case, irreducibility is at least plausible.

Ideally, we would like $x^{(0)}$ to be drawn directly from π, which is the goal of perfect MCMC algorithms (Section 3.8) but generally this is not viable. The usual fix is to ignore the output during a burn–in phase before collecting the sample $x^{(1)}, \dots, x^{(m)}$ for use in (5). There are no hard and fast rules for choosing the burn–in but assessment via formal analysis (e.g. autocorrelation times) and informal graphical methods (e.g. parallel box–and–whisker plots of the output) is usually adequate, though simple time–series plots can be misleading.

There are some contexts in which burn–in is a crucial issue; for example, with the Ising model in statistical physics and in some applications in genetics. It is then desirable to construct special purpose algorithms; see, among others, Sokal (1989), Marinari and Parisi (1992), Besag and Green (1993) and Geyer and Thompson (1995). Some keywords include *auxiliary variables*, *multigrid methods* and *simulated tempering* (which is related to but distinct from simulated annealing).

When X is very high–dimensional, storage of MCMC samples can become problematic. Storage can be minimized by calculating (5) on the fly for any given g, but often the g's of eventual interest are not known in advance. Because successive states $X^{(t)}$, $X^{(t+1)}$ usually have high positive autocorrelation, little is lost by *subsampling* the output. However, this has no intrinsic merit and it is not generally intended that the gaps be sufficiently large to produce in effect a random sample from π. No new theory is required for subsampling: if the gap length is r, then P is merely replaced by the new Markov t.p.m. P^r. Therefore, we can ignore this aspect in constructing appropriate P's, though eventually $x^{(1)}, \dots, x^{(m)}$ in (5) may refer to a subsample. Note also that burn–in and collection time are somewhat separate issues: the rate of convergence to π is enhanced if the second–largest eigenvalue of P is small *in modulus*, whereas a *large negative* eigenvalue can improve the efficiency of estimation. Indeed, one might use different samplers during the burn–in and collection phases. See, for example, Besag *et al.* (1995), especially the rejoinder, for some additional remarks and references.

Lastly here, we mention that the capabilities of MCMC have occasionally been undersold, in that the convergence of the Markov chain is not merely to the marginals of π but to its entire multivariate distribution. Corresponding functionals (3), whether involving a single component or many, can be evaluated with equal ease from a single run. Of course, there are some obvious limitations: for example, one cannot expect to approximate the probability of some very rare event with high relative precision without a possibly prohibitive run length.

3.2. Detailed balance. We need a method of constructing P_k's to satisfy (23). That is, we require P_k's such that

$$\sum_{x \in S} \pi(x)\, P_k(x, x') \;=\; \pi(x')\,, \tag{26}$$

for all $x' \in S$. However, we also need to avoid the generally intractable summation over the state space S. We can achieve this by demanding a much more stringent condition than general balance, namely *detailed balance*,

$$\pi(x)\, P_k(x, x') \;=\; \pi(x')\, P_k(x', x)\,, \tag{27}$$

for all $x, x' \in S$. Summing both sides of (27) over $x \in S$ implies that general balance is satisfied; moreover, detailed balance is much simpler to confirm, particularly if we insist that $P_k(x, x') = 0 = P_k(x', x)$ for the vast majority of $x, x' \in S$. Also note that (27) need only be checked for $x' \neq x$, which is helpful in practice because the diagonal elements of P_k are often quite complicated. The physical significance of (27) is that, if a stationary Markov chain $\ldots, X^{(-1)}, X^{(0)}, X^{(1)}, \ldots$ satisfies detailed balance, then it is *time reversible*, which means that it is impossible to tell whether a film of a sample path is being shown forwards or backwards.

It is clear that, if $P_1, \ldots, P_n$ individually satisfy detailed balance with respect to π, then so does P in (25). Time reversibility is not inherited in the same way by P in (24) but it can easily be resurrected by assembling the P_k's as a random rather than as a fixed permutation at each stage. The maintenance of time reversibility has some theoretical advantages (e.g. the Central Limit Theorem of Kipnis and Varadhan, 1986, and the Initial Sequence Estimators of Geyer, 1992) and is worthwhile in practice if it adds a negligible computational burden.

3.3. Hastings algorithms. Hastings (1970) provides a remarkably simple general construction of t.p.m.'s P_k satisfying detailed balance (27) with respect to π. Thus, let R_k be *any* Markov t.p.m. having state space S and elements $R_k(x, x')$, say. Now define the off–diagonal elements of P_k by

$$(28) \qquad P_k(x, x') = R_k(x, x') \, A_k(x, x'), \qquad x' \neq x \in S,$$

where $A_k(x, x') = 0$ if $R_k(x, x') = 0$ and otherwise

$$(29) \qquad A_k(x, x') = \min \left\{ 1 \, , \frac{\pi(x') \, R_k(x', x)}{\pi(x) \, R_k(x, x')} \right\},$$

with $P_k(x, x)$ obtained by subtraction to ensure that P_k has unit row sums, which is achievable since R_k is itself a t.p.m. Then, to verify that detailed balance (27) is satisfied for $x' \neq x$, either $P_k(x, x') = 0 = P_k(x', x)$ and there is nothing to prove or else direct substitution of (28) produces $\min \{\pi(x) \, R_k(x, x'), \, \pi(x') \, R_k(x', x)\}$ on both sides of the equation. Thus, π is a stationary distribution for P_k, despite the arbitrary choice of R_k, though note that we might as well have insisted that zeros in R_k occur symmetrically. Note also that P_k depends on π only through $h(x)$ in (1) and that the usually unknown and problematic normalizing constant c cancels out. Of course, that is not quite the end of the story: it is necessary to check that P, obtained via an amalgamation of different P_k's, is sufficiently rich to guarantee irreducibility with respect to π but usually this is simple to ensure in any particular case.

Operationally, any P_k is applied as follows. When in state x, a *proposal* x^* for the subsequent state x' is generated with probability $R_k(x, x^*)$. This requires calculating the non–zero elements in row x of R_k on the fly, rather than storing any matrices. Then either $x' = x^*$, with the *acceptance probability* $A_k(x, x^*)$, or else $x' = x$ is retained as the next state of the chain. Note that (28) does not apply to the diagonal elements of P: two successive states x and x' can be the same either because x happens to be proposed as the new state or because some other state x^* is proposed but is not accepted. Also note that the procedure differs from ordinary rejection sampling, where proposals x^* are made until one is accepted, which is not valid in MCMC.

3.4. Componentwise Hastings algorithms. In practice, we still need to choose a particular set of R_k's. It is important that proposals and decisions on their acceptance are simple and fast to make. We now openly acknowledge that X has many components and write $X = (X_1, \ldots, X_n)$, where each X_i is univariate (though this is not essential). Then, the most common approach is to devise an algorithm in which a proposal matrix R_i is assigned to each individual component X_i. That is, if x is the current state, then R_i proposes replacing the ith component x_i by x_i^*, while leaving the remainder x_{-i} of x unaltered. Note that we can also allow some continuous components: then the corresponding R_i's and P_i's become transition kernels rather than matrices and have elements that are conditional densities rather than probabilities. Although the underlying Markov chain theory must then be reworked in terms of general state spaces (e.g. Nummelin, 1984), the modifications in practice are entirely straightforward. For convenience here, we continue to adopt discrete state space terminology and notation.

In *componentwise* Hastings algorithms, the acceptance probability for x_i^* can be rewritten as

$$A_i(x, x^*) \;=\; \min\left\{1,\; \frac{\pi(x_i^*|x_{-i})\, R_i(x^*, x)}{\pi(x_i|x_{-i})\, R_i(x, x^*)}\right\}, \tag{30}$$

which identifies the crucial role played by the *full conditionals* $\pi(x_i|x_{-i})$. Note that these n univariate distributions comprise the basic building blocks of *Markov random field* formulations in spatial statistics (Besag, 1974), where formerly they were called the *local characteristics* of X.

The full conditionals for any particular $\pi(x)$ follow from the trivial but, at first sight, slightly strange–looking result,

$$\pi(x_i|x_{-i}) \;\propto\; \pi(x) \;\propto\; h(x)\,, \tag{31}$$

where the normalizing constant involves only a one–dimensional summation over x_i. Even this drops out in the ratio (30) and, usually, so do many other terms because likelihoods, priors and posteriors are typically formed from products and then only those factors in (31) that involve x_i itself need to be retained. Such cancelations imply enormous computational savings.

In terms of Markov random fields, the *neighbors* ∂i of i comprise the minimal subset of $-i$ such that $\pi(x_i|x_{-i}) = \pi(x_i|x_{\partial i})$. Under a mild positivity condition (see Section 3.5), it can be shown that, if $j \in \partial i$, then $i \in \partial j$, so that the n neighborhoods define an undirected graph in which there is an edge between i and j if they are neighbors. Similar considerations arise in *graphical models* (e.g. Lauritzen, 1996) and *Bayesian networks* (e.g. Pearl, 2000) in constructing the implied undirected graph from a directed acyclic graph or from a chain graph, for example. Note that conventional dynamic simulation makes use of directed graphs, whereas MCMC is based on undirected representations or a mix of the two, as in space–time (chain graph) models, for example. Generally speaking, dynamic simulation should be used and componentwise MCMC avoided wherever possible.

Example. Autologistic and related distributions. The autologistic distribution (Besag, 1974) is a pairwise–interaction Markov random field for dependent binary r.v.'s. It includes binary Markov chains, noisy binary channels and finite–lattice Ising models as special cases, so that simulation without MCMC can range from trivial to taxing to (as yet) impossible.

We define $X = (X_1, \ldots, X_n)$ to have an autologistic distribution if its p.m.f. is

$$\pi(x) \propto \exp\left(\sum_i \alpha_i x_i + \sum_{i<j} \beta_{ij} 1[x_i = x_j]\right), \qquad x \in S = \{0,1\}^n, \tag{32}$$

where the indices i and j run from 1 to n and the β_{ij}'s control the dependence in the system. The simplification with respect to a saturated binary model is that no terms involve interactions between three or more components. The autologistic model also appears under other names: in Cox and Wermuth (1994), as the quadratic exponential binary distribution and, in Jordan *et al.* (1998), as the Boltzmann distribution, after Hinton and Sejnowski (1986).

It is convenient to define $\beta_{ij} = \beta_{ji}$ for $i > j$. Then, as regards the special cases mentioned above, the r.v.'s $X_1, \ldots, X_n$ form a simple symmetric Markov chain if $\alpha_i = 0$ and $\beta_{ij} = \beta$ for $|i - j| = 1$, with $\beta_{ij} = 0$ otherwise. The noisy binary channel (17) is obtained when $\pi(x)$ becomes $\pi(x|y)$ with y fixed, $\alpha_i = (2y_i - 1)\alpha$ and $\beta_{ij} = \beta$ whenever $|i - j| = 1$, with $\beta_{ij} = 0$ otherwise. For the symmetric Ising model, the indices i are identified with the sites of a finite d–dimensional regular array, $\alpha_i = 0$ for all i and $\beta_{ij} = \beta$ for each pair of adjacent sites, with $\beta_{ij} = 0$ otherwise. In each case, the asymmetric version is also catered for by (32).

It follows from (31) that the full conditional of X_i in the distribution (32) is

$$(33) \qquad \pi(x_i|x_{-i}) \propto \exp\big(\alpha_i x_i + \sum_{j \neq i} \beta_{ij} 1[x_i = x_j]\big)\,, \qquad x_i = 0, 1,$$

so that i and j are neighbors if and only if $\beta_{ij} \neq 0$. i. For example, in the noisy binary channel (17)

$$(34) \quad \pi(x_i|x_{-i}, y) \propto \exp\left\{\alpha 1[x_i = y_i] + \beta(1[x_i = x_{i-1}] + 1[x_i = x_{i+1}])\right\},$$

where $x_0 = x_{n+1} = -1$ to accommodate the end points $i = 1$ and $i = n$; and correspondingly, $\partial i = \{i - 1, i + 1\}$, unless $i = 1$ or n for which i has a single neighbor.

Although it is trivial to evaluate the conditional probabilities in (33), because there are only two possible outcomes, we again emphasize that the normalizing constant is not required in the Hastings ratio, which is important in much more complicated formulations. We now describe the two most widely used componentwise algorithms, the Gibbs sampler and the Metropolis method.

3.5. Gibbs sampler. The term "Gibbs sampler" (Geman and Geman, 1984) is motivated by the simulation of *Gibbs distributions* in statistical physics, which correspond to Markov random fields in spatial statistics, the equivalence being established by the Hammersley–Clifford theorem (Besag, 1974). The Gibbs sampler can be interpreted as a componentwise Hastings algorithm in which proposals are made from the full conditionals themselves; that is,

$$(35) \qquad R_i(x, x^*) = \pi(x_i^*|x_{-i})\,,$$

so that the quotient in (30) has value 1 and the proposals are always accepted. The n individual P_i's can then be combined as in (24), producing a *systematic scan* of all n components, or as in (25), giving a *random scan*, or otherwise. Systematic and random scan Gibbs samplers are aperiodic, because $R_i(x,x) > 0$ for any $x \in S$; and they are irreducible under the positivity condition that the minimal support S of X is the Cartesian product of the minimal supports of the individual X_i's. Positivity holds in most practical applications and can be relaxed somewhat (Besag, 1994b) to cater for exceptions. To see its relevance, consider the trite example in which $X = (X_1, X_2)$ and $S = \{00, 11\}$, so that no movement is possible using a componentwise updating algorithm. On the other hand, if $S = \{00, 01, 11\}$, then positivity is violated but both the systematic and random scan Gibbs samplers are irreducible. Severe problems occur most frequently in constrained formulations and can be tackled by using *block* updates of more than one component at a time, to which we return in Section 3.7, or by augmenting S by dummy states (e.g. $\{01\}$ in the above example).

Although the validity of the Gibbs sampler is ensured by the general theory for Hastings algorithms, there is a more direct justification, which formalizes the argument that, if X has distribution π and any of its components is replaced by one sampled from the corresponding full conditional induced by π, to produce a new vector X', then X' must also have distribution π. That is, if x' differs from x in its ith component at most, so that $x'_{-i} = x_{-i}$, then

$$\Pr(X' = x') \;=\; \sum_{x_i} \pi(x)\, \pi(x'_i | x_{-i}) \;=\; \pi(x'_i | x_{-i})\, \pi(x_{-i}) \;=\; \pi(x') \,.$$

Example. Autologistic distribution. In the systematic scan Gibbs sampler for the distribution (32), every cycle addresses each component x_i in turn and immediately updates it according to its full conditional distribution (33).

3.6. Metropolis algorithms. The original MCMC algorithm is that of Metropolis et al. (1953), used for the Ising and other models in statistical physics. This is a componentwise Hastings algorithm, in which R_i is chosen to be a *symmetric* matrix, so that the acceptance probability (30) reduces to

$$A_i(x, x^*) \;=\; \min\{1,\, \pi(x^*_i | x_{-i}) / \pi(x_i | x_{-i})\}\,, \tag{36}$$

independent of R_i! For example, if X_i supports only a small number of values, then R_i might select x^*_i uniformly from these, usually excluding the current value x_i. If X_i is continuous, then it is common to choose x^*_i according to a uniform or Gaussian or some other easily–sampled symmetric distribution, centered on x_i and with a scale factor determined on the basis of a few pilot runs to give acceptance rates in the range 20 to 60%, say. A little care is needed here if X_i does not have unbounded support, so as to maintain symmetry near an endpoint; alternatively a Hastings correction can be applied.

The intention in Metropolis algorithms is to make proposals that can be generated and accepted or rejected very fast. Note that consideration of π arises only in calculating the ratio of the full conditionals in (36) and that this is generally a much simpler and faster task than sampling from a full conditional distribution, unless the latter happens to have a very convenient form. Thus, the processing time per step is generally much less for Metropolis than for Gibbs; and writing a program from scratch is much easier.

Example. Autologistic distribution. When updating x_i in (32), the obvious Metropolis proposal is deterministic, with $x_i^* = 1 - x_i$. This generally results in more mobility than in the corresponding Gibbs sampler, because $A_i(x, x^*) \geq \pi(x_i^* | x_{-i})$, and therefore increases statistical efficiency. The argument can be formalized (Peskun, 1973, and more generally, Liu, 1996) and provides one reason why physicists prefer Metropolis to Gibbs for the ferromagnetic Ising model, though these days they would usually adopt a cluster algorithm (e.g. Swendsen and Wang, 1987).

3.7. Gibbs sampling versus other Hastings algorithms. The Gibbs sampler has considerable intuitive appeal and one might assume from its ubiquity in the Bayesian literature that it represents a panacea among componentwise Hastings algorithms. However, we have just seen that this is not necessarily so. The most tangible advantage of the general Hastings formulation over the Gibbs sampler is that it can use the current value x_i of a component to guide the choice of the proposal x_i^* and improve mobility around the state space S. For some further discussion, see Besag *et al.* (1995, Section 2.3.4) and Liu (1996). Even when Gibbs is *statistically* more efficient, a simpler algorithm may be superior in practice if 10 or 20 times as many cycles can be executed in the same run time. Hastings also facilitates the use of block proposals, which are often desirable to increase mobility and indeed essential in constrained formulations. As one example, multivariate proposals are required in the Langevin–Hastings algorithm (Besag, 1994a).

Nevertheless, there are many applications where efficiency is relatively unimportant and componentwise Gibbs is quite adequate. In the continuous case, difficult full conditionals are often log–concave and so permit the use of *adaptive rejection sampling* (Gilks, 1992). Also, approximate histogram–based Gibbs samplers can be corrected by appropriate Hastings steps; see Tierney (1994) and, for related ideas involving *random* proposal distributions, Besag *et al.* (1995, Appendix 1).

Finally, smart block updates can sometimes be incorporated in Gibbs. For example, if $\pi(x_A | x_{-A})$, in a particular block A, is multivariate Gaussian, then updates of X_A can be made via Cholesky decomposition. Similarly, if $\pi(x_A|x_{-A})$ is a hidden Markov chain, then the Baum algorithm can be exploited when updating X_A. It may also be feasible to manufacture block Gibbs updates by using special recursions (Bartolucci and Besag, 2002).

3.8. Perfect MCMC simulation. We have noted that it may be difficult to determine a suitable burn–in period for MCMC. Perfect MCMC simulation solves the problem by producing an initial draw that is exactly from the target distribution π. The original method, called monotone coupling from the past (Propp and Wilson, 1996), in effect runs the chain from the infinite past and samples it at time zero, so that complete convergence is assured. This sounds bizarre but can be achieved in several important special cases, including the Ising model, even at its critical temperature on very large arrays (e.g. 2000×2000). Perfect MCMC simulation is a very active research area; see

`http://research.microsoft.com/~dbwilson/exact/`

Here we focus on coupling from the past (CFTP) and its implementation when π is the posterior distribution (17) for the noisy binary channel.

We assume that π has finite support S. We can interpret burn–in of fixed length m_0 as running our ergodic t.p.m. P forwards from time $-m_0$ and ignoring the output until time 0. Now imagine that, instead of doing this from a single state at time $-m_0$, we do it from *every* state in S, using the identical stream of random numbers in every case, with the consequence that, if any two paths ever enter the same state, then they coalesce permanently. In fact, since S is finite, we can be certain that, if m_0 is large enough, coalescence of *all* paths will occur by time 0 and we obtain the same state $x^{(0)}$ regardless of $x^{(-m_0)}$. It also ensures that we will obtain $x^{(0)}$ if we run the chain from any state arbitrarily far back in time, so long as we use the identical random number stream during the final m_0 steps. Hence $x^{(0)}$ is a random draw from π. It is crucial in this argument that the timing of the eventual draw is fixed. If we run the chain forwards from every possible initialization at time 0 and wait for all the paths to coalesce, we obtain a *random* stopping time and a corresponding bias in the eventual state. As an extreme example, suppose that $P(x', x'') = 1$ for two particular states x' and x'' but that $P(x, x'') = 0$ for all $x \neq x'$. Then $\pi(x'') = \pi(x')$ but the state at coalescence is never x''.

At first sight, useful implementation of the above idea seems hopeless. Unless S is tiny, it is not feasible to run the chain from every state in S even for $m_0 = 1$. However, we can sometimes find a monotonicity in the paths which allows us to conclude that coalescence from certain extremal states implies coalescence from everywhere. We discuss this here merely for the noisy binary channel but the reasoning is identical to that in Propp and Wilson (1996) for the ostensibly much harder Ising model and also extends immediately to the general autologistic model (32), provided the β_{ij}'s are non–negative.

Thus, again consider the posterior distribution (17), with α and $\beta > 0$ known. We have seen already that it is easy to implement a systematic scan Gibbs sampler based on (34). We presume that the usual inverse distribution function method is used at every stage: that is, when addressing component x_i, we generate a uniform deviate on the unit interval and, if its value exceeds the probability for $x_i = 0$, implied by (34), we set the new $x_i = 1$, else $x_i = 0$.

Now imagine that, using a single stream of random numbers, we run the chain as above from each of two states x' and $x'' \in S$ such that $x' \leq x''$ componentwise. Then the corresponding inequality is inherited by the new pair of states obtained at each iteration, because $\beta > 0$. Similarly, consider three initializations, 0 (all zeros), 1 (all ones) and any other $x \in S$. Because $0 \leq x \leq 1$ elementwise, it follows that the corresponding inequality holds at every subsequent stage and so all paths must coalesce by the time the two extreme ones do so. Hence, we need only monitor the two extremal paths. Note that coalescence occurs much faster than one might expect, because of the commonalities in the simulation method.

However, we must still determine how far back we need to go to ensure that coalescence occurs by time 0. A basic method is as follows. We begin by running simulations from time -1, initialized by $x^{(-1)} = 0$ and 1, respectively. If the paths do not coalesce at time 0, we repeat the procedure from time -2, ensuring that the previous random numbers are used again between times -1 and 0. If the paths do not coalesce by time 0, we repeat from time -3, ensuring that the previous random numbers are used between times -2 and 0; and so on. We terminate the process when coalescence by time 0 first occurs and take the corresponding $x^{(0)}$ as our random draw from π. We say coalescence "by" rather than "at" time 0 because, in the final run, this may occur before time 0. In practice, it is generally more efficient to use increasing increments between the starting times of successive runs, again with duplication of the random numbers during the common intervals. There is no need to identify the smallest m for which coalescence occurs by time zero.

For a numerical illustration, we again chose $\alpha = \ln 4$ and $\beta = \ln 3$ in (17), with $y = 111001110011100...$, a vector of length 100000. Thus, the state space has 2^{100000} elements. Moving back one step at a time, coalescence by time 0 first occurred when running from time -15, with an approximate halving of the discrepancies between each pair of paths, generation by generation, though not even a decrease is guaranteed. Coalescence itself occurred at time -2. There were 77759 matches between the CFTP sample $x^{(0)}$ and the MPM and MAP estimates, which recall are both equal to y in this case. Of course, the performance of CFTP becomes hopeless if β is too large but, in such cases, it may be possible to adopt algorithms that converge faster but still preserve monotonicity. Indeed, for the Ising model, Propp and Wilson (1996) use Sweeny's cluster algorithm rather than the Gibbs sampler. An alternative would be to use perfect *block* Gibbs sampling (Bartolucci and Besag, 2002). Fortunately, in most Bayesian formulations, convergence is relatively fast because the information in the likelihood dominates that in the prior.

4. Discussion. The most natural method of learning about a complex probability model is to generate random samples from it by ordinary Monte Carlo methods. However, this approach can only rarely be implemented. An alternative is to relax independence. Thus, in recent years, Markov chain Monte Carlo methods, originally devised for the analysis of complex stochastic systems in statistical physics, have attracted much wider attention. In particular, they have had an enormous impact on Bayesian inference, where they enable extremely complicated formulations to be analyzed with comparative ease, despite being computationally very intensive. For example, they are now applied extensively in Bayesian image analysis; for a recent review, see Hurn *et al.* (2003). There is also an expanding literature on particle filters, whose goal is to update inferences in real time as additional information is received. At the very least, MCMC encourages the investigator to experiment with models that are beyond the limits of more traditional numerical methods.

Acknowledgment. This research was supported by the Center for Statistics and the Social Sciences with funds from the University Initiatives Fund at the University of Washington.

REFERENCES

Bartolucci F. and Besag J.E. (2002). A recursive algorithm for Markov random fields. *Biometrika,* **89,** 724–730.

Baum L.E., Petrie T., Soules G., and Weiss N. (1970). A maximization technique occurring in the statistical analysis of probabilistic functions of Markov chains. *Annals of Mathematical Statistics,* **41,** 164–171.

Besag J.E. (1974). Spatial interaction and the statistical analysis of lattice systems (with Discussion). *Journal of the Royal Statistical Society B,* **36,** 192–236.

Besag J.E. (1994a). Discussion of paper by U. Grenander and M. I. Miller. *Journal of the Royal Statistical Society B,* **56,** 591–592.

Besag J.E. (1994b). Discussion of paper by L.J. Tierney. *Annals of Statistics,* **22,** 1734–1741.

Besag J.E. (2001). Markov chain Monte Carlo for statistical inference. *Working Paper No. 9,* Center for Statistics and the Social Sciences, University of Washington, pp. 67.

Besag J.E. (2003). Likelihood analysis of binary data in time and space. In *Highly Structured Stochastic Systems* (eds. P.J. Green, N.L. Hjort, and S. Richardson). Oxford University Press.

Besag J.E. and Clifford P. (1989). Generalized Monte Carlo significance tests. *Biometrika,* **76,** 633–642.

Besag J.E. and Clifford P. (1991). Sequential Monte Carlo p–values. *Biometrika,* **78,** 301–304.

Besag J.E. and Green P.J. (1993). Spatial statistics and Bayesian computation (with Discussion). *Journal of the Royal Statistical Society B,* **55,** 25–37.

Besag J.E., Green P.J., Higdon D.M., and Mengersen K.L. (1995). Bayesian computation and stochastic systems (with Discussion). *Statistical Science,* **10,** 3–66.

Chen M.-H., Shao Q.-M., and Ibrahim J.G. (2000). *Monte Carlo Methods in Bayesian Computation.* Springer: New York.

Cox D.R. and Wermuth N. (1994). A note on the quadratic exponential binary distribution. *Biometrika,* **81,** 403–408.

Doucet A., De Freitas N., and Gordon N. (2001). *Sequential Monte Carlo Methods in Practice.* Springer: New York.

Eddie S.R., Mitchison G., and Durbin R. (1995). Maximum discrimination hidden Markov models of sequence concensus. *Journal of Computational Biology,* **2,** 9–24.

Fishman G.S. (1996). *Monte Carlo: Concepts, Algorithms, and Applications.* Springer Verlag: New York.

Fredkin D.R. and Rice J.A. (1992). Maximum likelihood estimation and identification directly from single-channel recordings. *Proceedings of the Royal Society of London B,* **249,** 125–132.

Gelman A., Carlin J.B., Stern H.S., and Rubin D.B. (1995). *Bayesian Data Analysis.* Chapman and Hall/CRC: Boca Raton.

Geman S. and Geman D. (1984). Stochastic relaxation, Gibbs distributions and the Bayesian restoration of images. *Institute of Electrical and Electronics Engineers, Transactions on Pattern Analysis and Machine Intelligence,* **6,** 721–741.

Geman S. and Hwang C.-R. (1986). Diffusions for global optimization. *SIAM Journal on Control and Optimization,* **24,** 1031–1043.

Geyer C.J. (1991). Markov chain Monte Carlo maximum likelihood. In *Computing Science and Statistics: Proceedings of the 23rd Symposium on the Interface* (ed. E.M. Keramidas), 156–163. Interface Foundation of North America, Fairfax Station, VA.

Geyer C.J. (1992). Practical Markov chain Monte Carlo (with Discussion). *Statistical Science,* **7,** 473–511.

Geyer C.J. and Thompson E.A. (1992). Constrained Monte Carlo maximum likelihood for dependent data (with Discussion). *Journal of the Royal Statistical Society B,* **54,** 657–699.

Geyer C.J. and Thompson E.A. (1995). Annealing Markov chain Monte Carlo with applications to ancestral inference. *Journal of the American Statistical Association,* **90,** 909–920.

Gilks W.R. (1992). Derivative-free adaptive rejection sampling for Gibbs sampling. In *Bayesian Statistics 4* (eds. J.O. Berger, J.M. Bernardo, A.P. Dawid, and A.F.M. Smith), 641–649. Oxford University Press.

Gilks W.R., Richardson S., and Spiegelhalter D. (eds.) (1996). *Markov Chain Monte Carlo in Practice.* Chapman and Hall: London.

Green P.J. (1995). Reversible jump Markov chain Monte Carlo computation and Bayesian model determination. *Biometrika,* **82,** 711–732.

Hastings W.K. (1970). Monte Carlo sampling methods using Markov chains and their applications. *Biometrika,* **57,** 97–109.

Haussler D., Krogh A., Mian S., and Sjolander K. (1993). Protein modeling using hidden Markov models: analysis of globins. In *Proceedings of the Hawaii International Conference on System Sciences.* IEEE Computer Science Press: Los Alamitos, CA.

Hinton G.E. and Sejnowski T. (1986). Learning and relearning in Boltzmann machines. In *Parallel Distributed Processing* (eds. D.E. Rumelhart and J.L. McClelland). M.I.T Press.

Hughes J.P., Guttorp P., and Charles S.P. (1999). A nonhomogeneous hidden Markov model for precipitation. *Applied Statistics,* **48,** 15–30.

Hurn M., Husby O., and Rue H. (2003). Advances in Bayesian image analysis. In *Highly Structured Stochastic Systems* (eds. P.J. Green, N.L. Hjort, and S. Richardson). Oxford University Press.

Jordan M.I., Ghahramani Z., Jaakkola T.S., and Saul L.K. (1998). An introduction to variational methods for graphical models. In *Learning in Graphical Models* (ed. M.I. Jordan). Kluwer Academic Publishers.

Juang B.H. and Rabiner L.R. (1991). Hidden Markov models for speech recognition. *Technometrics,* **33,** 251–272.

Kipnis C. and Varadhan S.R.S. (1986). Central limit theorem for additive functionals of reversible Markov processes and applications to simple exclusions. *Communications in Mathematical Physics,* **104,** 1–19.

Kirkpatrick S., Gelatt C.D., and Vecchi M.P. (1983). Optimization by simulated annealing. *Science,* **220,** 671–680.

Lauritzen S.L. (1996). *Graphical Models.* Clarendon Press: Oxford.

Le Strat Y. and Carrat F. (1999). Monitoring epidemiologic surveillance data using hidden Markov models. *Statistics in Medicine,* **18,** 3463–3478.

Liu J.S. (1996). Peskun's theorem and a modified discrete-state Gibbs sampler. *Biometrika,* **83,** 681–682.

Liu J.S. (2001). *Monte Carlo Strategies in Scientific Computing.* Springer: New York.

Liu J.S., Neuwald A.F., and Lawrence C.E. (1995). Bayesian models for multiple local sequence alignment and Gibbs sampling strategies. *Journal of the American Statistical Association,* **90,** 1156–1170.

MacCormick J. (2002). *Stochastic Algorithms for Visual Tracking.* Springer: New York.

MacDonald I.L. and Zucchini W. (1997). *Hidden Markov and Other Models for Discrete-valued Time Series.* Chapman and Hall: London.

Marinari E. and Parisi G. (1992). Simulated tempering: a new Monte Carlo scheme. *Europhysics Letters,* **19,** 451–458.

Metropolis N., Rosenbluth A.W., Rosenbluth M.N., Teller A.H., and Teller E. (1953). Equations of state calculations by fast computing machines. *Journal of Chemical Physics,* **21,** 1087–1092.

Newman M.E.J. and Barkema G.T. (1999). *Monte Carlo Methods in Statistical Physics.* Clarendon Press: Oxford.

Nummelin E. (1984). *General Irreducible Markov Chains and Non-Negative Operators.* Cambridge University Press.

Pearl J. (2000).] *Causality.* Cambridge University Press.

Penttinen A. (1984). Modeling interaction in spatial point patterns: parameter estimation by the maximum likelihood method. *Jyväskylä Studies in Computer Science, Economics and Statistics,* **7.**

Peskun P.H. (1973). Optimum Monte Carlo sampling using Markov chains. *Biometrika,* **60,** 607–612.

Propp J.G. and Wilson B.M. (1996). Exact sampling with coupled Markov chains and applications to statistical mechanics. *Random Structures and Algorithms,* **9,** 223–252.

Rabiner L.R. (1989). A tutorial on hidden Markov models and selected applications in speech recognition. *Proceedings of the Institute of Electrical and Electronics Engineers,* **77,** 257–284.

Robert C.P. and Casella G. (1999). *Monte Carlo Statistical Methods.* Springer: New York.

Robert C.P., Rydén T., and Titterington D.M. (2000). Bayesian inference in hidden Markov models through the reversible jump Markov chain Monte Carlo method. *Journal of the Royal Statistical Society B,* **62,** 57–75.

Sokal A.D. (1989). Monte Carlo methods in statistical mechanics: foundations and new algorithms. *Cours de Troisième Cycle de la Physique en Suisse Romande,* Lausanne.

Swendsen R.H. and Wang J.-S. (1987). Non-universal critical dynamics in Monte Carlo simulations. *Physics Review Letters,* **58,** 86–88.

Tierney L.J. (1994). Markov chains for exploring posterior distributions (with Discussion). *Annals of Statistics,* **22,** 1701–1762.

Tjelmeland H. and Besag J. (1998). Markov random fields with higher-order interactions. *Scandinavian Journal of Statistics,* **25,** 415–433.

SEMIPARAMETRIC FILTERING IN SPEECH PROCESSING

BENJAMIN KEDEM* AND KONSTANTINOS FOKIANOS†

Abstract. We consider m data sets where the first $m-1$ are obtained by sampling from multiplicative exponential distortions of the mth distribution, it being a reference. The combined data from m samples, one from each distribution, are used in the semiparametric large sample problem of estimating each distortion and the reference distribution, and testing the hypothesis that the distributions are identical. Possible applications to speech processing are mentioned.

1. Introduction. Imagine the general problem of combining sources of information as follows. Suppose there are m related sources of data, of which the mth source, called the "reference", is the most reliable. Obviously, the characteristics of the reference source can be assessed from its own information or data. But since the sources are related, they all contain pertinent information that can be used collectively to improve the estimation of the reference characteristics. The problem is to *combine all* the sources, reference and non-reference together, to better estimate the reference characteristics and deviations of each source from the reference source.

Thus, throughout this paper the reader should have in mind a "reference" and deviations from it in some sense, and the idea of combining "good" and "bad" to improve the quality of the "good".

We can think of several ways of applying this general scheme to speech processing. The idea could potentially be useful in the combination of several classifiers of speech where it is known that one of the classifiers is more reliable than the rest. Conceptually, our scheme points to the possibility of improving the best classifier by taking into consideration also the output from the other classifiers.

The idea could possibly be useful also in speech processing to account for channel effects in different segments of the acoustic training data, for changes in genre or style in language modeling text, and in other situations where the assumption that the training material is temporally homogeneous is patently false, but the training data may be segmented into contiguous portions within which some homogeneity may be reasonable to assume.

Interestingly, the celebrated statistical problem of analysis of variance under normality is precisely a special case of our general scheme but without the burden of the normal assumption. We shall therefore provide a convenient mathematical framework formulated in terms of "reference" data or

*Department of Mathematics, University of Maryland, College Park, MD 20742, U.S.A (bnk@math.umd.edu).

†Department of Mathematics & Statistics, University of Cyprus, P.O. Box 20537, Nicosia 1678, Cyprus (fokianos@ucy.ac.cy).

their distribution and deviations from them in some sense. The theory will be illustrated in terms of autoregressive signals akin to speech.

The present formulation of the general scheme follows closely the recent development in Fokianos, et al. (2001) which extends Fokianos, et, al. (1998), and Qin and Zhang (1997). Qin and Lawless (1994) is the predecessor to all this work. Related references dealing with more general tilting or bias are the pioneering papers of Vardi (1982), (1986).

2. Mathematical formulation of source combination. In our formalism, "sources" are identified with "data". Deviations are formulated in terms of deviations from a reference distribution. Thus, a data set deviates from a reference set in the sense that its distribution is a distortion of a reference distribution.

To motivate this, consider the classical one-way analysis of variance with $m = q+1$ independent normal random samples,

$$x_{11}, \ldots, x_{1n_1} \sim g_1(x)$$

$$\vdots$$

$$x_{q1}, \ldots, x_{qn_q} \sim g_q(x)$$
$$x_{m1}, \ldots, x_{mn_m} \sim g_m(x)$$

where $g_j(x)$ is the probability density of $\mathrm{N}(\mu_j, \sigma^2)$, $j = 1, ..., m$. Then, holding $g_m(x)$ as a reference distribution, we can see that

$$\frac{g_j(x)}{g_m(x)} = \exp(\alpha_j + \beta_j x), \quad j = 1, ..., q \tag{1}$$

where

$$\alpha_j = \frac{\mu_m^2 - \mu_j^2}{2\sigma^2}, \qquad \beta_j = \frac{\mu_j - \mu_m}{\sigma^2}, \qquad j = 1, ..., q$$

It follows that the test $\mathrm{H}_0 : \mu_1 = \cdots = \mu_m$ is equivalent to $\mathrm{H}_0 : \beta_1 = \cdots = \beta_q = 0$. Clearly $\beta_j = 0$ implies $\alpha_j = 0$, $j = 1, ..., q$.

This sets the stage for the following generalization. With $g \equiv g_m$ denoting the reference distribution, we define deviations from g by the exponential tilt,

$$g_j(x) = \exp(\alpha_j + \beta_j h(x)) g(x), \; j = 1, ..., q \tag{2}$$

where α_j depends on β_j, and $h(x)$ is a known function. The data set $\mathbf{x}_j = (x_{j1}, ..., x_{jn_j})'$ corresponding to g_j deviates from the reference data set $\mathbf{x}_m = (x_{m1}, ..., x_{mn_m})'$ corresponding to $g \equiv g_m$ in the sense of (2). Our goal is to estimate g and all the α_j and β_j from the combined data $\mathbf{x}_1, ..., \mathbf{x}_q, \mathbf{x}_m$.

Expression (2) is what we mean here by filtering. It really is an operation applied to g to produce its filtered versions g_j, $j = 1, ..., q$.

The combined data set from the m samples is the vector $\mathbf{t}$,

$$\mathbf{t} = (t_1, ..., t_n)' = (\mathbf{x}'_1, ..., \mathbf{x}'_q, \mathbf{x}'_m)'$$

where $\mathbf{x}_j = (x_{j1}, ..., x_{jn_j})'$ is the jth sample of length n_j, and $n = n_1 + \cdots + n_q + n_m$.

The statistical semiparametric estimation/testing problems *using the combined data* $\mathbf{t}$ are.

1. Nonparametric estimation of $G(x)$, the cdf corresponding to $g(x)$.
2. Estimation of the parameters $\boldsymbol{\alpha} = (\alpha_1, ..., \alpha_q)'$, $\boldsymbol{\beta} = (\beta_1, ..., \beta_q)'$, and the study of the large sample properties of the estimators.
3. Testing of the hypothesis $\text{H}_0 : \beta_1 = \cdots = \beta_q = 0$.

Evidently, the general construction does not require normality or even symmetry of the distributions, the variances need not be the same, and the model does not require knowledge of the reference distribution. The main assumption is the form of the distortion of the reference distribution.

2.1. Estimation and large sample results. A maximum likelihood estimator of $G(x)$ can be obtained by maximizing the likelihood over the class of step cdf's with jumps at the observed values $t_1, ..., t_n$. Accordingly, if $p_i = dG(t_i)$, $i = 1, .., n$, the likelihood becomes,

$$\mathcal{L}(\boldsymbol{\alpha}, \boldsymbol{\beta}, G) = \prod_{i=1}^{n} p_i \prod_{j=1}^{n_1} \exp(\alpha_1 + \beta_1 h(x_{1j})) \cdots \prod_{j=1}^{n_q} \exp(\alpha_q + \beta_q h(x_{qj})). \tag{3}$$

We follow a profiling procedure whereby first we express each p_i in terms of $\boldsymbol{\alpha}, \boldsymbol{\beta}$ and then we substitute the p_i back into the likelihood to produce a function of $\boldsymbol{\alpha}, \boldsymbol{\beta}$ only. When $\boldsymbol{\alpha}, \boldsymbol{\beta}$ are fixed, (3) is maximized by maximizing only the the product term $\prod_{i=1}^{n} p_i$, subject to the m constraints

$$\sum_{i=1}^{n} p_i = 1, \ \sum_{i=1}^{n} p_i[w_1(t_i) - 1] = 0, ..., \ \sum_{i=1}^{n} p_i[w_q(t_i) - 1] = 0$$

where the summation is over *all* the t_i and

$$w_j(t) = \exp(\alpha_j + \beta_j h(t)), \ j = 1, ..., q.$$

We have

$$p_i = \frac{1}{n_m} \cdot \frac{1}{1 + \rho_1 w_1(t_i) + \cdots + \rho_q w_q(t_i)} \tag{4}$$

where $\rho_j = n_j/n_m$, $j = 1, ..., q$, and the value of the profile log-likelihood up to a constant as a function of $\boldsymbol{\alpha}, \boldsymbol{\beta}$ only is,

$$
\begin{aligned}
l = & -\sum_{i=1}^{n} \log[1+\rho_1 w_1(t_i)+\cdots+\rho_q w_q(t_i)] \\
& +\sum_{j=1}^{n_1}[\alpha_1+\beta_1 h(x_{1j})]+\cdots+\sum_{j=1}^{n_q}[\alpha_q+\beta_q h(x_{qj})].
\end{aligned} \tag{5}
$$

The score equations for $j = 1, ..., q$, are therefore,

$$
\begin{aligned}
\frac{\partial l}{\partial \alpha_j} &= -\sum_{i=1}^{n} \frac{\rho_j w_j(t_i)}{1+\rho_1 w_1(t_i)+\cdots+\rho_q w_q(t_i)} + n_j = 0 \\
\frac{\partial l}{\partial \beta_j} &= -\sum_{i=1}^{n} \frac{\rho_j h(t_i) w_j(t_i)}{1+\rho_1 w_1(t_i)+\cdots+\rho_q w_q(t_i)} + \sum_{i=1}^{n_j} h(x_{ji}) = 0.
\end{aligned} \tag{6}
$$

The solution of the score equations gives the maximum likelihood estimators $\hat{\boldsymbol{\alpha}}, \hat{\boldsymbol{\beta}}$, and consequently by substitution also the estimates

$$
\hat{p}_i = \frac{1}{n_m} \cdot \frac{1}{1+\rho_1 \exp(\hat{\alpha_1}+\hat{\beta}_1 h(t_i))+\cdots+\rho_q \exp(\hat{\alpha_q}+\hat{\beta}_q h(t_i))} \tag{7}
$$

and therefore, the estimate of $G(t)$ from the *combined data* is

$$
\hat{G}(t) = \frac{1}{n_m} \cdot \sum_{i=1}^{n} \frac{I(t_i \leq t)}{1+\rho_1 \exp(\hat{\alpha_1}+\hat{\beta}_1 h(t_i))+\cdots+\rho_q \exp(\hat{\alpha_q}+\hat{\beta}_q h(t_i))}. \tag{8}
$$

It can be shown that the estimators $\hat{\boldsymbol{\alpha}}, \hat{\boldsymbol{\beta}}$, are asymptotically normal,

$$
\sqrt{n} \begin{pmatrix} \hat{\boldsymbol{\alpha}} - \boldsymbol{\alpha}_0 \\ \hat{\boldsymbol{\beta}} - \boldsymbol{\beta}_0 \end{pmatrix} \Rightarrow \mathrm{N}(\mathbf{0}, \boldsymbol{\Sigma}) \tag{9}
$$

as $n \to \infty$. Here $\boldsymbol{\alpha}_0$ and $\boldsymbol{\beta}_0$ denote the true parameters and $\boldsymbol{\Sigma} = \mathbf{S}^{-1}\mathbf{V}\mathbf{S}^{-1}$, where the matrices $\mathbf{S}$ and $\mathbf{V}$ are defined in the appendix.

2.2. Hypothesis testing. We are now in a position to test the hypothesis $\mathrm{H}_0 : \boldsymbol{\beta} = \mathbf{0}$ that all the m populations are equidistributed.

We shall use the following notation for the moments of $h(t)$ with respect to the reference distribution:

$$
\begin{aligned}
E(t^k) &\equiv \int h^k(t) dG(t) \\
Var(t) &\equiv E(t^2) - E^2(t).
\end{aligned}
$$

2.2.1. The $\mathcal{X}_1$ test. Under $\mathrm{H}_0 : \boldsymbol{\beta} = \mathbf{0}$—so that all the moments of $h(t)$ are taken with respect to g–consider the $q \times q$ matrix $\mathbf{A}_{11}$ whose jth diagonal element is

$$
\frac{\rho_j[1+\sum_{k \neq j}^{q} \rho_k]}{[1+\sum_{k=1}^{q} \rho_k]^2}
$$

and otherwise for $j \neq j'$, the jj' element is

$$\frac{-\rho_j \rho_{j'}}{[1+\sum_{k=1}^q \rho_k]^2} .$$

For $m = 2, q = 1$, $\mathbf{A}_{11}$ reduces to a scalar $\rho_1/(1+\rho_1)^2$. For $m = 3, q = 2$,

$$\mathbf{A}_{11} = \frac{1}{(1+\rho_1+\rho_2)^2} \cdot \mathbf{M}_2$$

where

$$\mathbf{M}_2 = \begin{pmatrix} \rho_1(1+\rho_2) & -\rho_1\rho_2 \\ -\rho_1\rho_2 & \rho_2(1+\rho_1) \end{pmatrix} = \begin{pmatrix} \rho_1 & 0 \\ 0 & \rho_2 \end{pmatrix} \begin{pmatrix} 1+\rho_2 & -\rho_2 \\ -\rho_1 & 1+\rho_1 \end{pmatrix}$$

and the eigenvalues of the matrix on the right are $1, 1+\rho_1+\rho_2$.

The elements are bounded by 1 and the matrix is nonsingular,

$$|\mathbf{A}_{11}| = \frac{\prod_{k=1}^q \rho_k}{[1+\sum_{k=1}^q \rho_k]^m} > 0$$

and can be used to represent $\mathbf{S}$,

$$\mathbf{S} = \begin{pmatrix} 1 & E(t) \\ E(t) & E(t^2) \end{pmatrix} \otimes \mathbf{A}_{11}$$

with $\otimes$ denoting the Kronecker product. It follows that $\mathbf{S}$ is nonsingular,

$$|\mathbf{S}| = \{Var(t)\}^q |\mathbf{A}_{11}|^2$$

and,

$$\mathbf{S}^{-1} = \frac{1}{Var(t)} \begin{pmatrix} E(t^2) & -E(t) \\ -E(t) & 1 \end{pmatrix} \otimes \mathbf{A}_{11}^{-1} .$$

On the other hand, $\mathbf{V}$ is singular,

$$\mathbf{V} = Var(t) \begin{pmatrix} \mathbf{0} & \mathbf{0} \\ \mathbf{0} & \mathbf{A}_{11} \end{pmatrix}$$

as is

$$\mathbf{\Sigma} = \mathbf{S}^{-1}\mathbf{V}\mathbf{S}^{-1} = \frac{1}{Var(t)} \begin{pmatrix} E^2(t) & -E(t) \\ -E(t) & 1 \end{pmatrix} \otimes \mathbf{A}_{11}^{-1} .$$

Since $\mathbf{A}_{11}$ is nonsingular we have from (9),

$$\sqrt{n}\hat{\boldsymbol{\beta}} \Rightarrow \mathrm{N}\left(\mathbf{0}, \frac{1}{\mathrm{Var(t)}} \mathbf{A}_{11}^{-1}\right) \tag{10}$$

It follows under $\mathrm{H}_0 : \boldsymbol{\beta} = \mathbf{0}$

$$\mathcal{X}_1 = nVar(t)\hat{\boldsymbol{\beta}}'\mathbf{A}_{11}\hat{\boldsymbol{\beta}} \tag{11}$$

is approximately distributed as $\chi^2(q)$, and H_0 can be rejected for large values of $nVar(t)\hat{\boldsymbol{\beta}}'\mathbf{A}_{11}\hat{\boldsymbol{\beta}}$.

In practice, $Var(t)$ needed for χ_1, and defined above as the variance of $h(t)$ (not of t unless $h(t) = t$), is estimated from

$$\sum_{i=1}^{n} h^2(t_i)\hat{p}_i - \left(\sum_{i=1}^{n} h(t_i)\hat{p}_i\right)^2 .$$

2.2.2. A power study. In Fokianos et al. (2001) the power of $\mathcal{X}_1$ defined in (11) was compared via a computer simulation with the power of the two sample t and Wilcoxson rank sum tests for $m = 2$, and with the F and Kruskal–Wallis tests for $m = 3$.

The simulation study shows that the $\mathcal{X}_1$-test is more powerful than the t-test in lognormal and gamma cases, meaning that departures from the classical normal and variance equality assumptions can weaken the t-test considerably. In addition, the $\mathcal{X}_1$ test dominated the Wilcoxon rank sum and Kruskal–Wallis tests in all the cases considered. Interestingly, under normality and variance equality the $\mathcal{X}_1$-test is not dominated by the t-test and F tests.

2.2.3. Testing the linear hypothesis. We can further test the general linear hypothesis $\mathbf{H}\boldsymbol{\theta} = \mathbf{c}$ where $\mathbf{H}$ is a $p \times 2q$ predetermined matrix of rank p ($p < 2q$), $\boldsymbol{\theta} = (\alpha_1, \ldots, \alpha_q, \beta_1, \ldots, \beta_q)'$, and $\mathbf{c}$ is a vector in R^p. Then, using (9), we have under the hypothesis

$$\sqrt{n}(\mathbf{H}\hat{\boldsymbol{\theta}} - \mathbf{c}) \Rightarrow \mathrm{N}(\mathbf{0}, \mathbf{H\Sigma H}')$$

Thus, the random variable

$$\mathcal{X}_2 = n(\mathbf{H}\hat{\boldsymbol{\theta}} - \mathbf{c})'(\mathbf{H\Sigma H}')^{-1}(\mathbf{H}\hat{\boldsymbol{\theta}} - \mathbf{c}) \tag{12}$$

has an asymptotic chi-square distribution with p degrees of freedom provided the inverse exists (Sen and Singer 1993 p. 239). A consistent estimator of $\boldsymbol{\Sigma}$ can be obtained by replacing all the parameters by their maximum likelihood estimates.

Note that in general the results obtained from (11) and (12) are different, since in (11) we substitute the exact value $\boldsymbol{\beta} = \mathbf{0}$ in $\boldsymbol{\Sigma}$, while (12) requires the maximum likelihood estimate of $\boldsymbol{\theta}$ instead.

3. Illustration of the theory. We illustrate the semiparametric approach in terms of first order autoregressive (AR(1)) time series. Extensions to more complex models are straightforward. We have to distinguish between marginal and conditional models.

3.1. Marginal analysis. Suppose $m = 3$ and that the data come from

$$X_t = \phi_i X_{t-1} + \epsilon_t, \tag{13}$$

where ϕ_i is a parameter such that $|\phi_i| < 1$ for $i = 1, 2, 3$, and $\{\epsilon_t\}$ is a sequence of independent normal random variables with mean 0 and variance σ^2. Then it is well know that for sufficiently large t, X_t is normally distributed with mean 0 and variance $\sigma^2/1 - \phi_i^2$, that is

$$f_i(x_t) = \frac{\sqrt{1-\phi_i^2}}{\sigma\sqrt{2\pi}} \exp\left(-\frac{1-\phi_i^2}{\sigma^2} x_t^2\right), \quad i = 1, 2, 3. \tag{14}$$

With this (2) becomes,

$$\log\left[\frac{f_i(x_t)}{f_3(x_t)}\right] = \alpha_i + \beta_i x_t^2, \tag{15}$$

where

$$\begin{aligned} \alpha_i &= \frac{1}{2}\log\left[\frac{1-\phi_i^2}{1-\phi_3^2}\right], \\ \beta_i &= \frac{\phi_i^2 - \phi_3^2}{2\sigma^2} \end{aligned} \tag{16}$$

The last three equations show that the exponential distortion model (2) holds for Gaussian AR(1) data with $h(x) = x^2$. Thus we expect our method to produce consistent estimates for independent samples from sufficiently long time series. We show that this is indeed the case by resorting to the following limited simulation study.

We assume throughout that ϵ_t follows the normal distribution with mean 0 and variance 1, and that N_i denotes the length of time series i. Practically independent samples were obtained by sampling at random from long time series. Each entry in the Table 1 was obtained from 100 independent time series.

In Table 1 the reference $\phi_3 = 0.1$ is rather small meaning that the reference signal is close to being white noise. The table shows that there is a good agreement between the true parameter values and their estimates. To demonstrate the asymptotic normality, Figure 1 displays the Q–Q plots of the estimators when $N_1 = N_2 = N_3 = 200$. The plots indicate that the asymptotic distribution of the estimators is close to normal.

3.2. Conditional analysis. The exponential tilt model (2) also holds for the conditional one–step transition probability density function of the AR(1) model (13). First observe that the conditional density of X_t given X_{t-1} is given by

$$f_i(x_t|x_{t-1}) = \frac{1}{\sigma\sqrt{2\pi}} \exp\left[-\frac{(x_t - \phi_i x_{t-1})^2}{2\sigma^2}\right], \quad i = 1, 2, 3. \tag{17}$$

TABLE 1

Parameters estimates and their standard errors for different sample sizes for model (15). The data were generated according to (13) with $\phi_1 = -0.5$, $\phi_2 = -0.3$, and $\phi_3 = 0.1$, so that $\alpha_1 = -0.139$, $\beta_1 = 0.12$, $\alpha_2 = -0.042$, $\beta_2 = 0.04$.

Sample Sizes			Estimated Parameters and Standard Errors			
N_1	N_2	N_3	$\hat{\alpha}_1$	$\hat{\beta}_1$	$\hat{\alpha}_2$	$\hat{\beta}_2$
200	200	200	−0.120	0.118	−0.021	0.034
			0.067	0.059	0.073	0.071
200	500	200	−0.133	0.129	−0.037	0.049
			0.075	0.067	0.056	0.055
300	500	200	−0.136	0.131	−0.038	0.048
			0.067	0.063	0.069	0.067

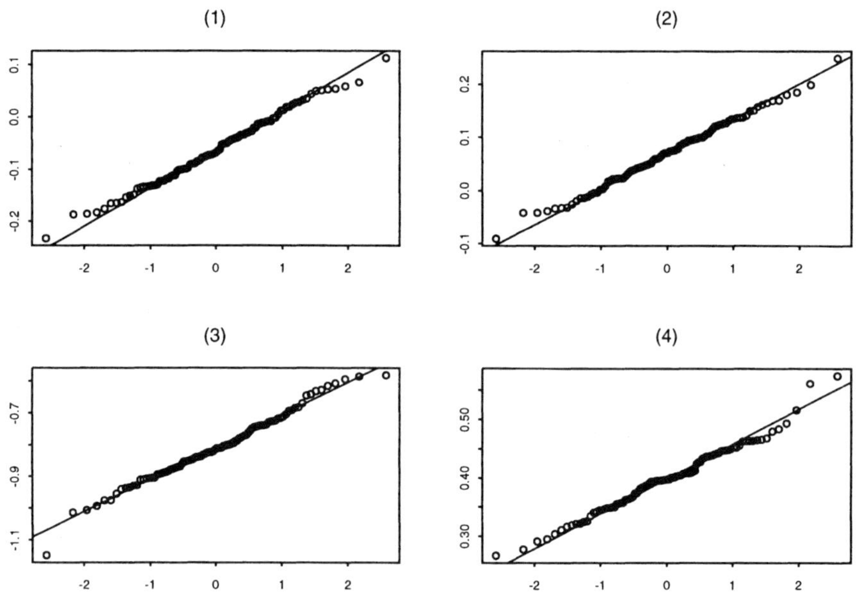

FIG. 1. *Q-Q plots of the parameter estimates under model (15). The data were generated according to (13) with $\phi_1 = -0.5$, $\phi_2 = -0.3$, $\phi_3 = 0.1$, and $N_1 = N_2 = N_3 = 200$. The number of runs is 100. (1) $\hat{\alpha}_1$ (2) $\hat{\beta}_1$ (3) $\hat{\alpha}_2$ (4) $\hat{\beta}_2$.*

Put $f_3(x_t|x_{t-1})$ for the reference distribution. Then a simple calculation shows that

$$\log\left[\frac{f_i(x_t|x_{t-1})}{f_3(x_t|x_{t-1})}\right] = \frac{\phi_i - \phi_3}{\sigma^2} x_t x_{t-1} + \frac{\phi_3^2 - \phi_i^2}{2\sigma^2} x_{t-1}^2, \tag{18}$$

for $i = 1, 2$. Thus, we are led to the following density ratio model

$$\log\left[\frac{f_i(x_t|x_{t-1})}{f_3(x_t|x_{t-1})}\right] = \alpha_i^c + \beta_i^c x_t x_{t-1} + \gamma_i^c x_{t-1}^2 \tag{19}$$

for $i = 1, 2$. Comparison of expressions (18) and (19) reveals that if the data are generated according to the Gaussian autoregressive process (13), then

$$\begin{aligned} \alpha_i^c &= 0 \\ \beta_i^c &= \frac{\phi_i - \phi_3}{\sigma^2} \\ \gamma_i^c &= \frac{\phi_3^2 - \phi_i^2}{2\sigma^2}, \end{aligned} \tag{20}$$

for $i = 1, 2$. A theory similar to the marginal case can be developed for the conditional case as well, but this requires further research.

4. Summary. We have outlined a method for combining data from several sources. This can be used in testing the similarity of m signals given a reference signal, and in the estimation of the marginal distribution of the reference signal from the combined data as in (8). The mathematical development assumes exponential distortions or deviations of the form (2), but the theory can be developed for other types of distortions. The analysis was illustrated in terms of Gaussian AR(1) data where $h(x) = x^2$. In general, $h(x)$ is not known and must be estimated. The choice/estimation of $h(x)$ can be approached in several ways. One possibility is to approximate $h(x)$ by a polynomial or by B-splines. Another possibility is to employ some type of kernel estimation applied to the log-ratio of probability densities. For skewed geophysical data $h(x) = \log(x)$ or even $h(x) = \log^2(x)$ may be helpful.

APPENDIX

We prove the asserted asymptotic normality (9). First define

$$\nabla \equiv \left(\frac{\partial}{\partial\alpha_1}, \ldots, \frac{\partial}{\partial\alpha_q}, \frac{\partial}{\partial\beta_1} \ldots, \frac{\partial}{\partial\beta_q}\right)'.$$

Then it is easy to see $E[\nabla l(\boldsymbol{\alpha}, \boldsymbol{\beta})] = \mathbf{0}$. To obtain the score second moments it is convenient to define $\rho_m \equiv 1,\ w_m(t) \equiv 1$,

$$E_j(t) \equiv \int h(t) w_j(t) dG(t),\ Var_j(t) \equiv \int h^2(t) w_j(t) dG(t) - E_j^2(t)$$

and,

$$A_0(j,j') \equiv \int \frac{w_j(t)w_{j'}(t)dG(t)}{1+\sum_{k=1}^q \rho_k w_k(t)}$$
$$A_1(j,j') \equiv \int \frac{h(t)w_j(t)w_{j'}(t)dG(t)}{1+\sum_{k=1}^q \rho_k w_k(t)}$$
$$A_2(j,j') \equiv \int \frac{h^2(t)w_j(t)w_{j'}(t)dG(t)}{1+\sum_{k=1}^q \rho_k w_k(t)}$$

for $j, j' = 1, ..., q$. Then, the entries in

$$\mathbf{V} \equiv Var\left[\frac{1}{\sqrt{n}}\nabla l(\boldsymbol{\alpha},\boldsymbol{\beta})\right]$$

are,

$$\frac{1}{n}Var\left(\frac{\partial l}{\partial \alpha_j}\right) = \frac{\rho_j^2}{1+\sum_{k=1}^q \rho_k}\left[A_0(j,j) - \sum_{r=1}^m \rho_r A_0^2(j,r)\right]$$
$$\frac{1}{n}Cov\left(\frac{\partial l}{\partial \alpha_j},\frac{\partial l}{\partial \alpha_{j'}}\right) = \frac{\rho_j\rho_{j'}}{1+\sum_{k=1}^q \rho_k}\left[A_0(j,j') - \sum_{r=1}^m \rho_r A_0(j,r)A_0(j',r)\right]$$
$$\frac{1}{n}Cov\left(\frac{\partial l}{\partial \alpha_j},\frac{\partial l}{\partial \beta_j}\right) = \frac{\rho_j^2}{1+\sum_{k=1}^q \rho_k}\left[A_0(j,j)E_j(t) - \sum_{r=1}^m \rho_r A_0(j,r)A_1(j,r)\right]$$
$$\frac{1}{n}Cov\left(\frac{\partial l}{\partial \alpha_j},\frac{\partial l}{\partial \beta_{j'}}\right) = \frac{\rho_j\rho_{j'}}{1+\sum_{k=1}^q \rho_k}\left[A_0(j,j')E_{j'}(t) - \sum_{r=1}^m \rho_r A_0(j,r)A_1(j',r)\right]$$
$$\frac{1}{n}Var\left(\frac{\partial l}{\partial \beta_j}\right) = \frac{\rho_j^2}{1+\sum_{k=1}^q \rho_k}\left[-A_2(j,j) + 2A_1(j,j)E_j(t) - \sum_{r=1}^m \rho_r A_1^2(j,r)\right] + \frac{\rho_j}{1+\sum_{k=1}^q \rho_k}Var_j(t)$$
$$\frac{1}{n}Cov\left(\frac{\partial l}{\partial \beta_j},\frac{\partial l}{\partial \beta_{j'}}\right) = \frac{\rho_j\rho_{j'}}{1+\sum_{k=1}^q \rho_k}\left[-A_2(j,j') + A_1(j,j')(E_j(t)+E_{j'}(t)) - \sum_{r=1}^m \rho_r A_1(j,r)A_1(j',r)\right].$$

Next, as $n \to \infty$,

$$-\frac{1}{n}\nabla\nabla' l(\boldsymbol{\alpha},\boldsymbol{\beta}) \to \mathbf{S}$$

where $\mathbf{S}$ is a $2q \times 2q$ matrix with entries corresponding to $j, j' = 1, ..., q$,

$$-\frac{1}{n}\frac{\partial^2 l}{\partial \alpha_j^2} \to \frac{\rho_j}{1+\sum_{k=1}^q \rho_k}\int \frac{[1+\sum_{k\neq j}^q \rho_k w_k(t)]w_j(t)}{1+\sum_{k=1}^q \rho_k w_k(t)}dG(t)$$

$$-\frac{1}{n}\frac{\partial^2 l}{\partial \alpha_j \alpha_{j'}} \to \frac{-\rho_j \rho_{j'}}{1+\sum_{k=1}^q \rho_k} \int \frac{w_j(t) w_{j'}(t)}{1+\sum_{k=1}^q \rho_k w_k(t)} dG(t)$$

$$-\frac{1}{n}\frac{\partial^2 l}{\partial \alpha_j \beta_j} \to \frac{\rho_j}{1+\sum_{k=1}^q \rho_k} \int \frac{[1+\sum_{k\neq j}^q \rho_k w_k(t)] h(t) w_j(t)}{1+\sum_{k=1}^q \rho_k w_k(t)} dG(t)$$

$$-\frac{1}{n}\frac{\partial^2 l}{\partial \alpha_j \beta_{j'}} \to \frac{-\rho_j \rho_{j'}}{1+\sum_{k=1}^q \rho_k} \int \frac{h(t) w_j(t) w_{j'}(t)}{1+\sum_{k=1}^q \rho_k w_k(t)} dG(t)$$

$$-\frac{1}{n}\frac{\partial^2 l}{\partial \beta_j^2} \to \frac{\rho_j}{1+\sum_{k=1}^q \rho_k} \int \frac{[1+\sum_{k\neq j}^q \rho_k w_k(t)] h^2(t) w_j(t)}{1+\sum_{k=1}^q \rho_k w_k(t)} dG(t)$$

$$-\frac{1}{n}\frac{\partial^2 l}{\partial \beta_j \beta_{j'}} \to \frac{-\rho_j \rho_{j'}}{1+\sum_{k=1}^q \rho_k} \int \frac{h^2(t) w_j(t) w_{j'}(t)}{1+\sum_{k=1}^q \rho_k w_k(t)} dG(t).$$

The entries are obtained by a repeated application of the facts $\int dG(t) = 1$ and $\int w_j(t)dG(t) = 1$, $j = 1, ..., q$. It should be noted that due to profiling, the matrix $\mathbf{S}$ is not the usual information matrix although it plays a similar role.

Thus, when (2) holds with true parameters $\boldsymbol{\alpha}_0, \boldsymbol{\beta}_0$, it follows under regularity conditions $\hat{\boldsymbol{\alpha}}, \hat{\boldsymbol{\beta}}$ are both consistent and asymptotically normal (See Sen and Singer 1993, Ch. 5),

$$\sqrt{n}\begin{pmatrix} \hat{\boldsymbol{\alpha}} - \boldsymbol{\alpha}_0 \\ \hat{\boldsymbol{\beta}} - \boldsymbol{\beta}_0 \end{pmatrix} \Rightarrow \mathrm{N}(\mathbf{0}, \boldsymbol{\Sigma}) \tag{A.1}$$

where $\boldsymbol{\Sigma} = \mathbf{S}^{-1}\mathbf{V}\mathbf{S}^{-1}$.

Acknowledgments. Research supported by NASA contract NAG52783. The authors are very grateful to the referee for a number of helpful suggestions which were incorporated in the paper.

REFERENCES

[1] Fokianos K., B. Kedem, J. Qin, and D.A. Short (2001), "A semiparametric approach to the one-way layout," *Technometrics*, **43**, 56–64.

[2] Fokianos K., B. Kedem, J. Qin, J. Haferman, and D.A. Short (1998), "On Combining Instruments." *Journal of Applied Meteorology*, **37**, 220–226.

[3] Qin J. and J.F. Lawless (1994), "Empirical likelihood and general estimating equations," *The Annals of Statistics*, **22**, 300–325.

[4] Qin J. and B. Zhang (1997), "A goodness of fit test for logistic regression models based on case-control data." in *Biometrika*, **84**, 609–618.

[5] Sen P.K. and J.M. Singer (1993), *Large Sample Methods in Statistics, An Introduction With Applications*, London: Chapman & Hall.

[6] Vardi Y. (1982), "Nonparametric estimation in the presence of length bias," *Annals of Statistics*, **10**, 616–620.

[7] Vardi Y. (1986), "Empirical distributions in selection bias models," *Annals of Statistics*, **13**, 178–203.

LIST OF WORKSHOP PARTICIPANTS

- Abeer Alwan, Department of Electrical Engineering, University of California, Los Angeles
- Steven Abney, AT&T Labs-Research
- Joan Bachenko, Linguistech Technologies
- James Baker
- Marian Barry
- Jerome Bellegarda, Spoken Language Group, Apple Computer, Inc.
- Julian E. Besag, Department of Statistics, University of Washington
- Peter Bickel, Department of Statistics, University of California, Berkeley
- Jeffrey A. Bilmes, Department of Electrical Engineering, University of Washington
- Daniel Boley, Department of Computer Science & Engineering, University of Minnesota
- Herve Bourlard, Dalle Molle Institute for Perceptual Artificial Intelligence (IDIAP), Swiss Federal Institute of Technology at Lausanne (EPFL)
- Mireille Boutin, School of Mathematics, University of Minnesota
- Jamylle Carter, Institute for Mathematics and its Applications
- David Castanon, ENG Electrical & Computer Engineering, Boston University
- Olivier Catoni, Probabilites et Modeles Aleatoires, C.N.R.S.
- Eugene Charniak, Department of Computer Science, Brown University
- Li-Tien Cheng, Institute for Mathematics and its Applications
- Zhiyi Chi, Department of Statistics, University of Chicago
- Jordan Cohen, Voice Signal Technologies, Inc.
- Michael Collins, AT&T Labs-Research
- Zoran Cvetkovic, Information Sciences Research, AT&T Shannon Laboratory
- Akira Date, Advanced Brain Signal Processing Lab., Riken Brain Science Institute
- Li Deng, Microsoft Research
- Mukund Deshpande, Department of Computer Science, University of Minnesota
- Fred Dulles, Institute for Mathematics and its Applications

- Jason Eisner, Department of Computer Science, University of Rochester
- Selim Esedoglu, Institute for Mathematics and its Applications
- Hollis Fitch, Institute for Defense Analysis
- Paul Garrett, School of Mathematics, University of Minnesota
- Stuart Geman, Applied Mathematics, Brown University
- Charles Geyer, School of Statistics, University of Minnesota
- Larry Gillick, Vice President of Research, Dragon Systems, Inc.
- Maria Gini, Computer Science and Engineering, University of Minnesota
- Herb Gish, BBN Systems and Technologies
- James Glass, Laboratory for Computer Science Massachusetts Institute of Technology
- Fikri Goksu, Department of Electrical Engineering, University of Minnesota
- Ramesh A. Gopinath, Speech Transcription Algoithms, HLT Department, IBM T.J. Watson Research Center
- Mitch Helle-Morrissey, Department of Research, Lernout and Hauspie
- Wendy J. Holmes, 20/20 Speech Limited
- Kirk Jackson, DIRNSA National Security Agency
- Frederick Jelinek, Center for Language, and Speech Processing, Johns Hopkins University
- Chuanshu Ji, Department of Statistics, University of North Carolina at Chapel Hill
- Mark Johnson, Department of Cognitive and Linguistic Sciences, Brown University
- Benjamin Kedem, Department of Mathematics, University of Maryland at College Park
- Irving Kessler, IDA-Center for Communications Research
- Sanjeev Khudanpur, Center for Language and Speech Processing, Johns Hopkins University
- Seongjai Kim, Department of Mathematics, University of Kentucky
- Dietrich Klakow, Lehrstuhl fr̈ Sprachsignalverarbeitung, Universität des Saarlandes
- Christopher Lang, Indiana University Southeast
- Chin-Hui Lee, Dialogue Systems Research, Lucent Technologies
- Lillian Lee, Department of Computer Science, Cornell University
- Zhijun Liu, Department of Mathematics and Statistics, Mississippi State University
- Elizabeth Lovance, Linguistic Technology

- Michael Macon, Center for Spoken Language Understanding, Oregon Graduate Institute of Science & Tech.
- Martin Maiers, Department of Computer Science, University of Minnesota
- Christopher Manning, Department of Computer Science, Stanford University
- David McKoskey, Research/Development, Linguistic Technologies, Inc.
- Dan Melamed, Westgroup
- Willard Miller, Institute for Mathematics and its Applications
- Mehryar Mohri, AT&T Labs-Research
- Alejandro Murua, Department of Statistics, University of Washington
- Partha Niyogi, Bell Laboratories, Lucent Technologies
- Alexei Novikov, Applied & Computational Mathematics, California Institute of Technology
- Peter Olver, School of Mathematics, University of Minnesota
- Mari Ostendorf, Department of Electrical Engineering, University of Washington
- Sergey Pakhomov, ILLASL (Linguistics Program), University of Minnesota
- Fernando Pereira, Whiz Bang! Labs
- Harry Printz, IBM T.J. Watson Research Center
- Yingyong Qi, Qualcomm Inc.
- Jianliang Qian, Institute for Mathematics and its Applications
- Kashif Riaz, Strategic Development, West Group
- Stefan Riezler, Institut fur Maschinelle Sprachverarbeitung, Universitat Stuttgart
- Ya'acov Ritov, Department of Statistics, The Hebrew University of Jerusalem
- Roni Rosenfeld, School of Computer Science, Carnegie Mellon University
- Yoshinori Sagisaka, ATR Interpreting Telecommunications Res. Lab
- Ananth Sankar, Nuance Communications
- Lawrence Saul, AT&T Labs-Research
- Guergana Savova, LTI/L&H, Lernout and Hauspie
- Judith Schlesinger, IDA Center for Computing Sciences
- Michael Schonwetter,
- Jackie Shen, School of Mathematics, University of Minnesota
- Kemal Sonmez, SRI International
- Richard Stern, Department of Engineering, Carnegie Mellon University

- Andreas Stolcke, SRI International
- Paul Thompson, West Group
- Jean-Philippe Vert, DMA, Ecole Normale Superieure
- Shaojun Wang, Center for Automated Learning and Discovery Carnegie Mellon University
- Ken Williams, University of Minnesota
- Jack Xin, Department of Mathematics, University of Texas at Austin
- Xiangsheng Xu, Department of Mathematics and Statistics, Mississippi State University
- Steve Young, Department of Engineering, Cambridge University

1999–2000 Reactive Flows and Transport Phenomena
2000–2001 Mathematics in Multimedia
2001–2002 Mathematics in the Geosciences
2002–2003 Optimization
2003–2004 Probability and Statistics in Complex Systems: Genomics, Networks, and Financial Engineering
2004–2005 Mathematics of Materials and Macromolecules: Multiple Scales, Disorder, and Singularities
2005-2006 Imaging

IMA SUMMER PROGRAMS

1987 Robotics
1988 Signal Processing
1989 Robust Statistics and Diagnostics
1990 Radar and Sonar (June 18–29)
New Directions in Time Series Analysis (July 2–27)
1991 Semiconductors
1992 Environmental Studies: Mathematical, Computational, and Statistical Analysis
1993 Modeling, Mesh Generation, and Adaptive Numerical Methods for Partial Differential Equations
1994 Molecular Biology
1995 Large Scale Optimizations with Applications to Inverse Problems, Optimal Control and Design, and Molecular and Structural Optimization
1996 Emerging Applications of Number Theory (July 15–26)
Theory of Random Sets (August 22–24)
1997 Statistics in the Health Sciences
1998 Coding and Cryptography (July 6–18)
Mathematical Modeling in Industry (July 22–31)
1999 Codes, Systems, and Graphical Models (August 2–13, 1999)
2000 Mathematical Modeling in Industry: A Workshop for Graduate Students (July 19–28)
2001 Geometric Methods in Inverse Problems and PDE Control (July 16–27)
2002 Special Functions in the Digital Age (July 22–August 2)
2003 Probability and Partial Differential Equations in Modern Applied Mathematics (July 21–August 1)
2004 n-Categories: Foundations and Applications (June 7–18)

IMA "HOT TOPICS" WORKSHOPS

- Challenges and Opportunities in Genomics: Production, Storage, Mining and Use, April 24–27, 1999

- Decision Making Under Uncertainty: Energy and Environmental Models, July 20–24, 1999
- Analysis and Modeling of Optical Devices, September 9–10, 1999
- Decision Making under Uncertainty: Assessment of the Reliability of Mathematical Models, September 16–17, 1999
- Scaling Phenomena in Communication Networks, October 22–24, 1999
- Text Mining, April 17–18, 2000
- Mathematical Challenges in Global Positioning Systems (GPS), August 16–18, 2000
- Modeling and Analysis of Noise in Integrated Circuits and Systems, August 29–30, 2000
- Mathematics of the Internet: E-Auction and Markets, December 3–5, 2000
- Analysis and Modeling of Industrial Jetting Processes, January 10–13, 2001
- Special Workshop: Mathematical Opportunities in Large-Scale Network Dynamics, August 6–7, 2001
- Wireless Networks, August 8–10 2001
- Numerical Relativity, June 24–29, 2002
- Operational Modeling and Biodefense: Problems, Techniques, and Opportunities, September 28, 2002
- Data-driven Control and Optimization, December 4–6, 2002
- Agent Based Modeling and Simulation, November 3–6, 2003

SPRINGER LECTURE NOTES FROM THE IMA:

The Mathematics and Physics of Disordered Media
Editors: Barry Hughes and Barry Ninham
(Lecture Notes in Math., Volume 1035, 1983)

Orienting Polymers
Editor: J.L. Ericksen
(Lecture Notes in Math., Volume 1063, 1984)

New Perspectives in Thermodynamics
Editor: James Serrin
(Springer-Verlag, 1986)

Models of Economic Dynamics
Editor: Hugo Sonnenschein
(Lecture Notes in Econ., Volume 264, 1986)

Forthcoming Volume:

Time Series Analysis and Applications to Geophysical Systems